Geology

McGraw-Hill Book Company
New York
St. Louis
San Francisco
Düsseldorf
Johannesburg
Kuala Lumpur
London
Mexico
Montreal
New Delhi
Panama
Paris
São Paulo
Singapore
Sydney
Tokyo
Toronto

Geology

Leon E. Long

Department of Geological Sciences
University of Texas at Austin

Geology

1 2 3 4 5 6 7 8 9 0 V H V H 7 9 8 7 6 5 4

This book was set in Helvetica Light by Progressive Typographers. The editors were Robert H. Summersgill and Laura Warner; the designer was Merrill Haber; the production supervisor was Joe Campanella. The drawings and cover were done by Felix Cooper.
Von Hoffmann Press, Inc., was printer and binder.

Library of Congress Cataloging in Publication Data

Long, Leon E   date
   Geology.

   Includes bibliographies.
   1. Geology. I. Title.
QE28.L85   550   73-20105
ISBN 0-07-038672-2

To Julie and Stephen, with affection

Contents

Preface

Our understanding of nature has always progressed very unevenly. Often while data are patiently gathered for years, even decades, the mysteries the data are supposed to explain seem only to grow darker. Then some new theory will quickly draw together the information into a pattern that is both reasonable and beautiful. Geology, the science of the composition, structure, and processes of the earth, has just entered its golden era. The turning point came in the 1950s, when geologic research suddenly flourished, handsomely supported by government funds. For the first time, the ocean basins, Antarctica, and other once inaccessible places were explored in detail. These investigations led in the 1960s to the theory of plate tectonics, a hypothesis that has given us a profound insight into the internal workings of the earth in space and time. Other dramatic findings have been announced throughout the present decade. Moon rocks, for example, have revealed clues to the earliest history of the earth, and the difficult question of the causes of the Ice Ages is being studied with renewed success.

Geology is vitally important for another reason. It is the science of the planet that is home for billions of human beings. Already, population pressure and the rapid consumption of materials and energy have reached a state of continual crisis. More resources must be found. The inevitable growth of cities must be carefully planned. Hurricane, earthquake, and pollution hazards must be clearly recognized, if not avoided altogether. Demand for geologists trained to cope with these ills of a technology-based civilization will increase in the future.

In contrast to the frenzied lives led by most of us modern men stands the unhurried tempo of changes in the earth. Although we may write poetry about the eternal hills, we know they are not truly eternal. Not only is geology concerned with the configuration and processes of the earth; it goes on to interpret the long and varied history of this planet. So strong is the historical flavor that the subject has traditionally been taught as separate courses in "physical" and "historical" geology. This book does not make such a distinction.

xi

Broadly speaking, geologic time is the unifying theme of the book. (Perhaps this choice reflects my own fascination with antiquity: my research specialty is in the use of radioactive decay as a means of telling the ages of ancient geologic events.) Chapters 1 and 2 examine the origins of the universe, the solar system, and the earth-moon system. Chapters 3 to 5 present the minerals and rocks (aggregates of minerals) of which the earth is made. Chapter 6, describing methods of telling geologic time, is preparation for the history of development of life (Chapters 7 to 9). In Chapters 10 to 12 we see how geophysical studies have culminated in the ideas of plate tectonics, a theory that perhaps some day, in better-developed form, will successfully describe the origin of every major feature of the face of the earth. The final three chapters focus on processes for which the more recent geologic record is by far the best preserved. These include deposition of sediments (Chapter 13), and climatic change (Chapter 14). In Chapter 15 we consider the occurrence and supply of earth resources, whose scarcity has become an acute problem in the geological here and now.

This book was written for a one-semester introductory course team-taught by my colleague Alan Scott and me at the University of Texas, Austin. Most of the several thousand students who over the years have taken the course did not intend to major in science, at least when they enrolled. I assume that my reader, who is probably a liberal arts major, can understand plain English, though he may have only a rudimentary acquaintance with biology, chemistry, or physics. The book includes only a few chemical formulas and almost no mathematics. It *does* contain hundreds of carefully selected illustrations that are tightly woven into the text. Neither the text nor the figures will make sense unless studied together. I have purposely designed much of the information into the illustrations to save space, in keeping with the Chinese proverb that one picture is worth more than ten thousand words. Felix Cooper and his assistants have drafted these illustrations with elegance.

While the writing was in progress, many persons commented on parts of the book from the viewpoint of a scientist, an educator, or a university student. I am deeply grateful to John Wilson, L. Jan Turk, James Sprinkle, Alan Scott, Peter Schultz, David Schramm, Suzanne Schoenner, William Rust, George Ridge, Ross Nicholson, John Maxwell, Fred McDowell, Ernest Lundelius, Mary Long, Donald Larson, Wann Langston, Lynton Land, Edward Jonas, F. Earl Ingerson, Christopher Henry, Charles Helsley, Robert Heller, William Donn, Stephen DeLong, Lou Deans, Dwight Deal, Arthur Cleaves, Stephen Clabaugh, Fred Bullard, L. Frank Brown, Alan Blaxland, Daniel Barker, and Victor Baker for steering me away from major blunders, and for gently suggesting better ideas, illustrations, and wordings. Lou Deans, Robert Heller, and my wife Mary read the entire manuscript, and their perceptive criticism was responsible for some major revisions in early stages of the writing. Alan Scott has been a constant source of inspiration and information to me. Joyce Best typed the manuscript impeccably. Laura Warner skillfully managed thousands of large and small decisions as the book took final form. And thank you, Mary, for sharing with me the strain of a task that took so long to accomplish.

Leon E. Long

Geology

chapter 1

Cosmic Beginnings

Heavy chemical elements were created when a star violently exploded in A.D. 1054. The Crab nebula is an expanding cloud of explosion debris. [*Hale Observatories photograph.*]

For at least 3000 years of recorded history, men have speculated about the universe. Their ancient questions, so simply posed and so difficult to solve, are largely unanswered today. How did the universe begin? Indeed, did it have a beginning, and will it come to an end? Is the universe finite, or unlimited? What is its structure? Are the laws that describe the behavior of material and energy in the laboratory equally valid on the universal scale of size?

Amongst the multitude of objects in the universe is located that infinitesimal speck of matter upon which we live. Clearly, the origin and history of the earth have been firmly linked to the development of the remainder of the universe. Great progress has been made in answering these profound, sometimes disturbing, questions of cosmic beginnings. And yet, few people are more keenly aware of human limitations than the scientist who is struggling to understand such a grand subject. He knows that we shall never see to the edge of the universe (if there is an edge), and that a human lifetime is but an instant compared with the age of the universe (if it has a finite age). Our perception is so narrow, so limited. Any attempt to define the universe is like trying to picture the vast underwater bulk of an iceberg when only the tip is showing. In view of these restrictions, how can we proceed?

ORIGIN OF THE UNIVERSE

One useful way to solve scientific problems is first to construct a *model*. A model railroad, for instance, is a familiar object. It is a mock-up that physically resembles the actual thing. Other models may be sets of symbols, and in fact many are strictly mathematical. Whatever the case, a model is a simplification that makes it possible to study part of reality as though it were isolated from the rest. Models are set up in order to make predictions about the behavior of the real world. Further observa-

FIGURE 1-1 Messier 101, a spiral galaxy turned face-on toward our own galaxy, is located in the constellation Ursa Major (Big Bear). [*Hale Observatories photograph.*]

tions of nature either confirm the prediction, or require the model upon which the prediction is based to be modified or discarded altogether For example, before the time of Copernicus a widely accepted model placed the earth at the center of the solar system. This model predicted the motions of the other planets and of the sun as seen from earth. Copernicus found that a simpler model, placing the sun at the center of the solar system, explained the observations just as well. Careful measurements over the following centuries have confirmed the Copernican model.

Structure of the Universe

Any model that explains the origin of the universe must begin with observations of its matter and energy. Early in this century it was discovered that the universe is populated, in all directions and as far as can be seen, with local concentrations of matter called *galaxies*. At least 1 billion galaxies are within the seeing range of large telescopes. Our own galaxy, the Milky Way, is a disk-shaped collection of about 100 billion stars, plus gas and dust that are not condensed into stars. In the thickened central region of the galaxy, stars are most densely crowded. Toward the edge, our galaxy is thinner, and possibly flung out into two broad spiral "arms" like those of a distant but similar galaxy (Figure 1-1). Between the arms are dark, relatively empty lanes.

Astronomers have devised clever ways to estimate sizes of galaxies and distances between them. The measurements show the Milky Way to be strongly flattened. The diameter of the star-populated disk proper is 100,000 light-years,* some 60 times greater than its thickness. Our galaxy is one of a local cluster of 26 galaxies typically spaced 1 to 2 million light-years apart. The local cluster is part of a super cluster numbering thousands of galaxies. We are reminded of Jonathan Swift's famous verse:

So, naturalists observe, a flea
Hath smaller fleas that on him prey;
And these have smaller still to bite 'em;
And so proceed ad infinitum.

De Morgan has written a fitting postscript:

And the great fleas themselves, in turn,
Have greater fleas to go on;
While these again have greater still,
And greater still, and so on.

Does the organization of galaxies, like the bigger and bigger fleas, proceed endlessly to larger groupings? We do not know.

Models of the Universe

Among the many important contributions of the physicist Albert Einstein is his theoretical model of a closed four-dimensional universe. (A fourth dimension, time multiplied by the speed of light, is introduced because an observer in one part of the heavens cannot recognize an event in a distant part until the message borne by light reaches him. The signal may have been traveling for a very long time. We see things not as they *are*, but as they *were*.) Einstein's model has some odd prop-

* One light-year, about 10 trillion kilometers, is the distance that light travels in a year. See Appendix A for a comparison of the metric system with other familiar units of measurement.

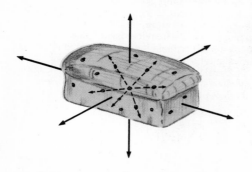

FIGURE 1-2 As the loaf of raisin bread rises (expands), every raisin, both on the surface and in the interior, moves farther from every other raisin.

erties not possible for us to imagine except by imperfect analogy. Picture a rising mass of raisin-bread dough (Figure 1-2). An insect crawling over the surface of the dough would soon return to the starting point of its local closed "universe." Similarly, a light beam could in principle circumnavigate the real universe because space is curved. Einstein's model of the universe was also static. In our bread-dough analogy, the raisins, representing galaxies, stay in fixed relative positions as long as the size of the loaf stays constant. When first proposed, the static aspect of the model seemed necessary, but it was a source of distress. Einstein knew that the attractive force of gravity should pull the galaxies together, causing the universe to collapse. Since the ultimate catastrophe did not seem in the making, his model required a mysterious, unknown repulsive force to counteract the gravitational force.

In the 1920s a series of astounding discoveries made the static model obsolete and set the stage for all modern cosmological thinking. Studies undertaken at the Mount Wilson Observatory, California, definitely established the existence of numberless galaxies outside the Milky Way. Other breakthroughs quickly followed when astronomers trained their newly refined spectrographs upon the stars. Spectrographs analyze the complex mixture of wavelengths of light (the light *spectrum*) reaching the earth from these distant sources. A typical spectrum is recorded on photographic film as a series of bright lines lying upon a dark background (Figure 1-3).

A particular set of lines, corresponding to particular wavelengths, can be identified with each chemical element. The astronomers promptly learned that stars are made of the same chemical elements familiar to us on earth. Moreover, the intensity and width of the spectral lines are related to the abundance of the element. They have revealed that the relative proportions of different elements in stars are greatly different from the abundances of these same elements on earth.

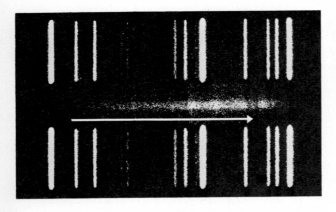

FIGURE 1-3 Upper and lower strips with vertical lines are reference spectra created by an incandescent lamp. They help to identify the spectral lines due to a distant cluster of galaxies (center strip). An arrow indicates the extent to which a faint line from the distant galaxies has been shifted right, in the direction of lengthened wavelengths (reddened light). The red shift in this case is equivalent to a recession of 60,000 kilometers per second. [*Hale Observatories photograph.*]

The Red Shift

Because of the great distances to even the closest galaxies, individual stars in them usually cannot be distinguished. For that reason, the spectra we receive from entire galaxies are essentially an average of the spectra of the individual stars. Galaxies in our local cluster exhibit "normal" spectra, but the same recognizable spectral lines from more distant galaxies are shifted somewhat toward longer wavelengths. Since the human eye interprets the lengthened wavelengths as a reddening of the light, the phenomenon was appropriately named the "red shift." The more distant a galaxy is, the more pronounced is the red shift of its light spectrum.

Some astronomers proposed that light is reddened as it passes through dust clouds. This is correct, but soon the red shift was noted in all directions in the sky, not just in regions where dust is present. A more plausible explanation makes use of the *Doppler effect*, which describes how an observer would receive radiation* coming from a moving source. Consider a source *A* that is sending out radiation in all directions (Figure 1-4). If the source is stationary with respect to ob-

* *Radiation:* the emission of waves or particles.

Stationary

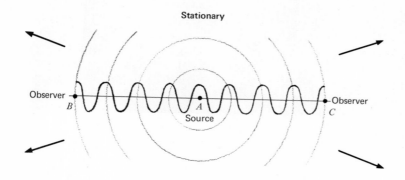

FIGURE 1-4 Radiation may be pictured as a series of circles constantly expanding away from the source. In the stationary situation, the circles are centered upon point *A*, whereas in the relative-motion situation, the circles are not concentric.

Relative motion between source and observers

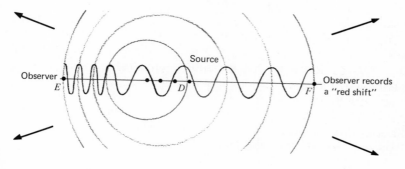

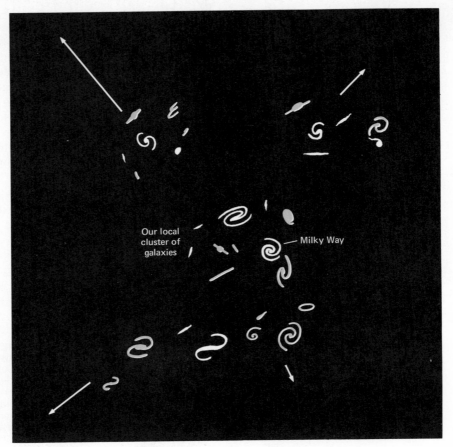

FIGURE 1-5 In an expanding universe, clusters of galaxies are receding from one another at a rate that increases as the distance between clusters increases.

servers at B and C, then the expanding circles, representing waves, are evenly spaced. On the other hand, if the source D is moving to the left, it partially catches up to its own radiation as it continues to give off more waves; consequently the waves received by observer E are crowded together (wavelength shortened). Conversely, from the viewpoint of observer F, the wavelengths are lengthened—they are redshifted. Thus the red shift signifies that the distant galaxies are traveling away from us; the substance of the universe is receding in all directions! This evidence for the expanding universe plays a prominent role in every modern cosmological model.

But why are the galaxies receding from *us*? Are we by chance at the center of the universe? The raisin-bread analogy helps at this point. When the loaf rises, every raisin grows more distant from every other raisin (Figure 1-5). Light from the Milky Way would appear red-shifted

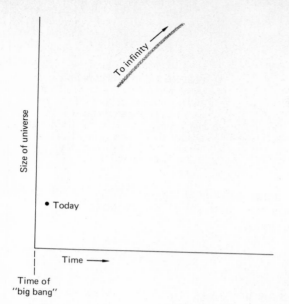

FIGURE 1-6 An open big-bang universe continues to get larger without limit in either space or time.

to a very distant observer, just as light from his galaxy appears red-shifted to us. There is nothing unique about either his or our position in the universe.

The Big Bang

In the light of this evidence, Einstein quickly abandoned his static model with its postulate of an antigravity force. Many other features of his original version, such as the curvature of space, have been retained in more recent models. One of these models takes the idea of an expanding universe to a logical conclusion. According to modern estimates, the "big bang" (the beginning of expansion) took place about 13 billion years ago at a time when the contents of the universe were compressed into a small, dense nucleus.

Will the expansion continue forever? If so, the universe is "open," or infinite (Figure 1-6). On the other hand, if the universe is actually "closed," or finite, the expansion will eventually slow to a stop. Gravity will pull the universe together back to a point from which it rebounds, and these gigantic oscillations, each about 100 billion years in duration, will repeat indefinitely (Figure 1-7). The choice between the open and closed models depends upon sophisticated observations of the present average density of matter in the universe—a datum that is not precisely known. If the density exceeds approximately one atom per cubic meter of space, the attractive force of gravity will eventually reverse the outward flight of the galaxies: the universe is closed. If the density is less, the universe is open. Preliminary measurements

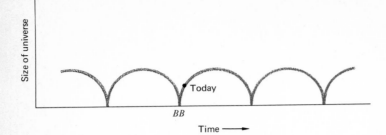

FIGURE 1-7 A closed big-bang universe repeatedly expands, halts, and collapses to a point. At present, we are in an expanding phase of development. Each collapse destroys all record of the previous existence of the universe. *BB* marks the time of only the *most recent* big bang.

suggest that the density is almost exactly at the balance point between these two alternatives. Surely each new set of observations will generate much interest and controversy.

ABOUT ATOMS

So far, we have considered the universe on the grandest scale of time and distance. We may gain an equally valuable insight by viewing nature on the ultramicroscopic scale of size. The "raw stuff" of the world is composed of 92 naturally occurring chemical elements, plus 13* additional elements that have been man-made. Their abundances and their manner of origin are now believed to be rather well understood. This information in turn provides important clues to the origin of the sun and the solar system. But first, let us briefly review some significant details of the structure of matter.

Structure of Matter

The early Greeks debated whether matter is continuous or discrete. If it is continuous, in principle it could be divided indefinitely into the tiniest pieces that are like large pieces in every respect except size. On the other hand, if matter is discrete, then it is composed of *elementary particles* that cannot be further divided. We now know the model of discrete particles to be the correct one. Ordinary matter is composed of *protons, neutrons,* and *electrons* (Table 1-1) that in turn are arranged into *atoms.* In the center of the atom is a *nucleus,* a composite of protons and neutrons bound together by powerful (but short-range) nuclear forces. A diffuse cloud of electrons outside the nucleus establishes the exterior size and shape of the atom. Although the nucleus occupies only about one-trillionth of the volume of an atom, it contains more than 99.95 percent of the total weight.

* This number occasionally must be revised as new elements are synthesized.

TABLE 1-1

Nuclear Particles

Particle	Mass	Electrical charge	Location
Proton	Heavy*	Positive	In nucleus
Neutron	Heavy	Zero	In nucleus
Electron	Light	Negative	Outside the nucleus

* A gram contains approximately a trillion trillion protons or neutrons. The mass of an electron is roughly $\frac{1}{2000}$ of the mass of one of the heavy particles.

The number of protons in a nucleus defines the particular chemical element. For example, all atoms of hydrogen have 1 proton each; a helium nucleus has 2 protons; carbon has 6; oxygen has 8; uranium has 92; and so on. Since the electrical charges of a proton and an electron are equal but of opposite sign, the charges cancel each other if the numbers of protons and electrons are equal, and a *neutral atom* results. In many situations the numbers of protons and electrons are not equal. If electrons are present in excess, a negative *ion* results. Conversely, if electrons have been removed, the unbalanced positive charge of the nucleus gives rise to a positive ion. In short, an ion is an electrically charged form of an atom.

Isotopes

What of the role of the neutrons, the other type of nuclear particle? Nuclei of a given element may contain various numbers of neutrons. For example, an atom of the element oxygen, with eight protons in the nucleus, may have eight, or nine, or ten neutrons in association with these protons (Table 1-2). Each of the proton-neutron combinations is an *isotope* of the element oxygen. To refer to the oxygen isotope with eight neutrons, we write ^{16}O (pronounced "oxygen 16"). The superscript designates the sum of the protons and neutrons.

Of the more than 1200 known isotopes, 327 are found in nature, the remainder having been created artificially. Twenty-two elements consist of only one isotope; for example, phosphorus in nature is entirely ^{31}P. All the other 70 or so naturally occurring elements are mixtures of isotopes. For instance, if we write the symbol "O" without a superscript, we refer to the natural mixture of atoms of ^{16}O, ^{17}O, and ^{18}O.

TABLE 1-2

Stable Isotopes of Oxygen

Isotope	Number of protons	Number of neutrons	Total
^{16}O	8	8	16
^{17}O	8	9	17
^{18}O	8	10	18

TABLE 1-3
Radioactive Decay Chain of Uranium 238

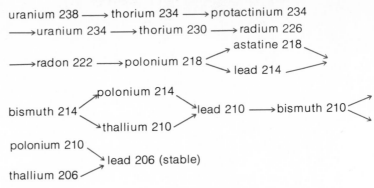

A distinctive property of each isotope is the degree of its stability—its tendency to remain unchanged, or to transform into something else. Most of the elements are represented by at least one stable isotope (the element tin has 10 of them). Atoms of unstable, or *radioactive*, isotopes have inherited an excess of energy that is released during the transformation. We say that in the process, a radioactive, or *parent*, atom decays into a *daughter* atom. A radioactive nucleus may decay into a daughter that is itself radioactive. In this case, the daughter is in fact the parent of yet a third species, and so on possibly down a long chain of decays until a stable daughter nucleus appears. For example, uranium 238 decays through a succession of 13 radioactive daughters to reach the ultimate stable daughter, lead 206 (Table 1-3). Accompanying the energy release, particles are shed from the decaying radioactive nuclei, as uranium with 238 nuclear particles eventually stabilizes as lead with only 206 nuclear particles. (A particular atom of polonium 218, or bismuth 214, or bismuth 210 decays along only one of the two possible routes available to it.)

We shall see that radioactive decay is a prominent source of the earth's internal energy. Radioactivity also provides our most accurate means to tell the ages of rocks.

ORIGIN OF THE ELEMENTS

It would seem that explaining the origin of the elements has grown very complicated. Far from having to account just for the elements as such, we must actually describe the origin of several hundred isotopes, each with unique properties. Here also, a model can help us to organize a wide variety of observations.

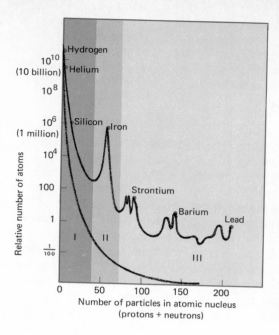

FIGURE 1-8 Cosmic abundances of the chemical elements. Note that because there is such an enormous variation in these abundances, the vertical scale is compressed in order to accommodate a wide range of data. A few of the elements are identified as points of reference. The lower curve shows the abundances created at the very outset by the big bang.

Cosmic Abundances

Among the basic data are the properties of each species of nucleus (its mass, whether stable or radioactive, and so on). Another clue to the origin of the elements (or, more exactly, to the isotopes of the elements) is found in the composition of the universe. What isotopes were created, and how much of each? A moment's reflection shows that the analysis of rocks would give a very false indication of these abundances. Unlike the stars, our planet is a dense, solid body accompanied by rather small quantities of water and the light gases. The earth is decidedly not a typical fragment of the universe. Instead, our estimates of cosmic abundances must be drawn in large part from the stars themselves. Intensities of spectral lines furnish the most representative data for chemical elements lighter than iron. Meteorites, which are the most ancient and primitive known solid objects, furnish the best abundance estimates for the heavy elements. Promptly it was discovered that every one of the stable isotopes is found in nature! Apparently, everything that *can* be created *was* created.

A graph (Figure 1-8) summarizes the results of the analyses. Suppose that we take 1 million silicon atoms as a standard of reference. Accordingly, hydrogen is about 100,000 times more abundant than silicon. The heaviest elements are about 1 million times less abundant than silicon. The abundance curve shows that 98 percent of the universe is hydrogen plus helium. (Our own solar system, dominated by the sun, is almost exactly of this composition.) Following these two elements is a very rapid, smooth decrease in the cosmic abundances of the light elements (region I), then a peculiar high-abundance peak in the vicinity of iron and nickel (region II). Elements heavier than iron are present in minute but fairly constant amounts (region III).

Synthesis of the Elements

In view of their atomic simplicity and their overwhelming abundance, hydrogen and helium were probably the starting points from which all other elements were synthesized. Only a little of the original hydrogen and helium has been "used up," yet significant quantities of even the heaviest elements were formed. What is the meaning of this odd situation? Suppose that protons, or heavier (proton-containing) nuclei, are approaching one another so as to fuse together. Acting to prevent the approach is a strong force of repulsion. (Why?) Consequently, fusion can take place only in a very hot environment, equivalent to high energies of particle motion. Extreme temperatures prevail today in the interiors of stars, and presumably once, long ago, during the big bang, when the contents of the universe were highly compressed. Could all the isotopes have been created at the very outset? Was the big bang big enough? Calculations suggest that the big bang accounted for most of the hydrogen and helium but for only a negligible amount of the heavier elements, as shown by the lower curve in Figure 1-8.

Suppose that we start with nothing but the lightest elements, provided by the big bang. First-generation stars that condense out of the primitive material could initiate the following series of reactions:

$$4 \, ^1H \xrightarrow{\text{(fusion taking place via several intermediate steps)}} {}^4He + \text{energy}$$

$$3 \, ^4He \xrightarrow{\text{(fusion)}} {}^{12}C + \text{energy}$$

$$^{12}C + {}^4He \xrightarrow{\text{(fusion)}} {}^{16}O + \text{energy}$$

Fusion of hydrogen into helium—the only process occurring in our sun—will release enough energy to maintain the sun at its present brightness for billions of years. Only stars much more massive than the sun can sustain the third reaction on the list. Each succeeding reaction fuses a ^4He nucleus into the nuclear "ashes" of the previous reaction. This process of helium addition has not progressed very far. That is, for any given heavier isotope, a much larger quantity of the lighter isotope that preceded it remains unreacted. Therefore the abundance curve in region I declines steeply.

Iron, cobalt, and nickel (region II) are the most stable of all elements. Because of this, the iron group elements are the "end of the line"—once formed, they simply stockpile. Or, to put it another way, if the universe had reached greatest stability, the condition toward which all reactions are directed, everything would be iron. Fortunately, only about 1 percent of the universe has so far become iron!

How, then, can elements that are even heavier (and less stable) than iron be created? Suppose that in a very massive star, an iron-rich central core has formed. Since no more energy can be released by fusion of iron with something else, the nuclear fuel supply has been used up. No longer is there an internal energy source to support the outer layers of the star from below by the pressure of radiation. Now the star must draw upon the force of gravity, the only source of energy left to it. Instantly the star collapses, then explodes as a titanic *supernova* that for a period of a few weeks may shine more brightly than all the remainder of the galaxy! Accompanying the explosion are numerous reactions that produce a flood of neutrons. Heavy elements of region III, and many isotopes of the lighter elements, are believed to be built up by a combination of neutron addition and radioactive decay during supernova explosions.

The sun and the solar system contain the heavy elements, yet the sun has not been, and never will become, a supernova. (Its interior will never reach the high temperature needed to synthesize the heavier elements.) If the heavy elements were not manufactured in the sun, they must have been present from the beginning; the sun cannot be a first-generation star. Early in the history of the Milky Way, there probably were quite frequent supernova explosions, estimated to be as many as one or two per year. Material ejected from these early explosions must have been mixed into the dust cloud that later condensed to form the sun and its solar system.

ORIGIN OF THE SOLAR SYSTEM

Models of the origin of the sun, the planets, their moons, and miscellaneous debris in the solar system are rather crude, though rapidly improving. For many years the only available data, aside from our direct knowledge of the earth, were certain types of astronomical measurements. Analyses of moon rocks and the space-probe explorations of Mars and Venus have already deeply influenced our ideas about early events in the solar system. Of course, basic to any model is the evidence that the solar system is indeed a *system* of related objects having a common origin. Take, for example, the motions of the planets. All 9 planets and their 32 satellites orbit in nearly the same plane, which is also close to the plane of the sun's equator. All the planets revolve about the sun in the same direction in nearly circular orbits. The sun itself rotates in this direction. Distances between planetary orbits,

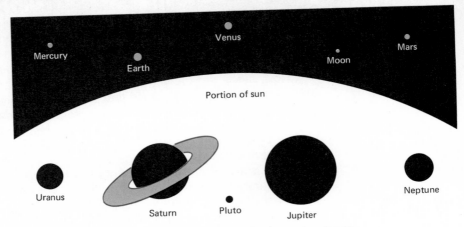

FIGURE 1-9 Inner planets are very much smaller than outer planets.

or between the orbits of neighboring moons, can be described by a simple mathematical formula. So accurate is this formula that it was used to guide successful searches for three major bodies, including the planet Uranus.

Chemical regularities in the solar system are also pronounced. The inner planets—Mercury, Venus, Earth, Mars—are small, dense, rocky bodies clothed in relatively tenuous atmospheres (Figure 1-9). According to the formula just mentioned, another planet should occupy the large gap beyond the orbit of Mars. This space is populated, not by a single planet, but by tens of thousands of small objects, the asteroids. Beyond the asteroid belt are the outer planets—Jupiter, Saturn, Uranus, Neptune, Pluto. Except for Pluto, they are large but of low density. In fact, the outer planets are almost nothing but the gases hydrogen and helium, with some methane (CH_4)* and ammonia (NH_3). Solid hydrogen, but little or no rocky material, may reside deep in their interiors.

Another significant though puzzling factor to be explained is the angular momentum of the sun and planets. Consider a planet in orbit about the sun (Figure 1-10). Angular momentum, a property of the orbital motion of the planet, is equal to its mass times its velocity of motion times its distance from the sun. Similarly, we could speak of the angular momentum of a single particle rotating about the spin axis of the sun or a planet. In this case, the total angular momentum is the sum of the angular momentums of all the particles contained in the spinning body. Surprisingly, about 98 percent of the angular momentum of the solar system resides in the motion of the planets. To put it another way, the sun's 27-day period of rotation is remarkably long. If the sun had

* The formula indicates that four atoms of hydrogen are chemically bonded to one atom of carbon.

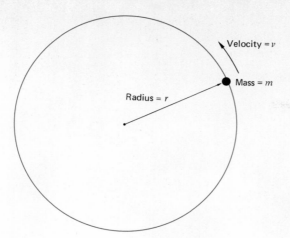

FIGURE 1-10 If a mass m is revolving about a center of motion with a velocity v at a distance r from the center, the angular momentum is defined as equal to mvr.

"spun off" the planets, we would expect the sun to be rotating rapidly today. More likely, the planets are sisters of the sun (formed independently, at about the same time), not its daughters (extracted from the body of an already condensed sun).

Yet another consideration, less well supported by data, is a certain confidence that the galaxy contains other solar systems. In part, this belief is wishful thinking; the idea of a solar system that formed in only one instance is distasteful to a scientist who is seeking a *general* explanation of reality. What if other stars do have planets? How could we detect these distant satellites? They would be invisible to the most powerful telescope. Barnard's Star, a rather dim neighbor about 6 light-years distant, gives some peculiar indications of an unseen orbiting companion. Gravitational attraction of the satellite, estimated to be about $1\frac{1}{2}$ times as massive as Jupiter, causes the star to wobble noticeably.

Toward a Reasonable Model

Let us attempt to draw this and other information together into a plausible model of the solar system. Today's great accumulation of data has put us in the uncomfortable position of knowing rather well what the questions are, but not the answers. Nevertheless, a good working hypothesis, even if speculative, is needed if we are to make further progress.

Scattered here and there in the spiral arms of the galaxy are large clouds of uncondensed gas and dust—regions such as the Lagoon nebula (Figure 1-11) where the density of matter may be up to 10,000 times greater than elsewhere. According to a current theory, a spiral arm signifies where a high-density wave, or shock wave, is passing

FIGURE 1-11 The Lagoon nebula is a diffuse cloud of dust and gas about 36 light-years across, located in the constellation Sagittarius. Small dark spots are local high concentrations of matter. [*Hale Observatories photograph.*]

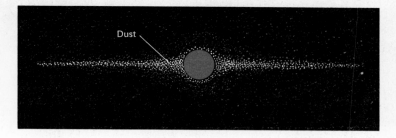

Dust

FIGURE 1-12 Presumably the evolving solar system once appeared as a flattened, rotating disk encircling a massive accumulation at the center. The disk became the system of planets; the central bulge became the sun.

through the galaxy. Temporarily, as the front of the shock wave reaches a given region, the diffuse material is highly compressed, perhaps enough to condense into more substantial clouds. Embedded within the Lagoon nebula are distinct dark globules in which matter is locally concentrated. These globules are actively shrinking so that, within 2 or 3 million years, by the time they are reduced to the size of the solar system, a new star will be born at the center almost literally overnight. In 1936, one such star suddenly began to shine in the midst of a concentration of dust and gas in the constellation Orion.

Our own solar system probably began as one of these clouds. During the shrinkage, parcels of cosmic material traveling in random directions began to interfere with one another, forcing adjustments of the motion. Slowly the infalling cloud was transformed into a more organized, rotating, flattened disk. At the center, a protosun* began to gather together somewhat tentatively (Figure 1-12). Already it was becoming hot as the force of gravity continued to drive the solar material together. The remainder of the disk, like the protosun, consisted mostly of gases, with a small amount of metal and rocky material mixed in. At this point the disk contained something between 2 and 10 times the combined mass of the modern planets.

Near the protosun, high temperatures prevented the solid material from condensing. Somewhat farther away, a frozen "snow" of methane and ammonia glued the dust particles into globs that eventually grew to about the size of basketballs. Larger bodies traveled faster and swept up the smaller objects. Once the bodies had become large enough, their gravitational attraction enabled them to draw in yet more particles. Detailed calculations suggest that at this time, events took place with catastrophic suddenness. The planets accumulated within 10,000 years, and most of the action took place in a few centuries. Occasionally the bodies suffered violent collisions that reduced them to rubble that was incorporated into other objects. (Some meteorites are made of fragments that were broken, recemented, rebroken, etc.) Eventually only a few large objects—today's planets, their satellites, and the larger asteroids—remained.

*Proto means "the earliest form," or "first in time."

Up to this time, the sun's heat was supplied by contraction due to the force of gravity. We can only surmise what happened next by observing young stars that are similar to the sun in composition and size. Apparently, when the interior becomes hot enough to ignite the nuclear furnace, a star experiences a time of instability. Newborn stars can be seen steadily streaming large quantities of their mass back into space. If the sun once experienced such vigorous turbulence, the frozen gases that made up the bulk of the nearby inner planets would have been stripped away. According to estimates, only about $\frac{1}{1200}$ of the earth's original body survived this process. The earth almost disappeared! That is why the inner planets are "cinders"—composed mainly of rock and metal that could not be entrained in the solar wind. The outer planets captured some of the solar gas, but even so, massive Jupiter probably was stripped of between one-half and nine-tenths of its original material. This matter, blasted far beyond the orbit of Pluto, may occasionally revisit us in the form of comets.

Perhaps magnetic fields also played an important role in the early solar system. According to the British astronomer Fred Hoyle, a strong magnetic field bound the ionized protosun and the surrounding disk together as though they were connected by tough, stretchable ropes. By way of the magnetic connection, most of the angular momentum of the once rapidly spinning sun was transferred to the planets. It has been noted that very young stars rotate rapidly, but that slightly more mature stars rotate much more slowly. Have the slowly rotating stars surrounded themselves with planets? Is our solar system unique, or is it one of many that have formed in the natural course of events?

FURTHER QUESTIONS

1 What is an application of the Doppler effect that is familiar to our everyday experience?

2 Refer to Figure 1-4. How would a light spectrum appear to be shifted to an observer located at point E?

3 Why can massive stars synthesize heavy elements, whereas less massive stars cannot? (There are several steps to the answer.)

4 Refer to Figure 1-12. As a solar nebula contracts, why does it assume a disk shape? Why does it not become a sphere? For that matter, why are Jupiter, Saturn, and other outer planets spherical, even though they are mostly gas? Why are they not disk-shaped, or irregular in shape?

READINGS

Listed in approximate order of technical difficulty.

Brandt, John C., and Stephen P. Maran, 1972: *New Horizons in Astronomy,* W. H. Freeman and Company, San Francisco, 496 pp.

Jastrow, Robert, and Malcolm H. Thompson, 1972: *Astronomy: Fundamentals and Frontiers,* John Wiley & Sons, Inc., New York, 404 pp.

Shklovskii, I. S., and Carl Sagan, 1966: *Intelligent Life in the Universe,* Holden-Day, Inc., Publisher, San Francisco, 509 pp.

chapter 2

The Double Planet

Full moon. [*Yerkes Observatory photograph.*]

Each of the planets is unique in some interesting way. Venus is permanently hidden beneath dense clouds; Mars is the other planet most likely to harbor life; Jupiter is the giant with many satellites; Saturn has rings, etc. One planet is accompanied by a single moon that is far larger in proportion to the primary body than any of the other satellites are in comparison to their planets. This "double planet" is, of course, the earth and the moon. Even at a distance of 380,000 kilometers, the moon exercises a considerable influence on the earth. We wish to learn more about the circumstances of the double planet. For example, did it originate as a pair, or did the earth capture the moon in a later, chance encounter?

Never before have exploratory expeditions so kindled our imaginations as the Apollo missions to the moon have done. Scientific findings from the Apollo voyages have firmly established our perspective of the earth-moon system, and they have raised new, unanticipated questions. This new information makes us recognize more than ever that the earth must be understood in relation to its kinship with the other bodies of the solar system. What are these relationships, and how do they help to explain the origin, structure, and history of the double planet? In this chapter we shall explore the basic features of the earth and the moon, with special attention to the latter.

METEORITES

A useful place to begin is with meteorites: solid objects that have accidentally collided with the earth while orbiting the sun. Although most meteorites underwent chemical and physical changes early in the history of the solar system, they are the closest approximation to original cosmic material that we have. Meteorites provide a good "first guess" upon which to build more elaborate models of the internal composition of the earth and the moon.

In ancient times, meteorites were revered as evidence of the working of great forces in nature. The Bible records* that the people of Ephesus, some 2000 years ago, were official keepers of a sacred stone that had fallen from the sky. During the following centuries, though, most people regarded the idea of stones from heaven as superstition, in spite of occasional eyewitness reports to the contrary. Then, on April 26, 1803, a shower of 2000 to 3000 meteorites fell in France near Paris. At last, this well-documented fall in a populated area convinced the gray-beards of the French Academy (at that time the prestige scientific society in Europe) of the reality of these visitors from outer space. Many questions have remained unanswered in spite of the vast amount of meteorite research accomplished since the early 1800s. John Wood, a meteorite expert, has commented, "Probably in no other branch of natural science is there such a wealth of observational data coupled with such a lack of unanimity in interpretation." One of the chief difficulties is that meteorites originate in an environment totally unfamiliar to our everyday experience.

Stony and Iron Meteorites

With few exceptions, meteorites can be simply classified as "stony" or as "iron." Since these contrasting types are so different, we must estimate their abundances before making up a chemical inventory of cosmic material. Most of the specimens on display in museums were found lying on the ground, but not actually seen to fall. As expected, nearly all of them are iron meteorites, which are so obviously different from ordinary rocks that almost anyone would recognize them. Iron meteorites are also likely to be noticed because they are extremely resistant to attack by weathering. Some iron meteorites have not rusted away even after lying on the earth's surface for hundreds of thousands of years. Stony meteorites, on the other hand, more quickly decompose into a nondescript loose pile of grains.

Clearly, only the meteorites that have been observed to fall are significant for a correct tabulation of meteorite types. On this more restricted basis, a census shows that stony meteorites are by far the most abundant:

Meteorite type	Percent abundance
Stony	92
Iron	6
Miscellaneous	2
	100

* Acts 19:35.

Catalogs list more than 1700 well authenticated occurrences of individual meteorites or swarms of fragments. Meteorites are named after the post office nearest the point where they were found. Exotic names like Abee, Cold Bokkeveld, and Sikhote-Alin abound in the literature about meteorites and are common vocabulary to the specialist. About 45 percent of known meteorites were seen to fall; the remainder were chanced upon afterward.

A Closer Look

Strictly speaking, the iron meteorites are not pure iron, but an intergrowth of two kinds of nickel-iron alloy. These alloys are distinguished by their nickel content (less than 8 percent nickel in one, more than 20 percent in the other), and by slightly different arrangements of nickel and iron atoms in their structures. A smoothly polished slab that has been lightly etched with acid shows a complicated network of the two alloys (Figure 2-1). This pattern may be continuous over a surface a meter or more across, indicating that iron meteorites at least as large as a meter are single crystals. Characteristic of a crystal is the regular, repeating pattern of its atoms. The perfection of the atomic pattern in iron meteorites suggests that very slow cooling permitted the atoms to settle into equilibrium positions in the crystal. Slow cooling implies that in the past, the iron meteorites must have existed in the interiors of much larger objects, which have come to be known as *parent bodies*.

How large were these parent bodies? A clever application of metallurgy comes to our aid here. The two iron-nickel alloys do not have fixed compositions. If they are in intimate contact (as in iron meteorites), the composition of each alloy depends on the temperature and, to a smaller extent, on pressure. It happens that as temperature decreases, nickel atoms slowly diffuse from the alloy with higher nickel

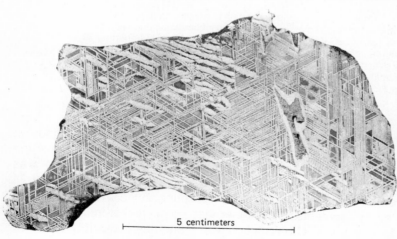

5 centimeters

FIGURE 2-1 A cut, polished, etched slice of the Edmonton nickel-iron meteorite. An inclusion of schreibersite $(Fe, Ni)_3P$ appears at the right center of the photograph. The external shape and size of a fallen meteorite are rarely the same as when the body was orbiting the sun. Like the nose cone of a space reentry vehicle, the forward surface of a meteorite is intensely heated and partly stripped away during its fiery plunge through the atmosphere. [*Smithsonian Astrophysical Observatory photograph.*]

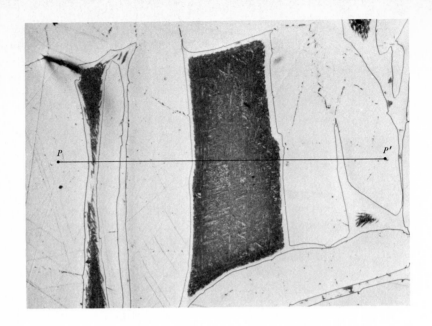

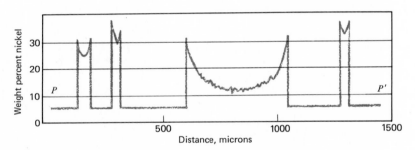

FIGURE 2-2 Under the microscope, platelets of nickel-iron alloy are clearly visible in the Anoka meteorite. Nickel concentration along the line *PP'* abruptly jumps between high and low values across the boundaries. Within the high-nickel platelets, the concentration has an M-shaped profile created by diffusion that ceased before full equilibrium was attained. (A micron is a millionth of a meter.) [*After J. A. Wood, "The Cooling Rates and Parent Planets of Several Iron Meteorites,"* Icarus, *vol. 3, pp. 429–459, 1964.*]

content to the low-nickel alloy. The movement of nickel atoms is very sluggish, however, and as the temperature continues to drop, diffusion practically ceases. Diffusion borders are left along the edges of the small metal platelets, indicating that the migration of atoms was "frozen" before chemical adjustment became complete (Figure 2-2).

Model calculations taking into account diffusion rates, border widths, nickel contents, etc., show that some nickel-iron meteorites cooled as slowly as 1°C per million years. Such an incredibly slow rate of cooling would take place deep in the insulated interior of a parent body some 200 to 400 kilometers in diameter. Parent objects were about the same size as the largest asteroids, but smaller than the moon. Probably the asteroids never did collect into a single planet. Rather, they are left-over cosmic "garbage." After a leisurely cooling period of at least several hundred million years, some asteroids with nickel-iron cores were violently disrupted by collisions. Iron meteorites are fragmental debris left from these shattering impacts in outer space.

Stony meteorites present their own, as yet unresolved, difficulties. Superficially, they look like ordinary rocks, but in many ways they are far different. Rocks and stony meteorites are both composed mostly of *silicate* minerals in which silicon and oxygen are the chief chemical elements. In stony meteorites, much of the iron and nickel is a free metal like that in iron meteorites. Iron and nickel in rocks are found in chemical combination with other elements, not as an alloy. Chemically, the largest class of stony meteorites strongly resembles the "condensible" fraction of the sun (the elements that would exist as solids at room temperature). These meteorites appear to be primitive cosmic material except that they are stripped of the abundant light gases. Terrestrial rocks, on the other hand, have been subjected to processes that have separated and concentrated different elements into different contrasting rock types.

Myriads of pinhead- to small, pea-sized, spherical grains found in many stony meteorites are a source of much puzzlement. Of a number of suggested origins, one of the earliest proposals may turn out to be the most accurate. H. C. Sorby, a British geologist who pioneered the study of rocks under the microscope, thought that the spherical grains are fused droplets of the original "fiery rain." Some of them are squashed together along flattened contacts, as though they had been thrown together while still molten. Eventually, the stony meteorite material accumulated into large parent bodies that may even be the same ones that gave birth to the iron meteorites.

Cosmic Cannonballs

Just as the very existence of meteorites was long disputed, not all geologists today agree upon the evidence for impact of one of these cosmic cannonballs. Like the circumstances of meteorite origin, the high pressure and velocity of a major impact are quite outside our daily experience. For example, it was once argued that if a large meteorite struck the earth at a low, glancing angle, why did it not simply dig a furrow? Why are supposed meteorite craters typically circular? To answer this objection, we must resort to model calculations of what probably happens during impact, as shown by the sequence in Figure 2-3. The energy of motion is abruptly converted to heat energy sufficient

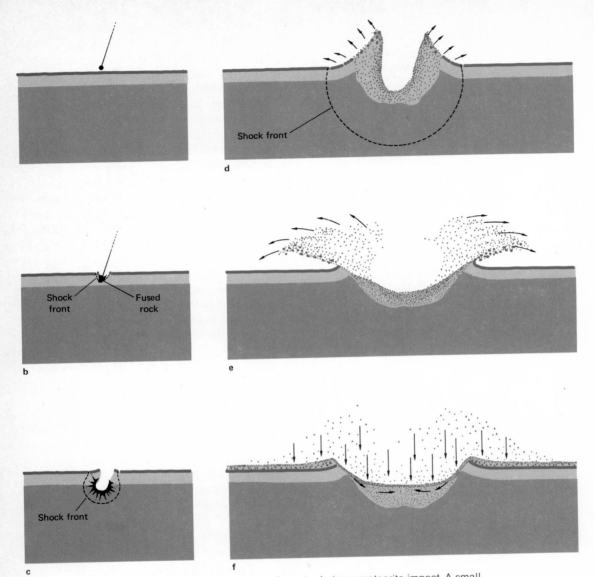

Shock front

Shock
front   Fused
             rock

Shock front

FIGURE 2-3 Sequence of events during a meteorite impact. A small meteorite can make a very large hole! The most deeply buried material disrupted by an impact explosion falls back into the crater or onto the crater rim. Target material that was originally just beneath the surface of the ground is ejected greater distances from the crater. This relationship was useful in identifying the depth of origin of material cast out of craters on the moon. [*After E. M. Shoemaker, "Penetration Mechanics of High Velocity Meteorites, Illustrated by Meteor Crater, Arizona," Report of the International Geological Congress, XXI session, part XVIII, Berlingske Forlag, Copenhagen, 1960.*]

FIGURE 2-4 Aerial photograph of the Barringer Crater, northern Arizona. The crater, about 1.2 kilometers across and 0.2 kilometer deep, was dug by an explosion equivalent to the detonation of 30 million tons of TNT. The meteorite itself weighed approximately 2 million metric tons. All that remains of it are scattered fragments on the crater rim and nearby plains. [*U.S. Geological Survey photograph by W. B. Hamilton.*]

to vaporize even an iron meteorite. In fact, the largest fragment likely to survive the crash in one piece would be only 3 meters or so in diameter. It is thus the explosion, not simply a plowing action, that excavates a circular crater regardless of the angle at which the meteorite has struck.

About a dozen meteorite impact craters are known, of which the most famous (but not the largest) is the Barringer Crater in northern Arizona (Figure 2-4). Curiously, every one of these craters was the work of an iron meteorite. Stony meteorites have possibly blasted some large craters that have gone unacknowledged because no distinctive pieces of iron are to be seen strewn about. Another possibility is that stony meteorites always break up into swarms of small fragments while still in the atmosphere; only the iron meteorites remain intact until they hit the ground. Meteorite craters are eventually eroded away or filled with sediment. The most ancient recognized crater is a little over 2 million years old.

On a larger scale are about 75 truly enormous circular structures, from several kilometers to as much as 90 kilometers in diameter. Not all circular areas of disturbed rock were created by impact. They may register the effect of volcanic eruptions, or of a broad updoming or downwarping of the bedrock, or other processes. How can we distinguish among these possibilities? Suppose the head of a very large comet* were to strike the earth. Fortunately, man has never witnessed this event, for the impact explosion would far outrival any other known type of catastrophe. The largest impact for which there is geologic evidence released an estimated 5 million times more energy than the biggest H-bomb ever exploded.

How would the earth absorb so much energy all at once? Because rock has a limited strength, there is also a limit to the size of the largest earthquake that can be set in motion. Impact energy that exceeds this limit is used, instead, to disrupt the target rock on the most intimate scale. Consider, for example, the common silicate mineral, quartz (SiO_2). In a quartz crystal, silicon and oxygen atoms are organized into a symmetrical pattern (Figure 2-5). Intense shock instantly converts the neat atomic pattern in a *crystal* into the disorganized array of atoms

* Comet heads are balls composed mostly of "ice" (frozen methane, ammonia, etc.) up to several kilometers in diameter.

FIGURE 2-5 (*a*) Oxygen and silicon atoms in quartz are precisely arranged in a regular network. (*b*) If a quartz crystal is severely shocked, the atoms are thrown into the confused disarray characteristic of a glassy substance.

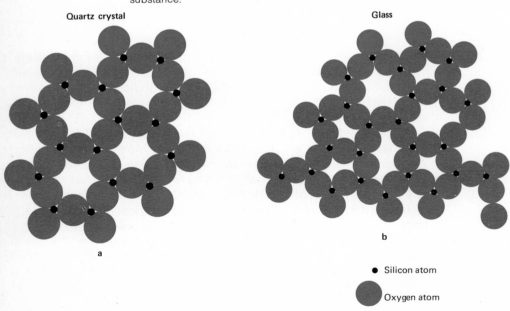

Quartz crystal

Glass

a

b

• Silicon atom

Oxygen atom

characteristic of a *glass*. Disruption in the most extreme cases went even further, as large volumes of the shocked rock actually melted.

Giant impact scars are less quickly effaced by erosion than are the smaller craters. The largest of them, the Vredefort Ring, in South Africa, is judged to be about 2 billion years old, which is nearly half the age of the earth. Still, it is not surprising that geologists continue to debate the origin of a mysterious upheaval alleged to be the work of a process never observed by man. Intense shock disturbance is the only convincing evidence of the impact origin in many doubtful cases. Impact scars are being discovered so rapidly that the final count may reach 200 or more.

THE MOON

In contrast to the earth (where circular depressions are few), the surface of the moon is populated by countless thousands of circular structures. What is their origin? Are they volcanic, created by an *internal* source of energy, or were they fashioned by the *external* energy of bombarding meteorites? Data from the Apollo flights have brilliantly confirmed the importance of both kinds of processes. Many of the lunar samples are found to be volcanic rocks that have been severely disturbed again and again by impact.

The Face of the Moon

But first, let us note some of the wealth of fascinating detail visible on the face of the moon (Figure 2-6). The light and dark areas apparent to the naked eye were first studied telescopically by Galileo in the early 1600s. He discovered the dark patches to be relatively smooth, depressed regions; he named them *maria*, which is Latin for "seas," though it is not certain that Galileo understood the areas literally to be covered by water. The more brightly reflecting regions are uplands which he named *terrae* (lands). Later and better telescopes revealed a variety of landforms that are important to our interpretation of lunar history. Among these are craters, some of which are centers of far-flung, radiating crater rays. Narrow trenches, the lunar rilles, are common in some areas. Everywhere, the moon is cut through by a network of intersecting fractures.

Lunar maria may be classed as circular or irregular in shape. Mare Imbrium ("Sea of Showers"), the grandest of the circular maria on the side of the moon turned toward the earth, is equal in area to Colorado plus New Mexico and Arizona (Figure 2-7). The Imbrium surface, some 3 to 6 kilometers lower than the surrounding uplands, is remarkably flat, or more accurately, it closely conforms to the overall spherical shape of the moon. Here and there, mountains rise above the plain, in part joining to form indistinct, partly buried concentric rings.

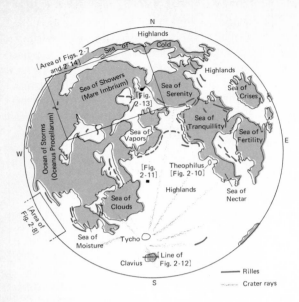

FIGURE 2-6 Compare this figure to the frontispiece of this chapter. Names given to the lunar maria hardly describe their true nature! Oceanus Procellarum, an irregular mare, is the largest of the flat plains. Its indistinct boundaries merge with Mare Imbrium and other low areas. Certain regions of the highlands have been named, but in general, these too are not sharply distinguished. Crater rays spread out from the crater Tycho over most of the moon's visible face. Because the Sea of Crises is so near the edge, it appears to be oval; actually, this lowland is circular.

FIGURE 2-7 Mare Imbrium, the "eye" of the Man in the Moon, is seen here in a closer view. [*Mt. Wilson Observatory photograph.*]

400 kilometers

FIGURE 2-8 Orientale, another circular mare, is much like Mare Imbrium except that its rugged, fractured face has not been flooded by lava. Surrounding Mare Orientale is a vast apron of loose material blasted outward from the central region. [*National Aeronautics and Space Administration photograph, National Space Sciences Data Center.*]

The near edge of Mare Orientale is barely visible from earth, the remainder being on the moon's far side. When orbiting cameras took photographs from a more advantageous position, it became evident that Mare Orientale is the most spectacular feature on the moon (Figure 2-8). This gigantic circular mare is 900 kilometers across (about the size of Mare Imbrium), but the broken concentric ring fractures of Mare Orientale have not been submerged beneath an upwelling of lava.

For an unknown reason, mare basins on the near side of the moon have been flooded with lava, but basins of maria on the back side of the moon have not. Peculiarities in the paths of orbiting lunar vehicles indicate that concentrations of unusually dense rock underlie the filled, smooth basins (Figure 2-9). Model calculations strongly suggest that the dense rock is in the form of a thin layer at the surface.

Highlands areas (*terrae*), covering two-thirds of the moon's visible face, are complex and little understood. Rock composing the highlands is less dense than the rock in the maria, suggesting that the highlands material long ago had separated and floated upward to the surface of the moon, like slag rising to the top of a pool of molten iron. The highlands terrain appears to be a fantastic jumble of craters that were partially destroyed by later, overlapping craters, then modified locally by volcanic outpourings. Probably the entire lunar surface was shaped into a rugged highlands during a period of intense bombardment as the

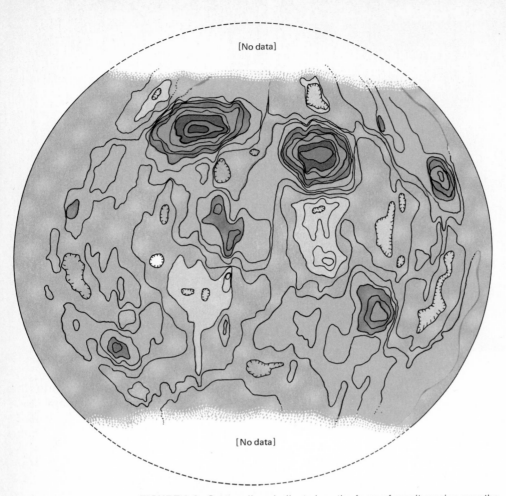

[No data]

[No data]

FIGURE 2-9 Contour lines indicate how the force of gravity varies over the moon's front face. "Bullseye" patterns correspond to unusually high gravity values due to dense material located not far beneath the lunar surface. A comparison with Figure 2-6 establishes that these dense rocks underlie the circular maria. [*After P. M. Muller and W. L. Sjogren, "Mascons: Lunar Mass Concentrations,"* Science, *vol. 161, pp. 680–684, 1968. Copyright 1968 by the American Association for the Advancement of Science.*]

moon was sweeping up the last bits of large debris that were incorporated into its bulk.

Craters were first noted by Galileo. Five craters on the visible face exceed 200 kilometers in diameter; 32 are between 100 and 20 kilometers, and yet smaller craters are present in rapidly increasing abundance down to tiny pits that are less than a millionth of a meter across. With the passage of time, the fresh, sharp-textured rims of young craters become rounded and subdued. Contours of the lunar surface

FIGURE 2-10 Geologists are still disputing the origin of the lunar craters. The elevated rim of Theophilus, photographed here by a lunar orbiter, suggests meteorite impact, but the origin of the central hills could be argued either way. They could be either an accumulation of volcanic rock or a mass of shattered lunar material that rebounded after a meteorite impact. Surrounding Theophilus and other lunar craters are thick blankets of ejected material. The volume of an ejecta blanket nicely matches the size of the crater cavity. [*National Aeronautics and Space Administration photograph, National Space Science Data Center.*]

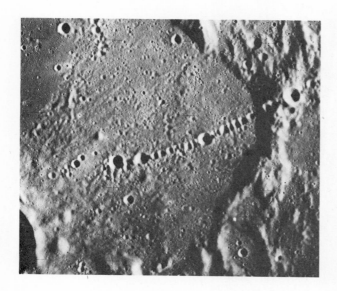

FIGURE 2-11 Are the lunar craters of impact origin or volcanic origin? Do you think it likely that a swarm of meteorites would have dug the neat chain of craterlets seen here? [*National Aeronautics and Space Administration photograph, National Space Science Data Center.*]

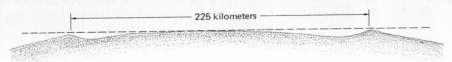

—————————— 225 kilometers ——————————

[Cross section]

FIGURE 2-12 The horizon of a lunar plain appears only 2 kilometers distant from a standing astronaut. So pronounced is the curvature of the moon's surface that a person in the center of the crater Clavius cannot see the rim!

are smoothed as loose material slides downhill. In addition, small meteorites pelting the moon ceaselessly overturn, or "garden" the uppermost few meters of material. (The astronauts found the surface to be pulverized into extremely fine dust.) This overturning helps to explain how crater rays (Figure 2-6) are eventually destroyed. Samples of ray material from the giant crater Copernicus proved to be fine particles of glass (created by shock?) splashed out of the crater in a radiating pattern. Crater rays are too thin to cast shadows of their own, and in fact, many are visible only under suitable illumination. A photograph taken at an unusual angle by the first orbiting astronauts showed some rayed craters not previously recognized as having rays.

The rilles are channels in the lunar surface (Figure 2-13). Straight rilles probably formed by fracturing followed by vertical displacements

FIGURE 2-13 Lunar rilles are not well understood. Many of them originate in a crater, meander across the moon for great distances (up to 700 kilometers), and mysteriously disappear. [*National Aeronautics and Space Administration photograph, National Space Science Data Center.*]

50 kilometers

of the rock, but the origin of meandering rilles that look like riverbeds is not well understood. Photographs taken on the rim of Hadley Rille show that layers of volcanic rock in the canyon walls have been cut through, but the pictures give no clue to the agent of erosion. There is not the slightest evidence for the former presence of water on the moon. Possibly a meandering rille was once an underground tube that carried lava. After draining, the roof collapsed, leaving an open valley. Nearly all the 200 or so known rilles are located at the edges of maria.

Lunar fractures, like the rays, are seen only under proper lighting conditions. Some fractures are subtle breaks oriented only in certain directions, imparting a crisscross grid pattern on a wide scale. Other profound systems of fractures radiate from the large circular maria out over the entire face of the moon. Impacts that created these maria were moon-shattering!

Lunar Stratigraphy

As historians of natural events, geologists are keenly interested in piecing together a connected story of the moon's origin and development. Working only from photographs, they were able to prepare detailed maps of much of the moon long before the first astronauts landed. These maps show the topography (the elevations and relief) and the appearance of the surface (smooth or rough, dark or light, etc.). They also interpret the events leading to the moon's present state. How can this be done? One way is to compare the freshness of appearance of the craters. Everything from sharp, young craters to old, subdued craters should be found in an ancient lava plain, whereas only younger craters should be found upon the surface of a more recent lava plain. (What would you predict about the density of population of craters in these two regions?)

Lunar *stratigraphy*—the arrangement of strata (layered deposits)—is an even more useful guide to the history of events. Take, for example, the region around Mare Imbrium, including the highlands and the craters Archimedes, Copernicus, and Eratosthenes (Figures 2-7 and 2-14). Which formed earliest: Mare Imbrium, or the rugged terrain surrounding it? The uplands and faint traces of interior rings in Mare Imbrium are similar in size to the distinct fractured rings of its sister mare, Orientale. In both cases, the uplands were probably cast up by stupendous impacts responsible for creating the mare basins. Archimedes obviously formed after the excavation of the mare basin upon which it rests, but note that the smooth floor of this crater is approximately at the same level as the mare material outside. In contrast, the rough crater floors of Copernicus and Eratosthenes are depressed below the level of the mare surface. Crater rays cast out of Copernicus overlie everything else, including the floor of Eratosthenes.

Thus from a single photograph we may deduce the following

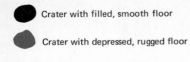

Crater with filled, smooth floor

Crater with depressed, rugged floor

Crater ray material

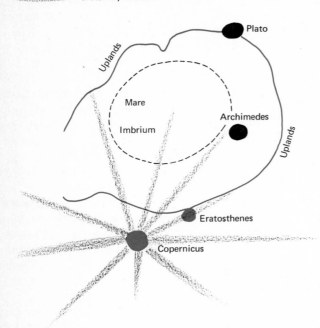

FIGURE 2-14 Compare this sketch with Figure 2-7. Geologists have interpreted a detailed history, not only of the Mare Imbrium region, but of much of the rest of the moon.

sequence of events: (1) creation of the Mare Imbrium basin and surrounding rim; (2) formation of the crater Archimedes; (3) flooding of the Imbrium basin and floor of Archimedes with mare fill material; (4) formation of the crater Eratosthenes, followed by (5) excavation of Copernicus. This commonsense reasoning makes use of the *law of superposition*, which states that *in an undisturbed series of strata, younger material lies on top of older material*. Obviously, we shall find this law to be just as useful in interpretations of layered rocks on earth.

EARTH AND MOON COMPARED

What are the similarities and differences between the moon and the earth? Are these two bodies so unalike that a common origin is improbable? We may explore these questions by comparing some basic geologic data from the two partners of the double planet.

Size

Although large as planetary satellites go, our moon is quite small, compared with the earth. The diameter of the moon is about 3500 kilo-

FIGURE 2-15 The moon superimposed on a map of the United States. Surface area of the moon slightly exceeds that of Africa.

meters (Figure 2-15), and its volume is only 2 percent of the earth's volume. The mass of the moon can be calculated from observations of its orbit. Once we know the mass and volume of the moon, we can calculate its overall density (mass per unit of volume). This value, 3.3 grams per cubic centimeter, is significantly less than the 5.5 grams per cubic centimeter mean density of the earth. Of course, the material buried deeply within the earth is under much higher pressure than it would be in the moon. (Why?) The greater compression of the earth material cannot account for all the density difference, however. In addition, the moon and earth must be made up of different mixtures of light and heavy substances. This fact immediately establishes that the origins of the two celestial bodies were somewhat different.

Internal Structure

In a following chapter we shall see how earthquake waves can be used as a "probe" of the inaccessible deep interior of our planet. They have revealed that the earth is a layered sphere consisting of three major divisions (Figure 2-16). A *core*, occupying the central 16 percent of the earth's volume, is made of very dense materials. Above the core lies the *mantle* which constitutes the great bulk of the earth, some 83 percent of its volume. Above the mantle, the outermost layer, the *crust*, rests like a thin scum; it contains the remaining 1 percent of the volume. The deepest drill hole has penetrated a trivial 0.15 percent of the earth's radius. Indeed, not even the mantle has ever been seen or sampled directly. Because of this, the chemical composition of the mantle and core is somewhat uncertain. Most geologists find an analogy with stony and nickel-iron meteorites to be rather persuasive. Could it be that the earth's core is a segregated mass of metal, whereas the mantle and crust are made of silicate and oxide minerals corresponding to the stony meteorite composition?

When the effect of the enormous pressure deep in the earth is taken into account, the known densities of the core and mantle rather nicely satisfy the meteorite model. Density variations are the reason for the earth's layering. The core is deepest and most dense; the mantle is of intermediate density, and the crust is the least dense part of the solid earth. Moreover, a layer of even less dense water lies above the crust in

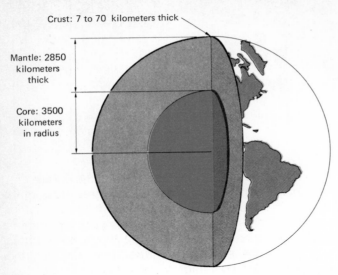

Crust: 7 to 70 kilometers thick

Mantle: 2850 kilometers thick

Core: 3500 kilometers in radius

FIGURE 2-16 A cutaway view of planet Earth shows its three major divisions. On the scale of this drawing, the crust is so thin that it is contained entirely within the thickness of the outer line.

the ocean basins, and the atmosphere rests on top of everything else. And so, in a broad sense, the earth is a series of concentric shells — spheres within spheres.

How did this layered condition come about? If the meteorite model is at all reasonable, we may assume that the original earth consisted of blebs of silicate material and of metal mixed together on a centimeter scale of distance. Segregation of the earth into the crust, mantle, and core must have taken place at a later time. Perhaps this process of *differentiation* is continuing even today, though at a sharply reduced pace.

Several sources of heat combined to drive the differentiation along. A considerable amount of *original heat* must have been generated by the countless collisions of particles as the earth accumulated. To this was added *heat from radioactivity*, which in the early days was produced roughly $4\frac{1}{2}$ times more rapidly than it is now. Long ago the moon was much nearer to the earth than it is presently, so that tides were higher. If the moon was already circling the earth from the very beginning (still an undecided question), the earth received a further quota of *heat from tidal stressing*. Once the temperature rose high enough to melt a little bit of the earth, the nickel-iron alloy (which melts at a lower temperature than silicates) would filter downward through the earth to its center. At the same time, less dense material, originally at the center, would be displaced upward. This exchange of positions released gravitational energy which was expressed as *heat of differentiation*. Intuitively, we may think of this heat as coming from friction as materials moved past each other.

The onset of differentiation was catastrophic. When added to heat from other sources, the heat of differentiation melted yet more material, thereby promptly furthering the process. Thus a chain reaction drove the differentiation faster and faster, eventually to engulf the entire earth. Whether or not all of the earth became molten at one time is not known; this condition would depend on how rapidly heat was produced, compared with how fast it could be transported to the surface and radiated into space.

Information about the internal structure of the moon is very sketchy because moonquakes have only just begun to be studied. (Moonquakes are vastly weaker and less frequent than earthquakes.) Already, we know that on the moon a porous outermost layer, churned by meteorite impacts, merges rather abruptly downward into a more solid crust. At a depth of about 65 kilometers, there is a sharp transition to a lunar mantle beneath. Perhaps other distinct divisions will some day be detected at yet greater depths.

The first recorded moonquake demonstrated the astonishing difference between the character of moonquakes and that of earthquakes. Once set to vibrating, the moon continues to reverberate, or "ring" like a bell for about 2 hours. That is, the moonquake energy is not quickly absorbed (converted into heat energy). A ringing action of the earth, though of long duration, is much less pronounced. In large part, these differences relate to the presence of water on earth and its total absence on the moon. On earth, water penetrates deep into the solid rock, where it coats individual grains as a thin film. The energy of earthquake vibration is literally dampened by the lubricating effect of water.

Atmosphere, Surface Water

Just as the divisions of the solid earth were a creation of later events, the earth's atmosphere and free water at the surface were probably not originally present. Contrary to popular opinion, the earth was "born naked and cold," devoid of an atmosphere but containing the ingredients of the ocean and the present-day atmosphere within it. What is the basis for this surprising assertion? One of the most cogent arguments is seen in the abundances of the "inert," or "rare," gases*—chemical elements that are so unreactive that they do not combine with any other atoms under natural conditions, not even with one another. A comparison of the terrestrial abundances of the inert gases with their "cosmic" abundances (Figure 2-17) shows that the supply of these gases on earth has been greatly depleted. That is why they are called rare gases. In view of this, why are generous amounts of water, nitrogen, carbon dioxide, and other chemically reactive gases still retained? These may be present to the extent of one-tenth or one-hundredth of their cosmic abundances, but the rare gases are present in only a millionth to a ten-billionth part of their cosmic abundances.

* Helium, neon, argon, krypton, xenon, radon.

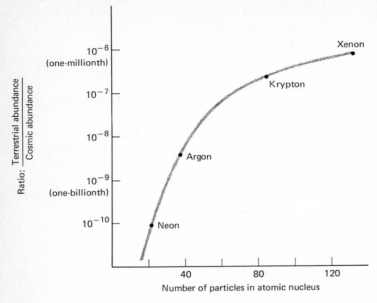

Ratio: Terrestrial abundance / Cosmic abundance

Number of particles in atomic nucleus

FIGURE 2-17 The earth's original endowment of inert gases has been severely depleted. Neon, a light gas, was depleted 10,000 times more than xenon, a heavy gas.

This odd situation can be explained if the primitive earth underwent a stripping process. The inert gases were not chemically combined and were lost; the reactive gases were held in chemical combination in the solid earth, later to be released to the surface. Possibly the stripping took place when the protosun first heated, as mentioned in the preceding chapter. At that time, nearly all of the mass of the inner planets was entrained and carried away by intense solar wind.

At first it might not seem possible for the great water mass of the ocean to be derived from the interior. However, solid rocks can contain a remarkable amount of water locked into *hydrous* minerals. For example, the mineral biotite (a variety of mica) contains hydrogen and oxygen. When biotite is decomposed, water is given off. Suppose that the original earth was an aggregate of meteorites that collected in the proportion of their "falls": 92 stony meteorites for every 6 iron meteorites. Iron meteorites contain practically no water, but the stony variety contains an average of about 0.5 percent H_2O in hydrous minerals. The amount of water incorporated into this hypothetical earth would be enough to fill the ocean 20 times over.

Popular accounts picture an early molten earth shrouded by a dense steam blanket that slowly condensed, forming the ocean. Even if this had been the situation, say at the time of differentiation, about 90 percent of the modern ocean would have remained dissolved in the depths. Throughout the ages, the water slowly emerged to the surface through volcanoes and hot springs. In fact, the amount of water delivered by this means is embarrassingly large. Estimates indicate that if hot springs had flowed as they do today during all of earth history, they

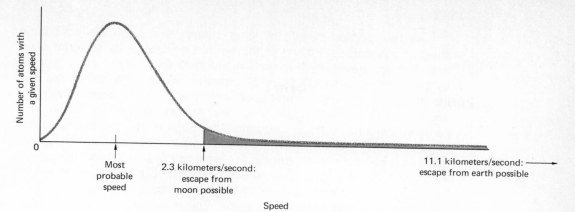

FIGURE 2-18 Collisions with neighboring atoms continually change the speed and direction of motion of an atom in a gas. If a series of chance collisions should knock an atom of the lunar atmosphere into an upward flight direction at a speed that exceeds 2.3 kilometers per second, it will never return. At any one instant, a considerable number of atoms may be escaping the moon, but only a trivial number of atoms in the earth's atmosphere are moving fast enough to leave the earth.

would have filled the ocean more than 100 times! Then why is the ocean as *small* as it is? Can you suggest a solution to the riddle?

Whatever liquids and gases the moon once contained would have moved upward toward the surface when the moon was differentiated. Yet the moon is an airless, waterless planet. (Today, most of the lunar atmosphere consists of gas from rocket exhaust.) Consider the constant motion of atoms in a gas (Figure 2-18). At a specified temperature, most of the atoms are traveling near a "most probable speed," but at any one moment, a small number of them are moving at high speeds. Calculations show that the moon's gravitational field is too weak to hold an atmosphere. If an atom happens to be speeding away from the moon faster than 2.3 kilometers per second, it will never return. Hydrogen and helium would have been lost in a few days, but heavier gases, such as krypton, would take somewhat longer to escape. The earth, with an escape velocity of 11.1 kilometers per second, constantly loses hydrogen and helium while retaining all the heavier gases.

Easily vaporized gases, including water, were expelled from the entire moon, not only from its surface. No hydrous minerals are present in lunar rocks. Even zinc, cadmium, lead, and other metals that form gaseous compounds at temperatures of hundreds of degrees have boiled away! Lunar rocks are rich in refractory elements such as aluminum, calcium, and titanium. Probably at one time the moon passed through a hot formative stage, much hotter than the earth ever experienced. Needless to say, the lunar environment has always been exceedingly hostile to life.

In Summary

An unexpected benefit of lunar studies has been the light they shed on the earliest history of the earth. Because of its superior size and stronger gravitational attraction, the earth must have been cratered by giant meteorite strikes about 50 times as frequently as the moon. But the face of the "lively" planet upon which we live has been destroyed and remade again and again. Such was not the case for the moon. Impacts on the lunar surface, frequent at the beginning, were dramatically reduced as the moon entered the present era of sweeping up a few last bits and pieces of cosmic debris. For reasons not understood, the moon long ago rather abruptly became a dead planet. Signs of internal activity (volcanism, etc.) ceased about the same time that the *oldest* rocks preserved on earth were being made. The moon records the hitherto missing first chapter in the story of the solar system.

ORIGIN OF THE EARTH-MOON SYSTEM

Before speculating on the origin of the earth-moon system, let us take note of one more source of data, the orbits of these two complementary bodies. Because both the earth and the moon contain mass, they attract each other gravitationally. The most obvious interaction is the rotation of the earth and moon about a mutual center of gravity (Figures 2-19 and 2-20). Superimposed upon their paths are a variety of minor wobbles, accelerations, and decelerations whose seemingly trivial effect may have profound significance over millions of years.

Tidal Interactions

Figure 2-21 shows, in exaggerated form, the effect of the lunar tides. The moon's gravitational pull slightly deforms the earth on the side facing the moon, and also on the opposite side. (Tides are·raised in the moon, too.) For the most part, the tidal bulges are a lifting of the oceanic water mass, but they also affect the solid earth. As the earth turns, the two bulges tend to remain directly beneath the moon, thus accounting for the approximately twice-daily ebb and flow so familiar to persons who live on the coast.

FIGURE 2-19 For calculation purposes, it is convenient to think of all the earth's mass as being concentrated at its center, and similarly, all the moon's mass at *its* center. The center of gravity (point C) is a balance point between a massive earth located at the end of a short lever arm EC, and a less massive moon located at the end of a long lever arm CM. Point C lies about 1500 kilometers beneath the earth's surface. (Earth-moon distance not to scale.)

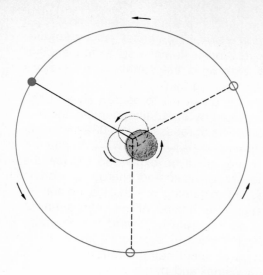

FIGURE 2-20 As the moon completes its orbit, the earth revolves about the common center of gravity. These motions are in turn superimposed upon the earth's orbit about the sun.

The upper diagram of Figure 2-21 corresponds to the "perfectly elastic" situation. If there were no friction, the high points of the tidal bulges and the centers of the moon and earth would all lie on the same straight line. Actually, friction prevents the crest of the tide from subsiding immediately as the earth's rotation carries it past the moon (lower diagram). Consequently, the crest "leads" the moon slightly (by about 2°) on the earth's near side, and on the far side the bulge falls slightly behind.

Now let us consider the attraction of the moon separately for each bulge. According to the law of universal gravitation, the attraction between two objects decreases in proportion to the distance squared; therefore the nearer tidal bulge dominates the interaction. Attraction of

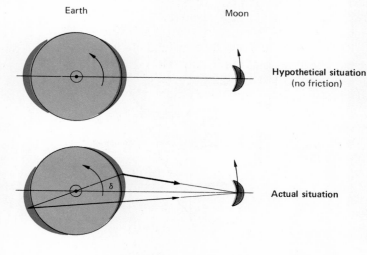

Earth

Moon

Hypothetical situation (no friction)

Actual situation

FIGURE 2-21 On the earth's near side (bottom figure), the tidal bulge slightly "leads" the moon. The (exaggerated) angle of misalignment, δ, is about 2°. A thickened arrow indicates that the moon's gravitational attraction to the near tidal bulge is stronger than its attraction to the more distant bulge.

45

this bulge tends to hold back the earth's rotation. In return, the earth's attraction pulls the moon forward in a sort of "slingshot" effect. Angular momentum of the earth's daily rotation is transferred to the angular momentum of the moon's orbit. With time, the earth's rotation slows down as the moon slowly recedes from the earth.

These changes take place much too slowly to be perceived. Each year, the length of the day is increased by about fifteen-millionths of a second on the average. The moon's rate of recession is 0.032 meter per year—the length of this line segment (├───────────┤) annually. It is obvious, furthermore, that energy must be expended to slow the earth's rotation. About two-thirds of the dissipated energy is used in moving the oceanic water mass; the remaining one-third becomes internal heat created by the constant tidal flexing of the solid earth. (A piece of wire gets hot when it is rapidly bent backward and forward.) The energy of tidal friction, accounting for 15 percent of the heat escaping from the earth, is produced at about half the rate at which all mankind consumes energy for industrial and personal uses. (Unfortunately, no more than a tiny fraction of the tidal energy could ever be made available to man.) Long ago, when the moon was closer to the earth, tidal heat was generated far more rapidly than it is today.

So minute are the changes happening today that some independent confirmation would be reassuring. Surprisingly, evidence has been provided on this matter by the study of fossil coral (Figure 2-22).

FIGURE 2-22 (*a*) Growth annulations on this specimen of an extinct coral suggest that the individual lived about 5 or 6 years. [*Judy Camps.*] (*b*) Superimposed upon the annulations are numerous much finer (and fainter) growth layers, believed to represent daily variations in the character of calcium carbonate deposition.

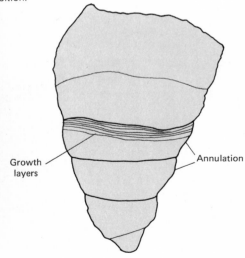

Growth layers

Annulation

a

b

Individuals of a genus of coral that lived about 400 million years ago slowly built their skeletons by adding one growth ring after another. Distinct larger ridges, called annulations, visible in the photograph, are believed to represent yearly variations in growth rate. High magnification of well-preserved specimens of the coral shows numerous fine layers that are interpreted to be daily growth additions. There are even intermediate-sized groups of layers that might correspond to lunar months.

Modern corals exhibit about 360 growth lines per annulation. The ancient corals present an average of 400 daily growth lines per annulation, as expected if the earth was rotating more rapidly then and if our interpretation of these faint layers is correct. It turns out that the number of days required for the moon to complete one revolution has hardly varied up to the present. As the earth's rotation slowed down, the moon's recession was accompanied by a longer period of orbit.

A Long Look into the Past

Tidal interactions are a key factor with which all theories of lunar origin must reckon. What would happen if, in our imagination, we were to "run the clock backward"? Where would the moon have come from? Calculations of the effect of all known forces, past and present, are intriguing, but they have not solved the problem. Hidden in the computations are some (as yet) poorly understood assumptions about tidal response. Even the most refined model ignores some of the complexities of nature. Three major hypotheses of the moon's origin are supported (skeptically!) by various researchers. The American geophysicist S. F. Singer aptly remarked: "Of the many proposed modes of origin of the moon, some violate physical laws; many are in conflict with observations; all are improbable."

Fission Certainly the moon was once closer to the earth. In 1898, Sir George Darwin, son of the Charles Darwin of biological evolution fame, proposed that the moon was actually spun off the earth (Figure 2-23). An appealing feature of the fission theory is the close match between the density of the earth's crust and upper mantle (which presumably gave birth to the moon) and the density of the moon. Later, it was suggested that the Pacific Ocean is the scar left by the moon's departure. Since the volume of the moon is 30,000 times greater than that of the Pacific basin, it scarcely matters whether the moon left the earth at the site of the Pacific or some other place. A serious objection to the fission mode of origin is that the moon literally never would have gotten off the ground. Calculations indicate that the newly separated moon would simply have fallen back on the earth. Still, the uncertainties of lunar origin are such that it seems premature at present to discard the fission theory.

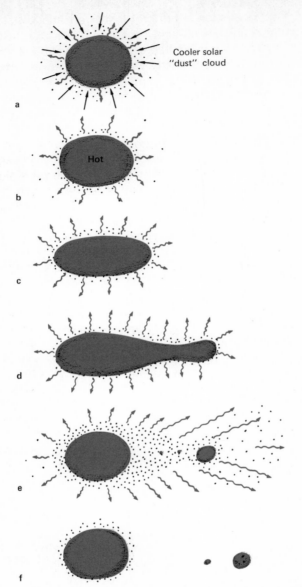

FIGURE 2-23 According to one version of the fission hypothesis, the initially cool, accumulating earth (*a*) is heated by the bombardment of dust and gas particles (*b*). When differentiation begins to concentrate the mass toward the center, the earth's rotation speeds up, just as a pivoting ice skater spins more rapidly when he pulls in his arms. Finally, the moon begins to separate (*d*) when the earth has a 2.65-hour day. Small moonlets, also cast off (*e*), eventually impact upon the near side of the moon to produce the circular mare basins (*f*). The fission hypothesis is beset by severe mechanical difficulties, and it definitely is not accepted by everyone. [*After Donald U. Wise, "Origin of the Moon from the Earth: Some New Mechanisms and Comparisons,"* Journal of Geophysical Research, *vol. 74, p. 6041, 1969. Copyright by American Geophysical Union.*]

Independent formation in earth-circling orbit If, as the majority of experts believe, the moon did *not* fission from the earth, then has the moon always been orbiting about the earth? This manner of origin would dispel some of the severe problems of accounting for the conservation of momentum, energy, etc., that plague the fission hypothesis. Regrettably, the *mechanical* difficulties of the fission model are replaced by a new set of *chemical* puzzles in the hypothesis of independent formation. For example, why does the moon seem to have had such a hot origin? Where did the materials go that were expelled from the moon? Proponents of the second hypothesis must devise an elaborate (and improbable) scheme to explain the pronounced chemical separation of the elements in the moon and earth, even though the two bodies accumulated near each other.

Formation elsewhere in the solar system A third hypothesis side-steps the chemical difficulties by permitting the moon to form in some other environment in a distant part of the solar system (nearer the sun). This theory substitutes a new collection of mechanical difficulties. Capture of the moon from an orbit greatly different from that of the earth is possible but unlikely. Elaborate model calculations support the capture hypothesis. They suggest that the moon approached the earth from a great distance, reached a minimum separation perhaps as close as 18,000 kilometers, and then has slowly receded ever since. A problem here is to evaluate the relative motions of the earth and moon over the past billions of years from a set of observations made over a period of only a few years. Laser reflectors placed on the moon's surface hold much promise for future studies. Precisely timed light beams, reflected back to earth from the moon, will make it possible to measure the distance to the moon to within less than 0.1 meter. At last, we shall have abundant, accurate information concerning these motions.

READINGS

Lowman, Paul D., Jr., 1969: *Lunar Panorama*, W. S. Heinman, New York, 105 pp.
Mutch, Thomas A., 1972: *Geology of the Moon*, Princeton University Press, Princeton, N. J., 324 pp.
Wood, John A., 1968: *Meteorites and the Origin of Planets*, McGraw-Hill Book Company, New York, 117 pp.

chapter 3

Plane Faces

Cluster of quartz crystals. [*Judy Camps*.]

The earth is made mostly of rocks. In turn, rocks are aggregates of minerals. And what is a mineral? At some time in the Middle Ages the word was coined to refer to a buried material of economic value, as that found in a mine. Long before that, perhaps as early as 6000 to 7000 years ago, mankind was already putting some 40 minerals to practical use. The science of mineralogy developed only very gradually. At first, minerals were classified with fossils, rocks, alloys, even with porcelain, glass, and pigments! Today the word *mineral* is defined in several ways that differ from the scientific meaning of the term. A corporation lawyer might refer to "mineral rights" in regard to coal or some other economic resource. In some situations even water is, legally speaking, a mineral.

This chapter, which considers the makeup of the solid earth, uses the word in a more restricted sense. A mineral is a *naturally occurring, inorganic* substance in which the atoms are *systematically arranged;* it is matter in the *crystalline* state. This concept took shape in two major scientific advances. The first came when analyses showed that a mineral species always has a restricted range of chemical composition. The second achievement was truly revolutionary. It took into account not only *what* atoms are present, but *how* they are positioned in a crystal. Let us explore some of these amazing developments.

CRYSTALS

Perfectly formed crystals have been prized as gemstones since ancient times. Someone has remarked that their beauty lies in the planeness of their faces. The very fact that crystals are bounded by flat surfaces calls for explanation. Straight lines and flat planes simply do not occur in nature purely by chance. Another regularity of crystals was discovered in the 1600s by Nicolaus Steno, a professor of anatomy who also was keenly interested in rocks and minerals. He pointed out that corre-

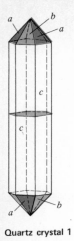

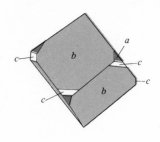

Quartz crystal 1

Quartz crystal 2

FIGURE 3-1 Quartz crystals may assume many forms. Here are two of them.

sponding faces of different crystals of a mineral always intersect at the same angle.

For example, consider two crystals of quartz, the mineral studied originally by Steno (Figure 3-1). Quartz crystals may vary in shape from long, slender needles to short, stubby forms. In spite of these differences, the angles between faces a and b, or between a and c, or between b and c are identical in both crystals. Suppose that crystal 1 were cut through the middle along the shaded plane. Viewed head-on in Figure 3-2a, this plane is seen to form a perfect hexagon. The corresponding plane through a less well formed quartz crystal might be a

FIGURE 3-2 The quartz crystals below have been sliced along a plane perpendicular to the faces labeled c in Figure 3-1, then turned cut-face up.

FIGURE 3-3 Building blocks in the Hauy model at the right are not necessarily miniature replicas of the entire crystal. In this example, a rather complex crystal form is constructed of numerous simple rectangular units. [*René Haüy, as appearing in Walter J. Moore,* Physical Chemistry, *Prentice-Hall, Inc., Englewood Cliffs, N.J., 1972.*]

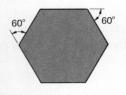

a

b

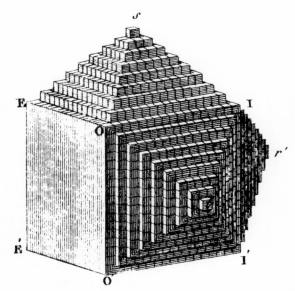

distorted hexagon (Figure 3-2b). Even so, the angle between any two adjacent sides in both crystals is 60°. Angles between crystal faces of other minerals differ from the angles of quartz. These diagnostic angles are a help in identifying minerals.

In 1784 a French crystallographer, R. Haüy (HAY-you-wee), proposed a model to explain the external regularity of crystals. He suggested that in a crystal, numbers of identical "building blocks" are stacked together as illustrated by one of his drawings (Figure 3-3). In hindsight, we recognize that the Haüy model was a major step forward in a day when understanding of chemistry was very confused. (A table published about the same time by a prominent chemist included heat and light among the chemical elements!)

Probing Crystals with X-rays

Still, no small cubelets like those suggested by Haüy's sketch are visible in crystals, even under high magnification. His building blocks, if they exist at all, must be submicroscopic. What are they? More than a century was to pass before discovery of x-rays would make possible further progress. At the outset, the nature of x-rays was just as obscure as the nature of Haüy's cubelets. Do x-rays consist of waves, or a stream of particles, or something else? (That is why they came to be known as *x*-rays, to denote an unknown quantity.) Then in 1912 the German physicist Max von Laue conceived the idea of training x-rays on crystals, and in so doing he solved both mysteries together. His hunch was that x-rays are similar to visible light, but more penetrating. Moreover, he reasoned, could it be that crystals have flat faces and constant interfacial angles because their atoms are arranged in an orderly pattern? The results of his experiments brilliantly confirmed both postulates. X-rays are waves, and crystals are internally ordered.

An obvious candidate for some of the earliest x-ray experiments were crystals of ordinary table salt, sodium chloride. As Laue predicted, x-rays blacken a photographic film in a symmetrical* pattern of spots after passing through this substance (Figure 3-4). One evident symmetry is the mirrored arrangement of spots on either side of a vertical line drawn through the center of the pattern. Can you pick out other types of symmetry?

Unlike the image of a person's ribs in a chest x-ray, these spots are not a picture of the positions of atoms in a crystal. Unfortunately, the information contained within the pattern is far less obvious than that. Let us see, at least in principle, how x-rays reveal the atomic structure of crystals.

A convenient way to represent wave motion is by a wavy line—a

---

* *Symmetry:* exact correspondence of a form on opposite sides of a dividing line or plane, or about a central point or axis.

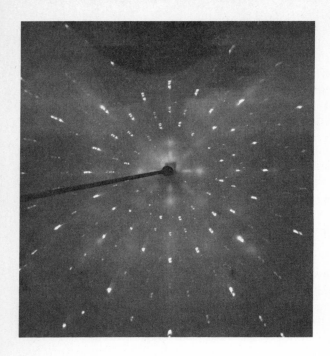

series of crests and troughs (Figure 3-5). The distance between adjacent crests, or between neighboring troughs, represents the *wavelength* of the x-rays. Suppose that we shine an x-ray beam, for which there is but a single value of the wavelength, upon a crystal (Figure 3-6). These x-rays are scattered (reflected), some from the topmost layer of atoms, and some from deeper layers that lie parallel to it. If reflected x-rays emerging from deeper layers are exactly "in step" with those reflected off the upper layers, the x-rays will reinforce one another.

As seen in Figure 3-6, the x-rays reflected off a deeper layer must travel distance $\overline{ABC}$ farther than x-rays reflected off the layer immediately above. If distance $\overline{ABC}$ is equal to some exact whole number of wavelengths, the entire crystal will reflect the x-rays. Otherwise, the waves cancel one another, and there is no reflection. Since the x-ray wavelength and the distance between layers of atoms are fixed, dis-

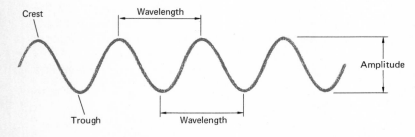

Crest

Wavelength

Trough

Wavelength

Amplitude

FIGURE 3-5 Uniform spacing of crests or troughs is equivalent to a single value of the wavelength.

54

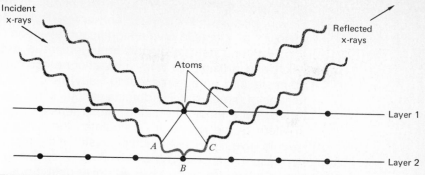

FIGURE 3-6 Distance $\overline{AB}$ is exactly one wavelength, as is distance $\overline{BC}$. Although the lower train of reflected x-rays emerges two wavelengths behind the upper train, the two are synchronized—they are "in phase" with one another.

tance $\overline{ABC}$ will be an integer number of wavelengths only if the incoming rays strike the crystal at certain angles. Or, to put it another way, we can learn the spacing between planes of atoms by observing the angles through which the crystal "flashes out" reflections.

But why are the x-rays not reflected in *every* direction? Would there not be atoms lying on a plane that passes through the crystal in any orientation whatsoever? The situation here is somewhat like a marching band on a football field (Figure 3-7). If the musicians know how to keep in step (which is not always so!), we should be able to sight along columns, along rows, or along various diagonals of marchers. Along diagonals they are more widely spaced than along rows or columns. Simi-

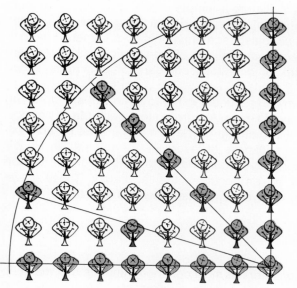

FIGURE 3-7 Evenly positioned marchers are analogous to layers of atoms in a crystal. Distances between marchers, or between atoms, vary according to direction. A row or column in this band contains eight marchers, but the same distance along the major diagonal includes only six musicians. Another diagonal contains only three. [*Courtesy Bill R. Deans.*]

larly, although there are nearly an infinite number of differently oriented layers of atoms in a crystal, only a few layers are populated densely enough by atoms to give a significant reflection.

Figuring out the atomic arrangement in a crystal calls for much skill and artistry. An x-ray crystallographer must know the chemical formula, and (in due respect to Haüy) he carefully studies the external shape of the crystal. He obtains various sorts of x-ray patterns. Then he proposes a model, in mathematical terms, describing the geometrical pattern of the atoms. He calculates the x-ray pattern that the model would give if it were an actual crystal. If the model calculation does not correspond accurately to the images on the photographic plate, he must refine the model over and over until it does agree acceptably. All the simple atomic structures have already been described. The millions of calculations necessary to explore the intricacies of complex crystal structures are done nowadays by computers.

X-rays can also be used to identify minerals. For example, some materials are so fine-grained that individual crystals are too small to see, even with a microscope. A chemical analysis would be difficult to interpret if a variety of different minerals were mixed together. On the other hand, an x-ray analysis is simple and accurate. We may easily match the angles and intensities of reflection of the sample of unknown composition with tables of x-ray data from pure samples of positively identified minerals.

Unit Cells

Now we may return to the question posed above: What *are* Haüy's building blocks? With the benefit of x-ray analyses, we may identify each block as a *unit cell*. Picture a simple crystal in which the pattern of the atoms extends indefinitely in all directions (Figure 3-8). A unit cell is the smallest and simplest fragment of this three-dimensional pattern that is necessary to convey the idea of the atomic positions. An entire crystal is made up of countless numbers of unit cells joined together.

We can also see why a true understanding of crystal structure had to wait until the discovery of x-rays. According to a principle of physics, the wavelength of radiation used to probe a material must be approximately the same as the thickness of layers in the material. In Haüy's day, only longer wavelengths corresponding to visible light or to radiant heat energy could be produced in the laboratory. These wavelengths are far too long to reveal significant detail in crystals.

MINERALS

Now we are ready to consider the most common minerals that make up the earth (feldspar, quartz, mica, calcite, gypsum, etc.). Most of the 2000

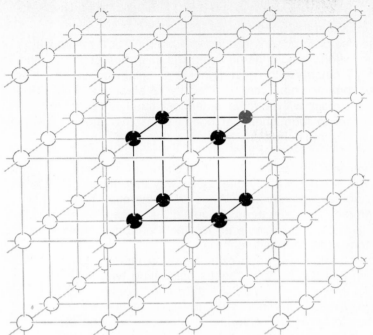

FIGURE 3-8 In this example, a crystal is assembled of cubic unit cells joined together in a framework extending in all directions. Focus on the colored atom. Not only does it occupy the corner of the boldly emphasized unit cell, but it participates equally in cells to the side, to the rear, and above it. How many unit cells share this atom? How many atoms does *one* unit cell contain?

known mineral species are exceedingly rare; only a single crystal is known of some of them. The great bulk of the earth is made of about 50 minerals or mineral families. Of these, only about one-half to one-quarter are likely to be encountered day in and day out by a professional geologist. Obviously, the minerals must be fashioned out of whatever atoms are at hand. Therefore, the chemical *abundances* play a key role in determining which of them are most common. Also important are the *size* and *charge* of the atoms. These factors work together to set limits upon what could be, in principle, an almost infinite number of possible combinations of chemical elements.

A Delicate Balance

What forces bind atoms together in a crystal? We saw in Chapter 1 that an atom consists of a positively charged nucleus surrounded by a cloud of negatively charged electrons. It is common knowledge that particles with like charges repel one another, whereas oppositely charged particles are mutually attracted. Since atoms contain *both* positive and negative charges, the result is a situation of delicately balanced forces. Suppose that two atoms are brought closer and closer together. As the electron clouds touch, then begin to overlap, they will experience a strong force of mutual repulsion. Because of this interplay of forces, the atoms in a crystal will settle into equilibrium positions that depend on the charges and sizes of the electron clouds and on the specific nature of all the other atoms in the vicinity. These positions are repeated over and over throughout the crystal, like the marchers who must stay in line in the band (Figure 3-7).

Another factor important in chemical bonding is the disposition of electrons in "shells" about the nucleus. The innermost shell, or zone, is occupied by 2 electrons; the next higher shell is fully occupied if it contains 8 electrons; the next shell is filled by 18 electrons, then 32 electrons, etc. An atom with a filled outermost shell is "saturated," and highly reluctant to form chemical bonds with other atoms. The inert gases exemplify this condition.

Ionic bonding In each of the other chemical elements, the outer electron shell is only partially filled. One way that atoms of these elements can achieve greater stability is by completion of the shell by donating or receiving electrons. Once this is done, the atom with a filled configuration has become an ion that can form chemical bonds with other ions of the opposite charge.

Sodium chloride is a good example of an ionic crystal. A neutral sodium atom has two "closed" shells containing two and eight electrons respectively, plus the start of a third shell populated by a single electron. (How many protons are in the sodium nucleus?) The third electron shell of a neutral chlorine atom lacks but one electron for completion. In a sodium chloride crystal the outermost, rather loosely bound electron of sodium is transferred to the vacant position in chlorine (Figure 3-9). Sodium has become a positively charged ion (a *cation*), and chlorine has become a negatively charged *anion*. Mutual attraction between the Na^+ and Cl^- holds the crystal together.

The structure of sodium chloride, the first to be determined by x-rays, is also one of the simplest (Figure 3-10). It is a three-dimensional checkerboard of alternating sodium and chloride ions. Each sodium ion is surrounded by six "nearest neighbor" chloride ions, and

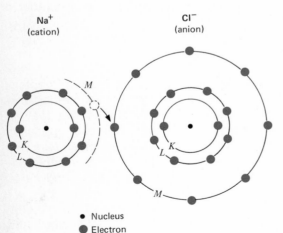

Na⁺
(cation)

Cl⁻
(anion)

• Nucleus
● Electron

FIGURE 3-9 K, L, and M designate the electron shells. When Na^+ and Cl^- are bonded together, the sodium ion donates its eleventh electron (which otherwise would be the sole occupant of the M shell) to fill a vacancy in the M shell of chlorine.

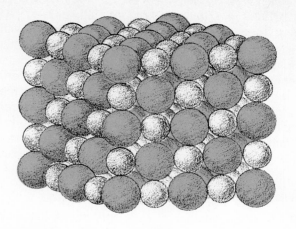
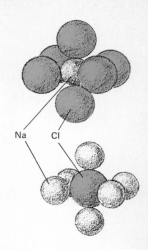

Na   Cl

FIGURE 3-10 Crystal structure of sodium chloride.

vice versa. Oppositely charged ions touch one another, whereas ions with the same charge are held apart at somewhat greater distances. Taken as a whole, a crystal of sodium chloride is electrically neutral.

Sodium chloride nicely illustrates one of a set of important principles known as Pauling's rules. The pertinent rule states that charges of ions are neutralized for the most part by adjacent, oppositely charged ions. These opposites must be intermixed on the most intimate possible scale.

Covalent bonding Another way in which the outer electron shell can be saturated is by participation in *covalent bonding*. Oxygen molecules (O_2) provide a splendid example of this bond type. An isolated oxygen atom lacks two electrons to complete a filled shell. By combining, each of the two atoms bound together in an oxygen molecule can achieve saturation, in effect, by *sharing* electrons with its partner (Figure 3-11). Covalent, or shared-electron bonding is the dominant type in the chemistry of carbon compounds (organic chemistry).

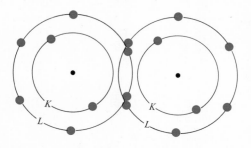

Oxygen molecule (O_2)

FIGURE 3-11 Covalently bonded oxygen atoms mutually share certain of the *L*-shell electrons.

59

Covalent and ionic bonds are similar and different in several respects. Both may be strong or weak. A mineral with strong chemical bonds is a hard substance, difficult to break or scratch because strong bonds are not easily ruptured. It so happens that carbon atoms in diamond, the hardest mineral, are covalently bonded, but some minerals in which ionic bonds are present are also extremely hard. Covalent bonds generally are strongly oriented toward just certain directions in space. In contrast, the charge of a cation or anion is felt equally in all directions. Finally, we note that every bond in a crystal is actually partly ionic and partly covalent. Nothing in nature is absolutely pure.

The Silicate Structure

By now, the number of rock specimens that have been analyzed chemically must be 100,000 or more. They include all the known rock types from a wide geographical distribution. Table 3-1, a grand average summary for the rocks in the earth's crust, shows that only eight elements make up nearly 99 percent of the total. If some rocks (particularly those beneath the sea) have not been adequately sampled, the averages might be somewhat in error, but not seriously enough to alter our conclusions. On a basis of weight or of numbers of atoms, oxygen and silicon are by far the most abundant elements. Moreover, oxygen atoms are large compared with the others, so that oxygen occupies almost 94 percent of the volume of the crust. The earth is essentially a packing of large but light oxygen atoms, among which the generally small but dense silicon and metal atoms are stuffed.

Oxygen and silicon, then, are the chemical backbone of the earth's crust. What governs the way these two species of atom are to combine?

TABLE 3-1
Abundances of Chemical Elements in the Earth's Crust

Element	% by weight	% by number of atoms	% by volume	Relative size of ions
Oxygen (O)	46.6	62.6	93.8	(O^{2-})
Silicon (Si)	27.7	21.2	0.9	(Si^{4+})
Aluminum (Al)	8.1	6.5	0.5	(Al^{3+})
Iron (Fe)	5.0	1.9	0.4	(Fe^{2+})
Calcium (Ca)	3.6	1.9	1.0	(Ca^{2+})
Sodium (Na)	2.8	2.6	1.3	(Na^{+})
Potassium (K)	2.6	1.4	1.8	(K^{+})
Magnesium (Mg)	2.1	1.9	0.3	(Mg^{2+})
All other elements	1.5	100.0*	100.0*	
	100.0			

* Includes only the first eight elements.

SOURCE: Brian Mason, *Principles of Geochemistry*, John Wiley & Sons, Inc., New York, 1966.

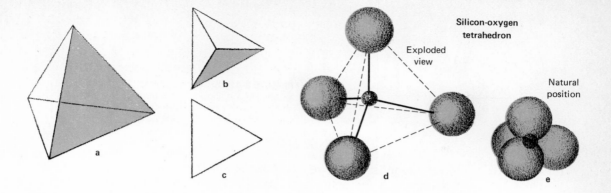

Silicon-oxygen
tetrahedron

Exploded
view

Natural
position

FIGURE 3-12 The tetrahedron, a solid figure, as viewed (*a*) in perspective, (*b*) from above, and (*c*) from below. Oxygen and silicon atoms fit together in a tetrahedral arrangement (*d* and *e*). In the natural position, the silicon is actually concealed from view, like a grape hidden in a pile of oranges.

Oxygen is a strongly nonmetallic element with an ionic charge of −2. Chemical properties of silicon are suggestive of both metals and non-metals. In the presence of oxygen, silicon behaves more as a metal, assuming an ionic charge of +4. The great size difference between ox-ygen and silicon (see Table 3-1) brings into focus another of Pauling's rules. It states: Negatively charged ions are grouped about positively charged ions in regular, geometric, three-dimensional patterns. For ex-ample, consider the four-sided object known as a *tetrahedron* (Figure 3-12). Large oxygen anions are spheres in mutual contact, located at the corners of a tetrahedron. Because of the curvature of the oxygen spheres, the interior of the tetrahedron has a small open space that can be occupied by a silicon atom. This arrangement of silicon and oxygen forms the basic framework of all silicate minerals.*

Since the proportion of silicon to oxygen atoms is 1:4, we may symbolize the association as SiO_4. What is the net electrical charge of a silicon-oxygen tetrahedron? Electrical neutrality can be restored to the overall crystal structure in two ways. In most silicates, cations such as K^+, Na^+, Ca^{2+}, and Al^{3+}† are located in spaces among the silicon-oxygen tetrahedra. The remainder of the charge imbalance is neutral-ized by the silicon-oxygen tetrahedra themselves. Tetrahedra may link together at the corners, forming chains, or sheets, or more complex structures. A negatively charged oxygen shared by two tetrahedra does double duty (Figure 3-13). By linking, the joined tetrahedra require

---

* All but one, that is. In stishovite, a silicate created under the extremely high pressures associated with the impact of a large meteorite, the oxygen and silicon atoms are organized somewhat differently.
† Al^{3+} can also substitute for Si^{4+} in the center of a silicon-oxygen tetrahedron.

61 MINERALS

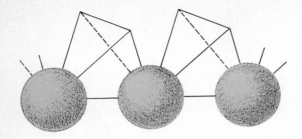

FIGURE 3-13 Skeleton outlines of linked tetrahedra are shown from a perspective view. A shared oxygen atom serves each of the two tetrahedra in which it participates equally well.

fewer oxygens for a given number of silicons; consequently the negative charge of the entire assemblage of atoms is reduced.

Chemical bonds joining oxygen and silicon in the tetrahedron are about equally covalent and ionic in character. We know that these bonds are very strong because the tetrahedra maintain a rigid, undistorted shape and size no matter what the remainder of the silicate structure may be like. Does this organization of silicon and oxygen satisfy the Pauling rule requiring intimate association of cations and anions?

Some Silicate Minerals

Long before the discovery of x-rays, mineralogists had worked out a classification of minerals based upon properties such as color, density, crystal shape, hardness, etc. X-ray analyses have enriched this classification by showing how the atomic structure of minerals can explain these other, more self-evident features. Here is a survey of the important families of silicate minerals. They are distinguished by the manner in which the silicon-oxygen tetrahedra are joined together.

Isolated tetrahedra *Olivine* (OL-i-veen) exemplifies a family of silicates in which the tetrahedra are not linked at all; no oxygens are shared between adjacent tetrahedra (Figure 3-14). Metal ions in olivine are either Mg^{2+} or Fe^{2+}, which may be present in any proportion. The same atomic structure is maintained throughout the entire range of variation, from pure magnesium olivine (Mg_2SiO_4) to pure iron olivine (Fe_2SiO_4). These two metal ions readily substitute for each other because of their similar sizes and identical charge. Ferrous iron and magnesium are characteristic, not only in olivine, but in many other minerals. As a group, these *ferromagnesian* minerals are quite dense, and in color they range from very dark green or brown to almost opaque black.

There is a striking relationship between the atomic structure and the physical properties of olivine. Oxygen atoms are closely packed, and they are connected to silicon and to magnesium (Mg) and iron (Fe) by strong chemical bonds. Therefore, olivine is one of the hardest, most dense minerals. Because the bonds are oriented in various direc-

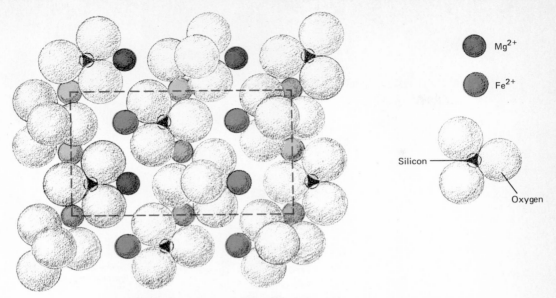

FIGURE 3-14 Silicon-oxygen tetrahedra in olivine form rows and columns in all three dimensions. Magnesium and ferrous iron ions occupy regular prescribed spaces amongst the tetrahedra, but whether any given cation site is filled by iron or magnesium is a matter of chance. The box encloses the atoms in one unit cell—the simplest repeating pattern.

tions in space, there are no especially preferred planes of atoms along which an olivine crystal might break. Rather, the crystal tends more to *fracture*—to break along irregular surfaces. Fire bricks made of olivine are used to line furnaces. What does this imply about the strength of the chemical bonds?

Single chains Another family of silicates is the pyroxene (PEER-oxeen) group, of which the mineral enstatite is a common example. Pyroxenes, the most abundant ferromagnesian minerals in the earth's crust, are prominent in a great variety of rocks of different origins. In these minerals, silicon-oxygen tetrahedra are linked corner-to-corner into long chains (Figure 3-15a). The net negative electrical charge of the chains is neutralized by layers of Mg^{2+} and Fe^{2+} ions that substitute freely for one another, just as in olivine (Figure 3-15b). The pyroxene group is much more complex than olivine, however. Cation sites in pyroxene can also accommodate calcium, sodium, aluminum, titanium, and other metal ions.

Double chains Linkage of tetrahedra into chains is the hallmark of yet another family of silicates, the *amphiboles*. In these minerals, two rows of silicon-oxygen tetrahedra join together, forming double chains

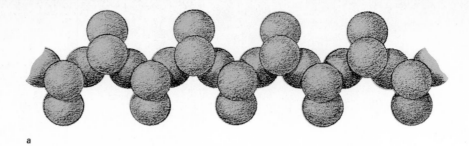

a

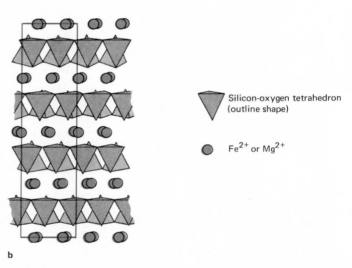

b

Silicon-oxygen tetrahedron
(outline shape)

Fe^{2+} or Mg^{2+}

FIGURE 3-15 (*a*) A short segment of the tetrahedron chain characteristic of the pyroxene group of silicate minerals. (Silicon atoms are not shown.) In enstatite, one of the pyroxene minerals (*b*), layers of these chains alternate between layers of Mg^{2+} and Fe^{2+} ions. The box encloses the atoms in one unit cell.

(Figure 3-16). Amphiboles outrival the pyroxenes in complexity. Both are ferromagnesians, but the cation sites in amphiboles can take up metal ions of almost any size or charge. This is especially true of *hornblende,* the most abundant species of amphibole. Hornblende is the "garbage can" of the mineral world, accommodating ions that are rejected by other, more discriminating mineral structures.

Because their atomic patterns are similar, pyroxenes and amphiboles share many other similarities. Members of both families are mostly rather hard, dense minerals. Chemical bonds within the silicon-oxygen chains are stronger than bonds connecting the chains to the remainder of the structure. Pyroxene or amphibole strongly tends to break along the chains, not across them—these minerals display ex-

FIGURE 3-16 The backbone of the amphibole family of minerals is the double chain of tetrahedra.

cellent *cleavage* along flat surfaces. Individual crystals may be blade-like or needlelike, and some varieties are even fibrous, like strands of hair. (Asbestos is a trade name for certain fibrous varieties of amphibole.) Angles between cleavage planes in pyroxene and amphibole are different. This detail, which shows particularly well under the microscope, is a handy way to identify and distinguish the two groups.

Sheets A natural extension of the linkage theme leads to minerals built around sheets of silicon-oxygen tetrahedra (Figure 3-17a). These sheets, which are basic to the *mica* family of minerals, display the most obvious connection between atomic structure and physical properties to be seen in any of the silicates. Mica splits into extremely thin, uniform flakes that may be either isolated, or stacked together in mica "books" (Figure 3-18).

In *biotite*, a very common, dark-colored mica, Mg^{2+} and Fe^{2+} ions are sandwiched between opposing sheets of silicon-oxygen tetrahedra (Figure 3-17b). *Muscovite,* almost as common as biotite, is a transparent mica that contains Al^{3+} ions in the place of biotite's magnesium and iron. (Muscovite is the first silicate mentioned thus far that is not a ferromagnesian.) Rather weakly bonded layers of K^+ ions separate the double sheets of tetrahedra in biotite and muscovite. The excellent cleavage of mica into flexible, springy flakes is due to breakage of bonds between the potassium layer and adjacent tetrahedra.

Another interesting family of silicates with sheet structure are the *clay minerals*. Because individual flakes of clay minerals are typically less than one-millionth of a meter across, extremely high magnification under an electron microscope is needed in order to study them. In fact, a single crystal may consist of only a few hundred to a few thousand unit cells. Microscopic and x-ray studies show that some varieties of

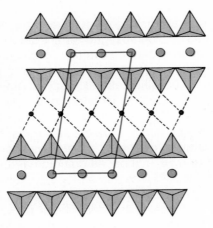

- ● Mg^{2+} or Fe^{2+}

- ● K^+

- ▲   Silicon-oxygen tetrahedron (skeleton outline)

FIGURE 3-17 (*a*) Tetrahedra are linked together in a continuous sheet in the mica family of silicates. (*b*) Viewed edgewise, the skeleton outlines of layers of tetrahedra, layers of potassium, and magnesium or iron ions are seen in biotite, a variety of mica. Biotite cleaves apart along the potassium layers. The box encloses the atoms in one unit cell.

FIGURE 3-18 Uniform, thin flakes are the leaves in thick "books" of mica. This mineral is an ingredient in wallboard cement, roofing material, and (because it is an electrical insulator) electronic devices. [*Judy Camps.*]

clay mineral are tiny, flat plates; others are fibrous bundles of hollow tubes.

The layers of tetrahedra in clays, imperfectly formed in the first place, are but feebly held together and easily sheared apart. Water and metal ions such as K^+, Na^+, and Mg^{2+} may enter and leave the "exchangeable" sites in clay minerals so readily that the chemical composition and unit-cell dimensions vary from day to day, depending upon humidity. Small wonder that wet clay is a slick, plastic substance, and that clay, with its ability to shrink and swell, is not a very desirable foundation material beneath a building. Clay is the chief ingredient in pottery and chinaware. It is used as a filler in paint, paper, and canned soup.

Framework silicates We have seen how the silicon-oxygen tetrahedra of silicate minerals may be isolated, or joined together as single chains, as double chains, or as continuous sheets. Finally, we note the *framework* silicates in which the linked tetrahedra are folded into complex three-dimensional patterns. Of the 10 or so groups of framework silicates, the *feldspars* and *silica minerals* are by far the most important. Neither of these two families includes a single representative of the true ferromagnesians.

Feldspars are chemically simple, but structurally they are frightfully complex. Metal ions in the abundant varieties of feldspar are potassium, sodium, and calcium. These three cations may substitute for one another, but not so freely as iron and magnesium do in the ferromagnesian minerals. Sodium and potassium both have a +1 charge, but

FIGURE 3-19 Cleavage is well developed in large feldspar crystals; indeed, two varieties of the mineral are named according to the angle between cleavage planes. Feldspar is an ingredient in porcelain and glass. [*Judy Camps.*]

quite different sizes (see Table 3-1). This moderate incompatibility is one of the reasons why the atomic structures of feldspar are so complicated.

The sizes of calcium and sodium ions are similar, but the charges do not match. If Ca^{2+} were to substitute for Na^+ in feldspar, how can the charge be balanced? Could half the metal ion positions perhaps be left unoccupied? If so many "holes" were left open, the atomic structure would rearrange into a more stable configuration; the mineral would no longer be a feldspar. Instead, what actually happens is a double substitution: calcium for sodium, and aluminum for tetrahedral silicon.

Metal Ion Positions

Between silicon-oxygen tetrahedra	*Within* silicon-oxygen tetrahedra	Sum of metal ion charge
Na^+ and	Si^{4+}	$+5$
	(simultaneous substitution)	
Ca^{2+} and	Al^{3+}	$+5$

Sodium and calcium are the diagnostic metal ions in the *plagioclase* feldspars.

We reach the ultimate in the cross-linkage of silicon-oxygen tetrahedra with the silica minerals. In the most common representative, *quartz*, every oxygen atom is shared between tetrahedra. All the negative charge of oxygen is satisfied by positively charged silicon. This being the case, what is the chemical formula of quartz? As usual, the tetrahedra are joined only at corners. Because of the rather open struc-

TABLE 3-2
Summary of Silicate Structures

Structure	Mineral group	Ferromagnesian minerals?	Si/O	Volume % in earth's crust	Density, g/cm^3	Volume of unit cell (cubic angstroms*)
Isolated tetrahedra	Olivine	Yes	SiO_4, or Si_4O_{16}	4	3.4	About 300
Single chains	Pyroxene	Yes	SiO_3, or Si_4O_{12}	17	3.1 to 3.3	About 430
Double chains	Amphibole	Yes	Si_4O_{11}			About 925
Sheets	Mica Clay minerals	Either yes or no	Si_2O_5, or Si_4O_{10}	4	2.8 to 3.3	About 980
Framework	Feldspar	No	$(Si,Al)_4O_8$	60	2.5 to 2.8	About 800 to 1500
	Silica	No	SiO_2, or Si_4O_8	12	2.6	
All other minerals (nonsilicates and miscellaneous silicates)				3		
				100		

* One angstrom is one ten-billionth of a meter.

ture of quartz and the varied orientations of tetrahedra in a number of directions, quartz is one of the least dense but one of the hardest of minerals. Would you expect quartz to fracture irregularly, or to break along smooth cleavage planes?*

Quartz and the other silica minerals are extremely pure substances with the same simple chemical composition (but different structures). Quartz is ordinarily transparent, but impurities present in the barest traces can impart to quartz beautiful tints of rose, yellow, smoky gray, etc.

Summary of silicates   Table 3-2 draws together some of the regular interrelationships among the silicate minerals. For each mineral family, the number of oxygen atoms is tabulated on the basis of four silicon atoms (the least common denominator). The proportions of silicon to oxygen systematically change down the list. Isolated tetrahedra of olivine can be packed tightly together, but the silicates in which the tetrahedra share corners are extravagant of space. Thus, density decreases and the volume of the unit cell increases.

* The clean-cut faces of quartz crystals shown in the frontispiece to this chapter do not necessarily signify that quartz will break along smooth flat planes.

And so, what is the composition of the earth's crust? Chemically it is mostly silicon and oxygen. From the viewpoint of crystal structure, it is a network of silicon-oxygen tetrahedra. As minerals go, feldspar is more abundant than all the other multiple hundreds of mineral species combined.

Some Nonsilicate Mineral Groups

With the discussion of silicates, we have noted all but a few of the common rock-forming minerals. Most of the remaining mineral families have relatively simple crystal structures and chemical formulas. Here are some of these minerals, classified according to their anions.

Carbonates These minerals are based upon the chemical bonding of various metal ions to the carbonate ion (CO_3^{2-}). Calcium and carbonate ions are arranged in alternating layers in *calcite* ($CaCO_3$), the most common carbonate mineral (Figure 3-20). As we might expect, calcite shows excellent cleavage along layers of atoms oriented in several directions in the crystal. Calcite crystals assume more varied shapes and sizes than do the crystals of any other mineral (Figure

FIGURE 3-20 In calcite, a layer containing calcium ions rests upon a layer of carbonate ions. Beneath the carbonate layer is another layer of Ca^{2+} (not shown), then a layer of CO_3^{2-}, etc.

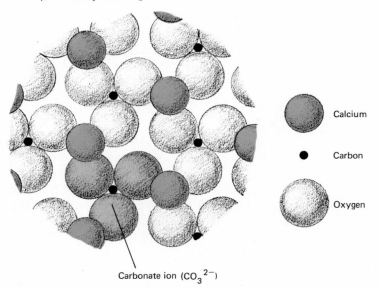

Calcium

Carbon

Oxygen

Carbonate ion (CO_3^{2-})

    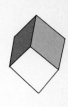

Formation under decreasing temperature ⟶

FIGURE 3-21 As a crystal is assembled, a plane oriented in one direction may accumulate atoms more rapidly than planes with other orientations. Since these relative growth rates can vary, the shapes of crystals may differ even though they are all the same mineral with the same atomic structure. More than 600 forms of calcite are known! Among the factors that govern these shapes during crystal growth are temperature and the presence of impurity ions.

3-21). Calcite may be deposited by animals as protective shell material, or it can precipitate from hot water solutions, or as dripstone in a cave, or in numerous other ways. A knife blade can scratch the surface of calcite; hydrochloric acid makes the mineral fizz violently (decompose, giving off carbon dioxide gas). These simple tests are handy for quick identification. Vast quantities of calcite are decomposed by heating to form quicklime (CaO), an ingredient of portland cement. More calcite goes into blast furnaces where it serves as a flux in the manufacture of steel.

Oxides In view of the abundance of oxygen, the importance of the oxide minerals comes as no surprise. *Hematite* (Fe_2O_3) and *magnetite* (Fe_3O_4) are prominent sources of iron. The structures of these two minerals are simple, but the atoms are organized into such dense patterns that they are difficult to diagram. Oxygen atoms in these iron oxides are arranged in a "closest packed" configuration (Figure 3-22). Iron atoms are concealed in certain of the small openings amongst the spheres of oxygen. The close packing, and the high density of iron, make hematite and magnetite the densest of the common minerals. Magnetite has no cleavage because all the layers of atoms are densely populated and knit together by strong chemical bonds. Magnetite is further distinguished by its magnetic properties. It can be attracted by a magnet, and may even act as a magnet.

Sulfides Copper, silver, zinc, lead, nickel, and other economically important metals are commonly found in nature bonded to the sulfide

Hematite   Magnetite

FIGURE 3-22 Two ways to pack equal-sized spheres as close together as possible are illustrated by the way oxygen atoms are positioned in hematite and magnetite.

(S^{2-}) ion. By far the most abundant sulfide mineral is *pyrite*: FeS_2. This opaque, dense, brassy yellow mineral ("fool's gold") is a source of both iron and sulfur where it occurs in large masses. The atomic structures of pyrite and of sodium chloride are similar in an interesting way (Figure 3-23).

Sulfates   In this family of minerals, the characteristic anion is SO_4^{2-}. Sulfur in this situation is about the same size as the Si^{4+} ion. Therefore, in what geometric pattern would you expect the four oxygen atoms to be grouped about sulfur in the sulfate ion? *Gypsum* ($CaSO_4 \cdot 2H_2O$), our chosen example of a sulfate mineral, is usually deposited by the intense evaporation of seawater. Water molecules occupy regular prescribed positions in the crystal structure; they "really belong," and are not present simply as an impurity. (Gypsum is a hydrous mineral.) Gypsum has excellent cleavage because the bonds between layers of water molecules and the neighboring atoms are easily broken. The mineral is so soft that a fingernail can scratch it. Commercial uses of gypsum include the manufacture of wallboard, cement, and fertilizer.

Halides   Another mineral group contains ions of the halogen elements: fluorine, chlorine, bromine, and iodine. Like gypsum, many of these halide minerals are quite soluble; they too form as a residue after seawater has evaporated. Easily the most common representative is rock salt, *halite* (Figures 3-10 and 3-23). Salt is used to season food, to melt ice from roads, and as a raw material in making chlorine gas and many other products.

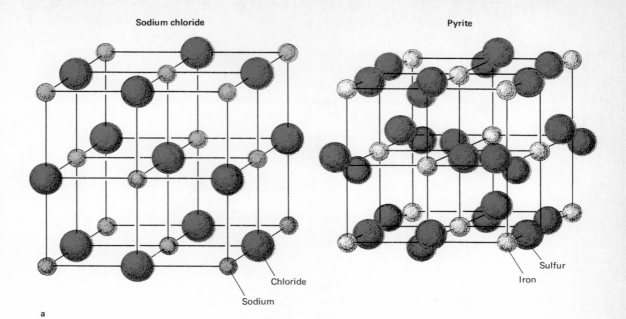

Sodium chloride

Pyrite

Chloride

Sodium

Sulfur

Iron

a

b

FIGURE 3-23 (*a*) Crystal structures of sodium chloride and pyrite are similar. In pyrite, we find iron instead of sodium, and dumbbell-shaped clusters of two sulfur atoms instead of chloride ions. (*b*) A common shape of pyrite crystals is the cube—the same as the shape of the unit cell.

73 MINERALS

Other mineral groups Anions in other minerals may contain vanadium, arsenic, phosphorus, uranium, boron, or bismuth as examples from a host of possibilities. Many of these minerals are vital sources of rare metals necessary to modern technology, but their abundance in the earth's crust is insignificant. Finally, some minerals are simply chemical elements in the free, or native, state. Prominent examples are gold, sulfur, and graphite and diamond (forms of carbon).

FURTHER QUESTIONS

1   Refer to Figure 3-10. The second Pauling rule states that anions (Cl^-, in this instance) are grouped about cations (Na^+) in a crystal. Why is it inconvenient to think about cations grouped about anions?

2   Refer to Figure 3-15*a*. What is the ratio of silicon to oxygen atoms in a single chain? (Recall that the silicon atoms are buried from view.) Count the oxygen atoms touching a silicon atom, but assign only half-credit to each oxygen that is shared between tetrahedra. The correct answer appears in Table 3-2.

3   Refer to Figure 3-16. What is the proportion Si/O in double chains? (*Caution:* There are both "interior" and "exterior" tetrahedra in this chain.) For the correct answer, consult Table 3-2.

4   Refer to Figure 3-17. What is the Si/O ratio in sheets? Correct answer: Table 3-2.

5   Refer to Figure 3-24, showing skeleton outlines of tetrahedra and silicon atoms contained within. A third Pauling rule observes that tetrahedra always link together at corners (*a*), *never* along edges (*b*) or faces (*c*). Why is this so? Note how the silicon-to-silicon distances decrease in going from *a* to *b* to *c*. Does this provide a clue to the answer?

FIGURE 3-24

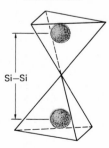

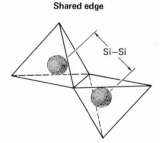

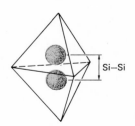

Shared corner	Shared edge	Shared face
One oxygen atom in common	Two oxygen atoms in common	Three oxygen atoms in common
a	b	c

READINGS

Listed in approximate order of technical difficulty.

Desautels, Paul E., 1968: *The Mineral Kingdom*, Madison Square Press, New York, 251 pp.

Hurlbut, Cornelius S., 1968: *Minerals and Man*, Random House, Inc., New York, 304 pp.

Sinkankas, John, 1964: *Mineralogy for Amateurs*, Van Nostrand Reinhold, New York, 585 pp.

Bragg, William L, 1937: *Atomic Structure of Minerals*, Cornell University Press, Ithaca, N. Y., 292 pp.

chapter 4

Born of Fire

Lightning display at night accompanying an
eruption of the new volcano Surtsey, on the
coast of Iceland. [*Courtesy Sigurgeir Jónasson.*]

The study of rocks is the major part of geology, though of course it is far from the whole subject. So varied, so immense, so complexly interrelated is the rocky substance of the earth that in a sense we can understand none of it without knowing all the other aspects. We might ask how the unique size, shape, and composition of any rock body came about. What role did living organisms play in its origin? What effect did energy sources—heat, magnetic, gravitational, seismic—exert on it? This chapter and the next take up the theme with a look at some of the general features of rocks. Following this we shall weave together yet more and more of the fabric of geology, finally bringing the entire earth into view.

We noted that rocks are composed of minerals, chiefly representatives of the several families described in Chapter 3. Rocks can be assigned into four groups whose origins differ. Like all other classifications, this one is artificial and somewhat indistinct; nevertheless, it is useful enough to have been accepted universally by geologists.

Originating as a clump of *cosmic matter*, the earth was probably converted wholly (or nearly so) into *igneous* rock when it melted and differentiated. At the same time, the beginnings of an ocean and atmosphere made their way to the surface. Immediately the water and air started to attack the igneous rock, weathering it into a variety of new minerals. Weathering broke up the rock mechanically, and also changed it chemically. Products of weathering were transported by wind, moving water, or ice, and deposited as *sedimentary* rocks. Not even these rocks are the end of the line, geologically speaking. Any rock can be transformed under high temperature and pressure into *metamorphic* rock, the final class. If subjected to extremely intense metamorphism, the rock may melt, thus returning the cycle to the igneous stage. Weathering, deposition, metamorphism, melting, solidification—these processes have been reenacted continually ever since those earliest times. Let us explore some aspects of the first two classes of rock in more detail.

77

COSMIC MATTER

Cosmic matter, the original condensate of the solar system, is found on earth today only in the form of meteorites. A rather trivial 100,000 tons or less of cosmic material are estimated to fall upon the earth each year. Once this substance has been weathered, there is no way it can be restored to its original state. On the other hand, igneous, sedimentary, and metamorphic rocks can be converted, one type into another, and back again.

IGNEOUS ROCKS

The second type, igneous* rock, solidified from melted material, or *magma*. The magma may have been wholly liquid, or perhaps a hot mush of crystals suspended in a silicate melt charged with dissolved H_2O and other fluids. Magma that breaks through the earth's surface is called *lava*.

Ironically, although igneous rocks comprise at least 80 percent or more of the volume of the crust, we have never been able to make direct observations of the environment in which magma is forming. (Magma is generated at depths in the earth of at least several kilometers, and of many tens of kilometers in extreme instances.) Therefore, studies of the origin of igneous rocks have necessarily had to proceed along indirect lines.

In the Field and in the Laboratory

An obvious place to begin with rocks of all kinds is out where they are—in what is familiarly called the "field." Systematic field studies began in the 1700s, and today, research in the field is the total occupation of thousands of geologists. Geologic fieldwork almost always includes preparing a map of a selected area and describing the various rocks that are found there (Figure 4-1). Additional studies may include a microscopic examination of specimens, chemical analyses, and identification of fossils. All these data must be blended together into a sensible interpretation of the geologic relationships.

Field investigations have provided truly staggering amounts of information. For one thing, they have shown us what is actually "out there" needing to be explained. They make it possible to piece together a sequence of past events. They have also pointed to generalities that are significant far beyond the confines of a local district. For example, field studies have established that certain types of igneous rock are as-

* The Latin word *ignis* means "fire."

a

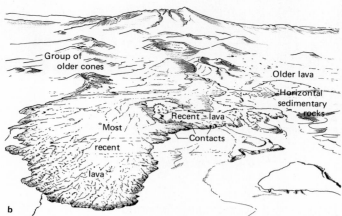

Group of
older cones

Older lava

Horizontal
sedimentary
rocks

Recent lava

Most

Contacts

recent

lava

b

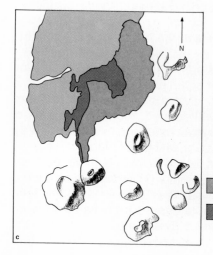

c

N

FIGURE 4-1 An oblique aerial photograph (*a*) southward across the San Franciscan volcanic field, near Flagstaff, Arizona, is interpreted in a drawing (*b*). Lava flows of different ages are easily distinguished from one another by superposition and by relative freshness of appearance. [*Photograph from* Geology Illustrated *by John S. Shelton and line drawing based upon* Geology Illustrated. *W. H. Freeman and Company. Copyright ©️ 1966.*] A geologic map of the foreground area (*c*) indicates four rock types and several of the volcanic cones. The map eliminates the foreshortened perspective of the aerial photograph. Moreover, the map is based on a careful examination of the rocks. In most areas, the resemblance between a photograph and a geologic map is much less obvious than in this example. [*Adapted from Harold S. Colton,* Cinder Cones and Lava Flows, *Northland Press, Flagstaff, Ariz., 1967.*]

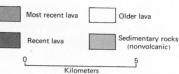

Most recent lava

Older lava

Recent lava

Sedimentary rocks
(nonvolcanic)

0   5

Kilometers

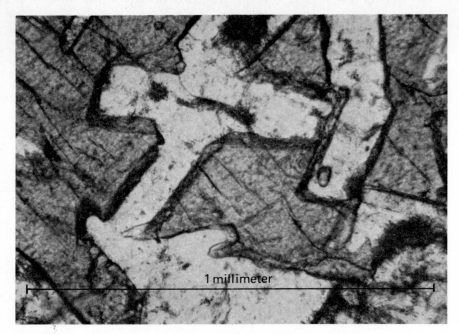

1 millimeter

FIGURE 4-2 A chip of rock that has been sliced to a thickness of 0.03 millimeter (much thinner than this page) can be examined under the microscope. When polarized light is passed through the slice, many otherwise dull-appearing minerals are clothed in brilliant colors that aid in their identification. In this microscopic view of an igneous rock, well-formed, bladelike crystals of plagioclase feldspar are surrounded by formless pyroxene crystals that fill the remaining space. [*Courtesy Douglas Smith.*]

sociated together over and over around the world, suggesting that in each place, the same processes must have created these distinctive rock assemblages.

Sometimes, however, the careful descriptions that are central to the field studies must be supplemented by geologic research that takes another line of approach. For example, consider the important question of the order in which different minerals begin to crystallize as a magma freezes. A relationship frequently seen in igneous rocks that crystallize at shallow depth is the presence of well-formed, bladelike crystals of feldspar set in the midst of irregular pyroxene crystals (Figure 4-2). Which formed earliest? Logically, it would seem that plagioclase crystallized first, then pyroxene, which was forced to solidify in all the open spaces that were left in the congealing liquid. Observations of cooling lava proved the situation quite otherwise. They showed that pyroxene and feldspar develop together, and that feldspar simply has a superior ability to form straight crystal boundaries. The example points up the need both to learn what the rocks *are* and, if possible, to establish how they *became* that way.

How can we accomplish the latter goal? Another useful technique is to simulate the melting and solidification of igneous rocks in the laboratory. Most of the early experiments were carried out at normal atmospheric pressure—an environment similar to that in a cooling lava flow. In order to reproduce conditions deep in the earth, we must use an apparatus that can apply a high pressure to the sample at a high temperature. (With increasing depth, both pressure and temperature increase.) Development of tough, heat-resistant alloys capable of withstanding such intense physical conditions made possible the first modern high-pressure experiments in the early 1950s. Today, some of these specialized machines can duplicate conditions 600 kilometers beneath the earth's surface!* To our knowledge, nearly all igneous rocks originate at depths shallower than 300 kilometers, under conditions that can easily be attained by the apparatus shown in Figure 4-3.

* A pressure of 200,000 atmospheres at a temperature of about 2000°C. Here, the term *atmosphere* means a unit of pressure equaling the air pressure at sea level, or about 14.7 pounds per square inch (1.03 kilograms per square centimeter).

FIGURE 4-3 In this massive high-pressure equipment, hydraulic rams converge on a sample assembly located near the center of the apparatus. This machine can exert a pressure of more than 50,000 atmospheres while the sample at the same time is being heated by an electrical resistance furnace. [*Courtesy Arthur Boettcher.*]

In a typical experiment, carefully measured amounts of simple, pure chemical compounds are reacted together at a known high temperature and pressure. After the reaction is complete, the sample is suddenly "quenched"—restored to room conditions. Often the starting materials have changed into an assemblage of minerals looking like some actual igneous rock. For example, a mixture of CaO, MgO, Al_2O_3, and SiO_2 might turn into a set of the minerals olivine plus pyroxene plus plagioclase feldspar.

Laboratory simulations have clarified many knotty problems that long have puzzled the field geologists. Among other things, they have shown why only certain minerals occur together in igneous rocks. Minerals that are known to be incompatible in nature were found to react together at high temperature. For example, quartz and olivine might react to form pyroxene. In doing so, one or both of the original minerals disappear. The laboratory studies have also spelled out many other details of igneous rocks: the probable composition of rocks in the interior of the earth, the temperatures and depths of origin, the changes that take place during crystallization, etc.

The very simplicity of the controlled experiments is at the same time a strength and a weakness of the method. It is much easier to begin a study by exploring the behavior of a simple mineral "system," but the simple system might not correspond very closely to reality. More elaborate experiments are being carried out, but probably another 50 years will go by before we can claim to have a thorough experimental understanding of igneous rocks.

Volcanic Rocks

Few natural events have ever terrified man more than do major volcanic eruptions. They are so powerful and so unpredictable! Legends of great eruptions can be traced back thousands of years to times when traditions were passed down only by word of mouth. Some of the early Greek and Roman writers described volcanic phenomena quite accurately. Other philosophers, convinced that volcanoes are sustained by the fiery breath of giants, or that they are gateways to Vulcan's underground forge, were carried away by enthusiasm for some ideas that a few observations would easily have dispelled. Systematic study of volcanoes began early in the eighteenth century. Like other sciences, volcanology developed in spurts as new interpretations and techniques made their appearance. For many decades, volcanologists have performed the necessary but preliminary task of describing and classifying volcanic phenomena—a worthy pursuit that will, of course, continue. Within the past 20 years or so, we have become aware of the importance of volcanism in shaping hitherto little explored regions such as the ocean basins and the surface of the moon. Laboratory studies of the physics and chemistry of volcanic rocks have added a new dimension to our understanding. Now at last a truly satisfying perspective of the role of volcanism in space and time is beginning to emerge.

Some comments   The earth today seems to be experiencing unusually intense volcanism (Figure 4-4). In Iceland, 200 young volcanoes have erupted within the past millennium. On a world basis, roughly 3500 separate eruptions have been recorded since A.D. 1700, many of which destroyed valuable agricultural or other property. About 520 volcanoes around the world are considered to be active; that is, they have erupted within historic times and appear likely to erupt again. Only a mere handful erupt constantly.

The list of active volcanoes might increase severalfold if submarine eruptions could be spotted more effectively. There is but a single recorded sighting of smoke rising from the ocean surface above a deepwater eruption, in spite of abundant evidence for frequent and copious outpourings of lava upon the ocean floor. (A number of volcanoes have erupted beneath shallow waters near the shores of Iceland, Japan, and elsewhere.) On the average, some 30 or 40 cubic kilometers of volcanic material are added to the top of the crust each year, mostly in underwater eruptions.

Active lifetimes of volcanoes may vary greatly. Undoubtedly, most volcanoes are like the famous Parícutin, in west central Mexico, which suddenly burst forth in a cornfield in 1943. Parícutin quickly built a cone to a respectable height of 400 meters, but within a brief 9 years after its birth, the volcano was extinguished, probably forever. In contrast, the active life of the giant volcanoes may be 10 to 20 million years. An example of long-lived volcanoes are the Hawaiian Islands, which (including the vast undersea portion of their bulk) easily rank as the largest single mountain peaks on earth.

Variations in composition and eruptive style among volcanoes are impressive. An eruption may bring forth a quietly flowing liquid or an explosive outpouring of already solidified particles. Lava may issue from a central crater atop a volcanic cone, or from an elongated fissure in the ocean floor. The congealed lava may be rich in dark ferromagnesian minerals, or consist of feldspar and quartz. In any case, the igneous rock emerges through the earth's crust: it is *extrusive* igneous material.

Measurements show that, in different environments, a magma may contain from 3 or 4 percent to as much as 17 percent of its weight in dissolved water. (Just as water can dissolve other substances, it can be dissolved *in* other substances.) What would happen when pressure is suddenly released as the magma reaches the surface? The outcome of this radical change may be awesome indeed. In 1883, the volcano Krakatoa nearly disappeared in a series of violent explosions that have been called "the loudest noise on earth" (Figure 4-5). Some 20 cubic

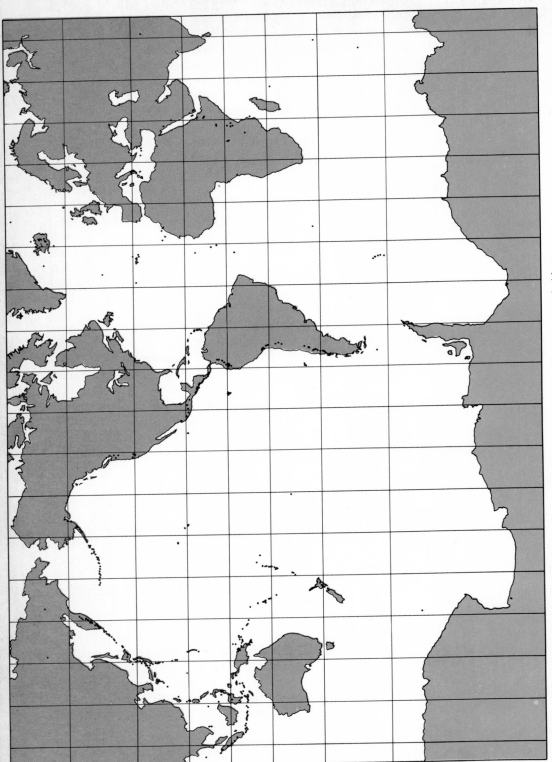

FIGURE 4-4 Many, perhaps most, of the world's active volcanoes are unknown, concealed from view beneath the ocean waters. The "ring of fire" bordering the Pacific Ocean is especially prominent. Volcanoes are abundant in the Mediterranean region and in Africa, but are seldom found in the interiors of the other continents. [*After Gordon A. Macdonald, Volcanoes, Prentice-Hall, Inc., Englewood Cliffs, N.J., 1972.*]

kilometers of material were blasted skyward by energy equivalent to the explosion of 200 million tons of TNT. In other, more quiet eruptions, the importance of volcanic gases may be overlooked as they continue to escape practically unnoticed. According to one study of an eruption in 1906, Mount Vesuvius (Italy) expelled a mass of gases greater than the combined total of lava and volcanic dust.

The solid rock remaining as the only visible evidence of a volcanic explosion also bears the imprint of the departed gases. Two common types of fragmental material are light frothy glass (pumice) and deposits of pulverized fine particles (volcanic ash: not to be confused with

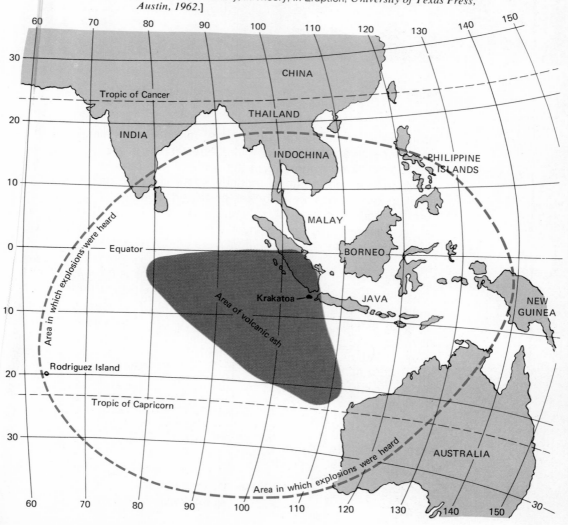

FIGURE 4-5 When superheated water flashes into steam, its volume increases at least a thousandfold. Effects of the catastrophic disruption of Krakatoa, both the noise and the spewed-out volcanic ash, spread outward great distances. Sounds of the explosion traveled about 4 hours before reaching the outer fringe of the area of detection! [*After Fred M. Bullard, Volcanoes: In History, in Theory, in Eruption, University of Texas Press, Austin, 1962.*]

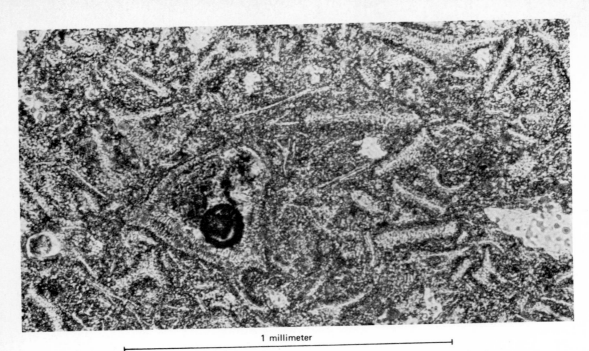

1 millimeter

FIGURE 4-6 Sharp, curved fragments in this volcanic ash are shards of glass, set in a matrix of extremely fine particles of glass and crystalline material. [*Courtesy Eric Swanson.*]

products of combustion). Under the microscope, minute delicate shards, or curved platelets of glass, are visible in the volcanic ash. They are shattered fragments of the walls of gas bubbles (Figure 4-6).

Much has been learned recently about volcanic gases, once the sampling techniques were further perfected. Methods had to be devised to collect and analyze the intensely heated, often furiously rushing gases on the spot without burning up the experimenter. One must not wander too close to nature's version of Hades!

Basalt   By far the most abundant volcanic rock is *basalt* (ba-SALT), a product of a generally quiet style of eruption. Basalt underlies the ocean basins and builds volcanic edifices such as Iceland and the Hawaiian Islands upon the ocean floor. On the continents, flows of basalt cover millions of square kilometers in the northwestern United States, in India, in Brazil, and elsewhere. We see evidence in the geologic record that basalt eruptions have been ·common throughout earth history.

Basalt is made of pyroxene, plagioclase feldspar in which calcium is the chief cation, and minor amounts of magnetite or other members of the iron-titanium oxide family. Some basalts also contain olivine or representatives of framework silicates similar to feldspar. Consequently, basalt is a dark, dense rock.

FIGURE 4-7 The front of an approaching basalt flow, here about 4 meters thick, advanced typically at a rate of about 400 meters per hour during the 1955 eruption of Kilauea (Hawaii). Contrary to popular opinion, lava cannot "flow like water." Measurements prove that even the most fluid lava is at least 100,000 times more viscous than water. [*Photograph by Gordon A. Macdonald, in* Basalts: The Poldervaart Treatise on Rocks of Basaltic Composition, *vol. 1, Interscience Publishers, a division of John Wiley & Sons, Inc., New York, 1967.*]

Basaltic lava is vented to the surface at about 1200°C; by the time it has cooled to 900°C, within a matter of minutes to at most a few weeks, the basalt is completely solid. (It may remain too hot to touch for many years.) When the rock begins to solidify, innumerable tiny seed crystals appear throughout the mass. As each crystal grows larger, it competes for more material with about the same degree of success as its neighbors. The end result is a rock so uniformly fine-grained that individual crystals can be distinguished only under a microscope. Some basalts contain large crystals of olivine embedded in the fine matrix. Can you suggest how this somewhat more complicated texture could have been created?

A notable property of liquid basalt is its low viscosity, which permits the lava to flow long distances before solidification brings it to a halt* (Figure 4-7). Viscosity is a measure of the rate that a substance will deform under the influence of shearing forces. In more descriptive terms, viscosity is related to the "stiffness" of a liquid. You will recall that silicon-oxygen tetrahedra are not very strongly linked in pyroxene, and not linked together at all in olivine. Tetrahedral fragments, or perhaps short segments of chains, are present in a disorganized state, able to slip past one another easily in the runny basalt liquid.

Types of eruption The viscosity of a magma may vary between such extremes that this factor becomes a natural basis on which to classify volcanic eruptions. For example, the *Icelandic* type of eruption does not even form volcanic peaks.† Instead, the lava wells up out of fissures to flow as sheets across the countryside, flooding stream

---

* Single flows extending 300 kilometers or more are known.

† Iceland does contain many conventional volcanoes, however.

[Cross section]

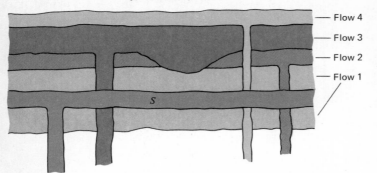

— Flow 4
— Flow 3
— Flow 2
— Flow 1

S

FIGURE 4-8 Many volcanic fields are complicated mazes of lava flows laced by dikes that were once conduits that fed lava upward to flows higher in the sequence. This cross section suggests that a long time interval elapsed after flow 2 was deposited. During this period, erosion carved a stream drainage on the land surface. Later the stream was disrupted by being filled with lava.

valleys and plains and devastating the works of mankind. The Icelandic style probably accounts for a greater volume of volcanic rock than all other types combined.

From time to time the conduits shift to new positions as later sheets are erupted (Figure 4-8). Solidified lava in the abandoned fissures forms tabular (book-shaped) bodies that cut across the previously deposited lava flows. These transecting bodies, or *dikes*, illustrate another useful way to tell the sequence of geologic events. The *law of crosscutting relationships* states that *a rock body must be older than another rock that has been emplaced into it.* Are the relative ages of the flows and feeder dikes in Figure 4-8 consistent with the laws of superposition and of crosscutting relationships?

During its ascent, the magma might encounter a buried surface such as the contact between two strata. If the magma is injected along such a contact, it forms another type of sheetlike igneous body, a *sill.* Dikes and sills are similar except that a dike crosscuts the local structure, whereas a sill runs parallel to it. A sill may look deceptively like just another lava flow in a layered sequence, whereas it is actually younger than the lava flows both above and below it. Does this relationship violate the law of superposition? How can you tell that rock *S* (Figure 4-8) is in fact a sill?

The *Hawaiian* type of eruption differs considerably from the Icelandic type. Here too the basalt erupts from fissures, but the same fractures continue to pour forth lava for a very long time. Thick, lens-shaped piles of lava whose gentle slopes are no steeper than a few degrees accumulate as *shield volcanoes* (Figure 4-9). Eruptions of the active Hawaiian shield volcanoes, Kilauea and Mauna Loa, come from long continuous rifts, sometimes from a part of the rift near the summit, and sometimes from a section of the rift situated farther down on the flanks. These volcanoes are built up of thousands of individual flows, each only a few meters thick.

Kilauea has become a classic locality for volcanic studies, thanks to the ongoing activities of the Hawaiian Volcano Observatory operated

5 kilometers
0
−5 kilometers

Mauna Loa

Kilauea

Ocean

Ocean

FIGURE 4-9 A profile through Hawaii, without vertical exaggeration, shows the gentle slopes of shield volcanoes that have merged together to form the island. Most of the volcanic mass is below sea level. [*After Fred M. Bullard*, Volcanoes: In History, in Theory, in Eruption, *University of Texas Press, Austin, 1962.*]

by the U.S. Geological Survey. At the summit of the volcano is located an oval-shaped depression about 3 by 4 kilometers across and up to 400 meters deep (Figure 4-10). This large *caldera** is evidently a collapse feature formed by the withdrawal of magma from below. At times the caldera of Kilauea holds a seething, rising and falling, lake of lava. Occasionally the lava spurts high as fountains or "curtains of fire." In spite of their spectacular appearance, the lava fountains do not pose an explosion hazard.

Careful measurements show that Kilauea inflates gradually before an eruption, then deflates rapidly when the lava drains from the summit caldera, or during a flank eruption. The top of Kilauea is a surveyor's nightmare; reference marks may shift horizontally and vertically by as much as 0.1 meter from year to year. The observations can be described by a mathematical model that postulates a reservoir located within the shield volcano about 3 kilometers beneath the summit. Emptying and filling of this reservoir cause the volcano to shrink and swell. The reservoir, which probably is a dense network of intersecting dikes

*The Spanish word *caldera* means "kettle."

FIGURE 4-10 Incandescent lava spurts upward through a lava lake in Halemaumau, a deep pit within the caldera of Kilauea. The wall in the background is constructed of numerous thin, solidified lava flows. [*From Willie T. Kinoshita, Robert Y. Koyanagi, Thomas L. Wright, and Richard S. Fiske, "Kilauea Volcano: The 1967–68 Summit Eruption,"* Science, *vol. 166, no. 3904, 1969. Copyright 1969 by the American Association for the Advancement of Science.*]

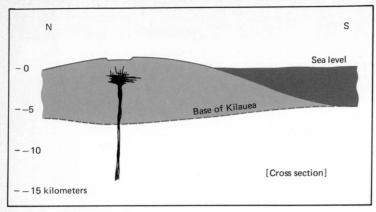

a

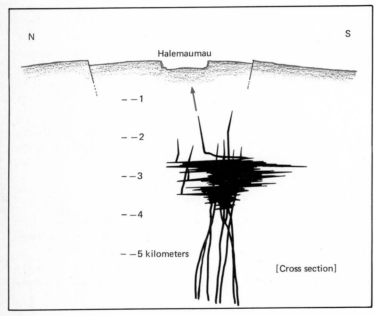

b

FIGURE 4-11 (*a*) A secondary magma chamber, fed from below, is located within the lens-shaped mass of Kilauea. (*b*) In detail, the chamber probably consists of a dense network of mutually intersecting dikes and sills. [*After Richard S. Fiske and Willie T. Kinoshita, "Inflation of Kilauea Volcano Prior to Its 1967–1968 Eruption," Science, vol. 165, no. 3891, 1969. Copyright 1969 by the American Association for the Advancement of Science.*]

and sills (Figure 4-11), is only a collecting spot, not the point of origin of the magma. Rather indirect evidence suggests that the magma is "sweated out" of rocks located at a much deeper 30 to 50 kilometers below the surface.

A third eruptive type, the *Strombolian*, is named after Stromboli volcano, which is located between Sicily and mainland Italy (Figure 4-12). Stromboli is one of the few volcanoes that sustains a constant, mild eruption. Apparently it has continued for at least the past 25 centuries. Strombolian lava is puffed out as showers of thick, pasty clots

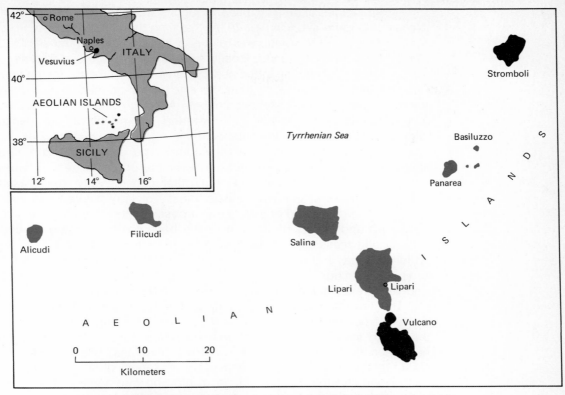

FIGURE 4-12 Locations of Stromboli, Vulcano, and Vesuvius. [*After Fred M. Bullard*, Volcanoes: In History, in Theory, in Eruption, *University of Texas Press, Austin, 1962.*]

accompanied by a dense cloud of white steam. At the summit of Stromboli is a *crater*, a closed volcanic depression that differs in size and origin from a caldera. A crater is simply the vent through which the material is erupted, and as such it is rarely more than 1 kilometer across, in contrast with a caldera, which may be up to 30 kilometers or more in diameter.

The *Vulcanian* type of eruption represents yet another long step in the direction of explosive violence. Named for the Roman god of fire, Vulcano is an island about 50 kilometers from Stromboli (Figure 4-12). A Vulcanian eruption is more infrequent but more destructive than the Strombolian type. At the beginning, pressure builds up in the blocked throat of the volcano until a series of major explosions hurls aloft ominous black clouds of hot but already solidified particles—dust, volcanic ash, and chunks of pumice. Torrential thunderstorms often accompany a Vulcanian eruption, probably because the fine dust particles "seed" the condensing raindrops. Once cleared, the vent may send forth a great volume of lava.

91 IGNEOUS ROCKS

Most famous of all Vulcanian eruptions was the outburst of Mount Vesuvius (Figure 4-12) in A.D. 79. At that time the coastal town of Pompeii at the foot of Vesuvius was plunged into complete darkness for 3 days. An avalanche of mud, locally exceeding 20 meters in thickness, caused more destruction in the nearby community of Herculaneum than did the fall of ash directly from the sky. A fine-grained rock, *tuff*, is deposited by Vulcanian-type eruptions. Although directly of volcanic origin, tuff is usually reworked (converted into mudflows by intense storms) just as it was at Herculaneum.

Fortunately the final, most destructive type of eruption is a rare event. For years, geologists puzzled over the origin of certain widespread layers of volcanic rock that differ considerably from basalt flows. The rocks in question consist mostly of light-colored minerals, especially feldspar and quartz. Many deposits contain flattened pumice fragments and shards of glass that have been compacted and welded into a solid mass. Up to that time, the only known extrusive rocks in which feldspar and quartz predominate were extremely viscous domes or spinelike plugs. How, then, could a single sheet like one in Mexico and west Texas come to extend across 10,000 square kilometers?

The first clue to this intriguing problem came with totally unexpected ferocity in Martinique, a Caribbean island. At the northern end of the island stands Mount Pelée, whose slumber for hundreds of years had been punctuated by only two minor eruptions. Then, on May 8, 1902, renewed eruption culminated in a series of four deafening explosions that hurled out a cloud of glowing fragments entrained in superheated steam. Traveling downslope at 150 kilometers per hour, the cloud instantly wiped out 30,000 lives in a town at the base. In these *Peléan*-type eruptions, a surge of "fluidized" material keeps close to the earth (Figure 4-13). As the hot fragments rush along, they continue to release dissolved steam in which they remain suspended. Unless high topography intervenes, the dense avalanche may travel on and on for many kilometers. The great mountainous backbone of western Mexico is built up of the world's largest continuous exposure of *ignimbrite*, the product of Peléan-type eruptions. No catastrophic event in the history of man remotely approaches the intensity that must have marked these eruptions.

From the Icelandic type to the Peléan type of eruption we note a steady increase of explosive power related to the viscosity of the magma, hence its ability to trap steam. The ferromagnesian minerals of low viscosity, Icelandic-type lava, stand in contrast to the framework silicates—feldspar and quartz—which crystallize in long, entangled chains of silicon-oxygen tetrahedra. A volcano may switch from one type of eruption to another. For example, most of the time Parícutin behaved in the Vulcanian style, but occasionally it acted in a Strombolian manner. Can you postulate how changing conditions beneath the vent could bring this about?

FIGURE 4-13 In 1968, Mayon Volcano, Philippines, one of the world's most symmetrical cones, sustained a series of explosive Peléan-type eruptions. The vertically rising cloud is steam; the angry dark cloud beneath is a tempestuous avalanche of hot fragments surging downslope at speeds up to 50 meters per second. As usual, these eruptions were accompanied by spectacular lightning strikes and violent thunderstorms. [*Courtesy Bernardo Tolentino. Photograph by SIX-SIS Studio, Legaspi City, Albay, Philippines.*]

Intrusive Igneous Rocks

Not all magma reaches the surface. A second category, *intrusive* igneous rocks, crystallized at depths of several hundred meters to as much as about 20 kilometers. Rocks that formed at even greater depths are only extremely rare "foreign" pieces carried up in volcanic pipes by liquid lava. Naturally, intrusive rocks are more difficult to study because they have formed in the depths of the earth.

Granite and granodiorite *Granite* and a closely related rock, *granodiorite*, are the commonest intrusive igneous rocks. In contrast to basalt, most intrusive rocks are medium- to coarse-grained. Granite has cooled deep beneath an overburden that conducts away the heat extremely slowly. Not only did the first-appearing seed crystals have opportunity to grow large, but since they were kept at high temperature for perhaps millions of years, they tended to anneal together into fewer but larger crystals.

Granite consists of quartz, potassium feldspar, and somewhat smaller amounts of sodium-rich plagioclase feldspar. Scattered crystals of biotite, muscovite, or hornblende may be present too. Grano-

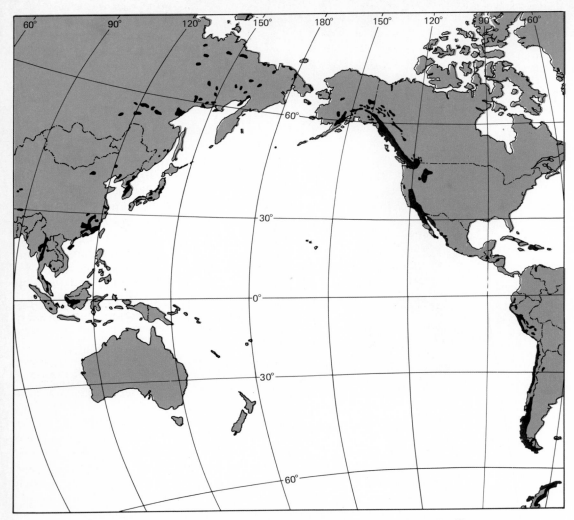

FIGURE 4-14 The batholiths (in black) that are marginal to the Pacific Ocean originated within the past 200 million years. Compare the distribution of batholiths to the distribution of active volcanoes (Figure 4-4). Could the batholiths, the volcanoes, and the Pacific itself be united in a common origin? [*After Paul C. Bateman and Jerry P. Eaton, "Sierra Nevada Batholith," Science, vol. 158, no. 3807, 1967. Copyright 1967 by the American Association for the Advancement of Science.*]

diorite differs only slightly; it has more plagioclase than potassium feldspar. Granite and granodiorite are rather light-colored, and of only moderate density (about 2.7 grams per cubic centimeter).

Though not so widespread as basalt, these two rock types underlie major portions of the continents, especially in regions where deep-seated forces have crumpled the rocks into mountain ranges. Enormous masses of granite and granodiorite are scattered around the margins of the Pacific Ocean (Figure 4-14). These *batholiths* are the largest nearly homogeneous bodies of rock accessible to man. The Canadian Coast Range batholith alone is the size of Kansas or Minnesota. Batholiths are found not only in mountainous regions. Vast areas of low elevation in each of the continents are floored by granite. These regions too were once mountainous, but they have been reduced to lowlands through the courtesy of millions of years of erosion. Ocean basins are practically devoid of granite and kindred intrusive rocks.

Granites and granites   What is the origin of granite and of its emplacement as batholiths? After two centuries of investigation and debate, geologists still cannot agree on the answer. Perhaps that is why they cannot—there is no one answer. Let us see how the field and laboratory studies suggest various possibilities of the origin of granite.

First, the batholiths must have formed at a considerable depth. In some regions it can be established that thousands of meters of rock have already been eroded away, and that some batholiths extend downward at least several additional kilometers. Many batholiths cut abruptly across the enclosing, or "host," rock (Figure 4-15). The sharpness of the "transgressive" or "discordant" contact suggests that the host rock, though strongly heated by the invading magma, nonetheless remained rigid enough to be punched through cleanly. In fact, the appearance of the host rock may not have been changed at all.

Other batholiths are not so easy to interpret. Indistinct margins of these bodies seem to grade by degrees from host rock to rock that appears increasingly granitelike and finally into unmistakable granite. In some places the rock has become a curious "half-and-half" mixture of contorted host rock saturated with streaks of granite (Figure 4-16). The field geologists rightly insisted that this granite must have stayed more or less in the same spot where it formed. Somehow the distinction here between invading magma and the original terrain has become rather blurred.

FIGURE 4-15 This intrusive, light-colored granite in Manitoba exhibits a sharp contact where it has invaded the dark-colored host rock (basalt).

FIGURE 4-16 Pods or thin layers of granite (light-colored) are intimately mixed with the host rock in this outcrop in Manitoba. Both rock types are contorted, as though squeezed like toothpaste while in a plastic condition. The helmet indicates the scale of the photograph.

After the debates of the 1930s and 1940s, geologists slowly came to recognize that there must be "granites and granites." How can we account for the different styles of emplacement? Could it be that the contrasting environments even once existed in the same batholith? At great depth, the host rock partly melted, while the remainder of it weakened into a softened, plastic mass. Some of the granite magma extracted from this zone moved upward as an intrusion into colder, more brittle host rock. Which type of granite we see today depends on how deeply erosion has cut down into the batholith at that place (Figure 4-17).

Laboratory simulations have also helped us to understand the origin of granite. Suppose we perform melting and crystallization experiments on the minerals that make up granite. It was early discovered that these minerals react with one another in complicated ways. The melting

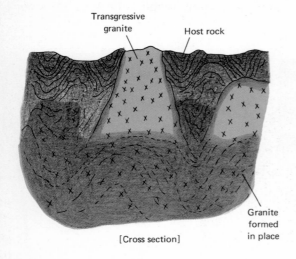

Transgressive granite

Host rock

Granite formed in place

[Cross section]

FIGURE 4-17 A cross section through part of the earth's crust shows a granite batholith blurring into its surroundings at great depth, but assuming ever more distinct identity toward the top. Host rock near the surface was forcibly shoved aside and distorted by the rising granite. At depth, the host rock was recrystallized (metamorphosed) and soaked by stringers of granite liquid. Long after the entire terrain had cooled and solidified, erosion laid bare the uppermost level of the batholith.

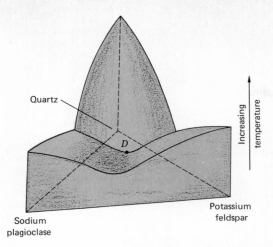

Quartz

Sodium
plagioclase

Potassium
feldspar

Increasing
temperature

D

FIGURE 4-18 A three-dimensional block diagram shows the temperatures at which heating would convert all possible combinations of quartz, sodium plagioclase feldspar, and potassium feldspar into liquid. This diagram is valid only if water vapor under high pressure is also present — a situation common in the depths of the earth.

temperature of a pure mineral may be far different from its melting or decomposition temperature in the presence of other minerals.

Nevertheless, experiments have already pointed to some important conclusions. Consider the rather odd-looking block diagram in Figure 4-18. The corners of this three-dimensional figure represent pure quartz, or potassium feldspar, or sodium-rich plagioclase feldspar, the principal minerals in granite. A point anywhere inside the figure represents some combination of these three minerals. The curved surface corresponds to the temperature (for each combination of these minerals) at which crystals begin to appear in the liquid. Above this surface, the temperature is too high for the liquid to contain crystals. This figure must not be taken for any physical reality; it is an abstraction that is useful in visualizing what happens as a granite magma crystallizes.

One of the findings of the experiments is that when a solid rock of *any combination whatever* of quartz and the two kinds of feldspar is heated, the first liquid to appear has a composition shown by point D, the lowest point on the surface of the figure. Analyses of many granite specimens from around the world plot close to this very point (Figure 4-19).

Whence granite? With the field and laboratory data at hand, we can now propose the origin of granite. In the mountain-making process, great masses of rock are crumpled together by deep-seated forces. At the base of the pile, the temperature is high enough to cause slight melting, perhaps just enough to form a microscopically thin film of silicate liquid coating the otherwise solid mineral grains. This "first-melting" film of granitic composition percolates upward, collecting into ever larger streams. In part the low-density liquid is carried up by its own buoyancy, but the streaming motion is greatly assisted by small quantities of water dissolved in the melt. Eventually, tongues and sheets of granitic magma combine into balloonlike masses (batholiths)

97 IGNEOUS ROCKS

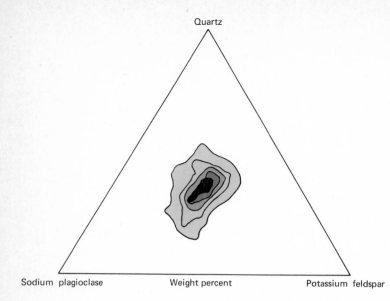

Quartz

Sodium plagioclase   Weight percent   Potassium feldspar

FIGURE 4-19 If the "first melting liquid" of composition *D* (Figure 4-18) were squeezed out of a heated rock and emplaced somewhere else, the magma would crystallize into granite having about equal proportions of quartz, sodium plagioclase, and potassium feldspar. Chemical compositions of 571 granites from around the world were recalculated to equivalent proportions of these three minerals. Compositions of 99 percent of these granites lie within the outer contour line in this figure; 94 percent of the granites lie in the black area. [*After O. F. Tuttle and N. L. Bowen, "Origin of Granite in the Light of Experimental Studies in the System* $NaAlSi_3O_8 - KAlSi_3O_8 - SiO_2 - H_2O$*,"* Geological Society of America Memoir 74, *Geological Society of America, Boulder, Colo., 1958.*]

that rise high into the earth's crust (Figure 4-17). The laboratory simulations also suggest that in most instances the first-melting liquid can be only a small fraction of the original rock. Therefore the batholiths, large as they are, must have been extracted from even much larger volumes of rock at depth.

FURTHER QUESTIONS

1   Volcanic eruptions bring material from the depths of the earth only *up* to the surface. Speculate on what process might take the rocks of the crust back down again. Or could there be some other way to "close up the hole" left when magma ascends from a deep source?

2   How would the following help to distinguish a buried lava flow from an intrusive sill: (*a*) A zone of very fine-grained igneous rock that signifies that the *base* of the igneous body was quickly chilled when it came in contact with cold host rock? (*b*) Chilled rock in a zone at the *top* of the igneous body? (*c*) Fragments that have been torn loose from the host rock, transported by the magma (or lava), and frozen into the igneous mass? (*d*) An upper zone in the igneous rock that is peppered with holes created by expanding steam? (*e*) Baked and hardened host rock at the contact with the *base* of the igneous body? (*f*) Baked host rock in contact with the *top* of the igneous mass? Does each of these features indicate unambiguously whether the igneous rock is a flow or a sill?

READINGS

Listed in approximate order of technical difficulty.

Gass, I. G., P. J. Smith, and R. C. L. Wilson (eds.), 1971: *Understanding the Earth: A Reader in Earth Sciences*, The M.I.T. Press, Cambridge, Mass., 355 pp.

Bullard, Fred M., 1962: *Volcanoes: In History, in Theory, in Eruption*, University of Texas Press, Austin, 441 pp.

Macdonald, Gordon A., 1972: *Volcanoes*, Prentice-Hall, Inc., Englewood Cliffs, N.J., 510 pp.

Green, Jack, and Nicholas M. Short (eds.), 1971: *Volcanic Landforms and Surface Features: A Photographic Atlas and Glossary*, Springer-Verlag OHG, Berlin, 519 pp.

chapter 5

Recycled Rocks

Sedimentary strata, deposited horizontally, have been tilted and eroded in a mountainous region in Yugoslavia. Deformation of rocks here and elsewhere has permitted study of great thicknesses that otherwise could be reached (if at all) only by drilling. [*Courtesy Robert Folk.*]

Among the inner planets, Mercury and the Moon are naked solids; Venus is shrouded in a dense, hot atmosphere devoid of liquid H_2O; no more than a trace of water is present in the rarefied air of Mars. Only the earth contains a well-developed ocean and atmosphere, and abundant life. We may be sure that the weathering of rocks and the deposition of sediments on earth have followed a richly varied course not possible in the neighboring planets. Even the processes of metamorphism are remarkably influenced by the presence of water. This chapter introduces the sedimentary and metamorphic rocks, the final two classes. These rock types demonstrate yet other means by which the substance of the earth's crust is continually recycled.

WEATHERING: A TRANSITIONAL STAGE

Weathered rock occupies a special niche in the total scheme of things. In places such as eastern Canada where ice has scoured the bedrock clean, the weathered layer is absent. Where intense tropical weathering has been prolonged, the layer may be hundreds of meters thick. This material has not been transported—it is not sedimentary rock—but neither does it look much like the rock beneath, from which it was derived. As the weathered zone is removed by erosion, it constantly renews itself by encroaching down into the bedrock. Like water in a waterfall, the weathering zone maintains the same general shape while the material of which it is made is constantly passing into it, then out again.

Weathering of Granite

Weathering of minerals in granite is especially striking because this rock forms in an environment so very different from the surface environment where weathering occurs. At great depth, many reactions proceed in a direction opposite to the pathway of the reaction at the

101

surface. For example, consider the differences in density and grain size. Moderate- to high-density minerals, with closely packed crystal structures, tend to form under high pressure. Low-density minerals are strongly favored at the surface. During weathering, the coarse mineral grains that are the rule in granite are reduced to smaller and smaller dimensions. A freshly exposed outcrop of granite merely decomposes into a mass of loose-fitting grains at first, but later on, new minerals appear whose dimensions range down to as little as a hundred-millionth of a meter.*

Accompanying these changes is a spectacular increase of the surface area of the material. For example, if a cube 1 centimeter on edge were finely divided into cubelets having the tiny dimensions mentioned earlier, the surface area would multiply a millionfold. Water, carbon dioxide, and oxygen can gain intimate access into every part of the rock. This condition too is unlike that in fresh granite, in which only a trace of water is found, locked away unreactively in the hydrous minerals biotite, muscovite, and hornblende.

Mechanical weathering   As erosion strips away the overburden, granite responds by popping apart into sweeping curved sheets that end up in a pile of rubble at the base of the slope. Bald granite hills, *exfoliation domes*,† are left as testimony to this process (Figure 5-1). Ice wedging and growing tree roots may also help to pry the rock apart. So does the increase in volume that occurs because chemical weathering has created a new set of minerals that are less dense than the original minerals. (Chemical alterations lead to physical changes as a side effect.) Once the mineral grains are separated from one another, they are highly subject to chemical disruption. Quartz is a striking exception to the rule, however. Quartz grains can be washed or blown about, but only a little quartz dissolves in the weathering zone.

* Individual crystals this small are visible only under an electron microscope.
† *Exfoliation* comes from the Latin meaning "to strip of leaves."

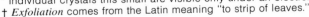

FIGURE 5-1   One of the most spectacular of all exfoliation domes is Half Dome, in Yosemite National Park, California. Its summit towers about 150 meters above camera level. Curved exfoliation slabs of this homogeneous igneous rock are 3 or 4 meters thick down to just a few centimeters thick. In the foreground, a boulder has partially decomposed into loose mineral grains. [*U.S. Geological Survey photograph by F. C. Calkins.*]

Chemical weathering In most environments, water mediates a chemical attack on rock that is much more profound than mechanical weathering. Consider the role of the hydrogen ion (H^+) which forms by dissociation of a small proportion of the water molecules:

$$H_2O \rightleftharpoons H^+ + OH^- \tag{5-1}$$

Water   Hydrogen   Hydroxyl
           ion   ion

Additional hydrogen ions appear when atmospheric carbon dioxide dissolves in water:

$$H_2O + CO_2 \rightleftharpoons H^+ + HCO_3^- \tag{5-2}$$

Water   Carbon   Hydrogen   Bicarbonate
      dioxide   ion   ion

Unique in its small size and highly concentrated electrical charge, the hydrogen ion is a potent wrecker of a silicate structure. For example, it participates in the destruction of feldspar along these lines:

$$2KAlSi_3O_8 + 2H^+ + 9H_2O \rightleftharpoons Al_2Si_2O_5(OH)_4 + 4H_4SiO_4 + 2K^+ \tag{5-3}$$

Potassium   Hydrogen   Water   Kaolinite, a   Silicic acid   Potassium
feldspar   ion [con-   clay mineral   (in solution)   ion (in
(framework   tributed by   (sheet structure)   solution)
silicate)   reactions
              (5-1) and (5-2)]

Since feldspar is the most abundant mineral, the most common of all chemical weathering reactions is the breakdown of feldspar into clay. Although both minerals are silicates, far more is required than a simple rearrangement of framework silicon-oxygen tetrahedra of feldspar into the sheet structures of the clay minerals. Details of the feldspar-clay conversion are not well understood. In any case, the silicon-oxygen tetrahedra are so extraordinarily stable that even after being torn completely apart, they are able to reassemble themselves.

Through weathering, the metal ions in feldspar (K^+ in the example just given, but also Na^+ and Ca^{2+}) have been *replaced* by hydrogen ions (H^+). (Therefore the weathering product, clay, is a hydrous mineral.) The released potassium, sodium, and calcium ions can enter into *solution* in water. For a while, vegetation may trap these ions, which are necessary to plant growth, but eventually they enter the ocean.

What happens to the ferrous iron that characterizes biotite and the other ferromagnesian minerals? When the rather soluble Fe^{2+} is met by air or by water containing dissolved oxygen, it immediately oxidizes to form Fe_2O_3 (hematite) or other iron oxide minerals. These iron compounds are well known as being some of the most *in*soluble of natural substances. Iron oxides can precipitate as a crust or penetrating stain that remains behind after most of the rock has disappeared. They are intensely colored in warm hues of red and yellow; even when present in trace amounts, they contribute much to the brilliant scenery of deserts.

TABLE 5-1
Weathering Series of Silicate Minerals, in Increasing Order of Stability

Discontinuous series*	Continuous series
Olivine	Calcium plagioclase
Augite (a pyroxene mineral)	(Intermediate plagioclase)
Hornblende	Sodium plagioclase
Biotite	
Potassium feldspar	
Muscovite	
Quartz	

* Crystal structures of minerals in the discontinuous series differ greatly, whereas a total reorganization of the atoms is not required as sodium replaces calcium in plagioclase. "Continuous" does not imply that weathering converts Ca plagioclase into Na plagioclase.
SOURCE: After Samuel S. Goldich, "A Study in Rock-weathering," *Journal of Geology*, vol. 46, pp. 17–58, 1938.

Weathering Series

The American geologist S. Goldich pioneered in studies of the relative susceptibility to breakdown of different silicate minerals (Table 5-1). To a remarkable degree, the order shown by the table is just the opposite of the order of stability in a magma. In a magma rich in magnesium and iron, olivine would tend to crystallize at a higher temperature than pyroxene or amphibole would. (Olivine is more stable at high temperature.) Similarly, calcium plagioclase freezes out of a cooling silicate melt before sodium plagioclase. The less stable mineral quartz crystallizes late, at a lower temperature. Yet quartz is one of the most persistent minerals against weathering, while olivine and calcium plagioclase are readily destroyed.

Soils

Much confusion has surrounded the description of soils, which are regarded differently by farmers, engineers, and geologists. Most maps of soil types have been made by agriculturalists, to whom a soil is the layer (typically about 2 meters thick) that can be penetrated by plant roots. To an engineer, soil is surface material that can be scooped out; it is contrasted with hard bedrock lying below. A geologist considers soil to be the end product of weathering. He is interested in the conversion of minerals into other groups of minerals, and the movement of dissolved ions or tiny clay particles in the weathered zone.

Soils are so complex that they are difficult to classify. Shall we group them on the basis of grain size, parent material, degree of maturity, color, chemical composition, fertility, thickness, slope of the land surface, or some other variable? The decision here may not be easy, for a good classification should give insight into the origin or history of whatever is classified. Often a useful classification can emerge only late, after the essential outline of a subject has been well mapped out.

Studies suggest that two variables, climate and the nature of the bedrock, dominate the evolution of soils. Immature soils naturally reflect mostly the differences in the bedrock from place to place. With greater maturity, the effect of life and climatically controlled chemical reactions begins to appear. Organisms leave their accumulated dead remains in the soil, and they excrete waste products such as CO_2, NH_3 (ammonia), and CH_4 (methane), all of them highly reactive chemicals. Indeed, the organic content of soil, plus its development in place (not transported), is what distinguishes soil from most sedimentary rock.

Soil scientists gradually came to recognize that local climate, not the character of the bedrock, governs the nature of a fully mature soil. (How would you test this hypothesis?) Maturity is marked by the development of soil "horizons" (layers). Although generally not all of the five horizons recognized by soil scientists are well developed in a given soil profile, these zones are shaped by basically the same processes everywhere. An uppermost, O horizon (often missing) is dark-colored, saturated with decayed plant debris. Below it lies the A horizon, from which percolating water has leached away the soluble material. Downward-seeping water has slowly enriched the underlying B horizon in deposited soluble or colloidal (extremely finely divided) material. (In most places the B horizon marks the deepest penetration of plant roots.) At still greater depth lies the C horizon, in which new soil is beginning to form. This zone is made up of fragments of partly decomposed bedrock merging down into the R horizon: fresh, hard bedrock (which is, strictly speaking, not part of the soil).

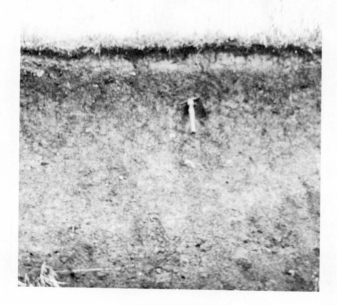

FIGURE 5-2 The leached A horizon in this soil profile in Minnesota is twofold: an uppermost zone darkened by decayed grass roots, and a lower, bleached horizon consisting of fine grains of quartz. Below it lies the dark B horizon impregnated with colloidal clay particles. Unweathered cobbles of bedrock (glacial debris) are visible in the C horizon at the base of the photograph. [*Courtesy Dwight Deal.*]

105 WEATHERING: A TRANSITIONAL STAGE

Variations on the theme of soil horizons seem to be almost endless. About 40 soil groups, each with related subgroups, are recognized in the United States alone. A useful classification takes note of various degrees of soil leaching, which in turn are related to temperature and rainfall patterns. In the semiarid Great Plains and farther west, the deeper soil horizons are enriched in vast accumulations of caliche (calcium carbonate), which in extreme cases is cemented into a solid layer. Measurements show that even where the bedrock does not contain calcium carbonate, windblown dust alone is able to account for the caliche deposit. Soils whose *B* horizon is enriched in *cal*cium carbonate are aptly named *pedocals*.

In more humid regions, especially east of the Mississippi River, the soluble Ca^{2+} ions are leached entirely from the soil. The *B* horizon in this area is enriched in particles of insoluble iron oxide and clay (aluminum-rich minerals). Hence these soils are nicknamed *pedalfers* (a play on *al* for aluminum and *fer* for ferric iron).

Another important soil group is called *laterite*, after the Latin word for "brick." Like most brick, laterites are colored yellow to rusty red by a high concentration of iron oxides. Commonly the *B* and *C* horizons are missing, and the severely leached *A* horizon may be almost nothing but insoluble oxides. Lush tropical jungles that grow on laterites conceal the unhappy fact that they are among the most impoverished soils on

FIGURE 5-3 Lateritic soils (black) are confined mainly to the tropics. [*After R. Ganssen and F. Hädrich,* Atlas zur Bodenkunde, *Bibliographisches Institut AG, Mannheim, 1965.*]

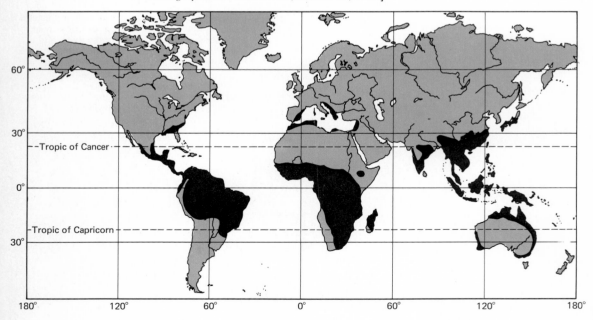

earth. Sustained heat and alternate wetting and drying encourage the soil bacteria to scavenge the organic matter out of this soil with the utmost efficiency. Dense vegetation grows in the tropics because of high rainfall and *in spite of* the infertile soil. Lateritic soil underlies a vast region inhabited by perhaps a third of the world's population (Figure 5-3). Nature has played a cruel joke on our efforts to carve productive farmland out of the tropical rain forests. After a few years of cultivation, fields lying open to the sun turn quite literally into pavements of solid brick!

A key reaction in the formation of laterite is the union of atmospheric oxygen with ferrous iron in rock, producing ferric iron oxides. The earth's crust contains nearly 300 times more atoms of ferrous (oxidizable) iron than the number of oxygen atoms in the atmosphere. What would happen if this iron were permitted to react freely with the atmosphere? Luckily for all of us, laterites are confined to the tropical and semitropical latitudes.

SEDIMENTARY ROCKS

Sedimentary comes from the Latin word for "settling," which is what gravity causes dense particles to do when they are immersed in a fluid. Fine sand, being moved along by wind, skips along, keeping close to the earth. Fine mud, stirred up into the water of a shallow bay by a storm, will slowly settle out of suspension. In these and most other situations, the sediment gravitates into roughly horizontal strata. Sedimentary rocks almost everywhere are stratified, but so are some igneous rocks, such as basalt flows or sheets of ignimbrite. However, sediments are deposited under "cool" (surface) temperature conditions. About 75 percent of the continents, and probably 90 percent or more of the ocean floor, is veneered by sedimentary rock. In volume, though, this rock type occupies only about 5 percent of the crust and a trivial fraction of the earth's total bulk.

If all the world's sedimentary rock were distributed uniformly, it would cover the surface of the earth to a depth somewhat less than a kilometer. Actually, there are enormous variations in thickness from place to place. Over much of the ocean floor the thickness is several hundred meters or less. Where streams have brought sediment to the edge of the continent for millions of years, the thickness may exceed 15,000 meters.* Rates of deposition vary too. A sediment thickness of 100 meters has accumulated within the past 2000 years at the mouth of the Mississippi River. Yet at Galveston, Texas, a coastal city some dis-

---

* At present, the rivers of southeast Asia are pouring forth 4 times as much sand and mud as all the rest of the world's rivers combined! Arid Australia, though not far from Asia, is being eroded less rapidly than any other continent.

tance to the west, only 0.3 meter has been deposited in the same length of time, and in the mid-ocean, the equivalent thickness is roughly 1 centimeter (0.01 meter).

From Weathered Granite to Sediment

We noted that when granite is weathered, some of the rock remains as solid fragments, and some of it goes into solution. A third part forms an intermediate dispersed state called a colloid. What kind of sedimentary rock might each of these materials become?

Terrigenous clastics Let us take first the solid fragments. A glance at Table 5-1 reminds us that quartz, muscovite, and potassium feldspar are best able to survive chemical weathering. These minerals, especially the very durable quartz, are concentrated in *terrigenous clastic* sediments. (*Terrigenous* means "land-derived," and *clastic* refers to sedimentary particles obtained from some preexisting rock.) These sediments may be further classified according to grain size. If the fragments are large—boulders or cobbles, down to pea-sized—the rock is a *conglomerate*. Progressively smaller particles are characteristic of *sandstone*, then *siltstone*, and finally mudstone or *shale*. A summary of particle sizes in terrigenous sediment shows three rather clearly defined groups (Figure 5-4). Shale is roughly twice as abundant as all other types combined. This fact is frequently not appreciated because lowlands overgrown by vegetation (in temperate climates) are commonly underlain by shale, an easily eroded rock, whereas sandstone may announce its presence as bold cliffs. There is also a relationship between particle size and mineral composition. Could the three groups have had separate origins?

Boulders and cobbles are large fragments of still recognizable "whole rock." On the other hand, the particles in sandstone and siltstone were broken apart along grain boundaries. These grains have been knocked together by wind, wave action, or streams which have abraded them into finer and finer material.

Nature's mill can grind only so far, however. The fluid medium that transports the sediment also surrounds and cushions the jostling particles so effectively that the very small grains seem unable to damage one another any more. Simulation experiments in which quartz sand was propelled by water around a circular tank showed that a 0.5-millimeter cube would have to be rolled a distance equivalent to 50 times around the equator to become rounded into a sphere. This suggests that stream abrasion is ineffectual in producing rounded sand grains. Estimates are that windblown sand is rounded about 1000 times faster than sand in streams. Wind speeds are 10 to 100 times higher than stream velocities, and the viscosity of air, which cushions the impacts, is one-fiftieth of that of water.

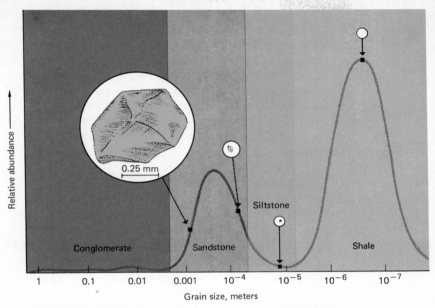

FIGURE 5-4 An abundance diagram for terrigenous clastic sediments also shows the relative sizes of typical grains of coarse sand, fine sand, and silt. An individual clay particle is invisible at this magnification.

FIGURE 5-5 Ripples are preserved in this thin-bedded sandstone in Illinois. Was this sedimentary rock deposited in still water or in moving water? Was the water disturbed by the back-and-forth motion of waves, or by a steady current flowing in one direction? If the latter, what was the direction? [*Illinois Geological Survey photograph, in F. J. Pettijohn and P. E. Potter*, Atlas and Glossary of Primary Sedimentary Structures, *Springer-Verlag New York Inc., New York, 1964.*]

109 SEDIMENTARY ROCKS

Since it is difficult to shatter very tiny grains, thus making them even smaller, how can it be that shale, with the smallest particles of all, is the most abundant sedimentary rock (Figure 5-4)? The answer to the paradox is that the mineral grains in shale mostly are not pulverized, *original* bedrock; they are particles that have been *reconstituted* by chemical weathering. We might expect that the abundant feldspar would supply much of the material to shale, the most common sediment type. When feldspar is chemically weathered, what does it become? Is your answer consistent with the fact that many of the shale particles are ultramicroscopic?

Colloids Clay minerals in water behave as a colloid—a suspension of minute particles that settle out only under special circumstances. Within a crystal, the charges of positive and negative ions are balanced, but at the surface a myriad of "unsatisfied" chemical bonds reach out into the suspending medium. Colloidal particles are so small that these surface effects are quite pronounced. For example, hydroxyl and oxygen ions occupy the surfaces of certain clay minerals (Figure 5-6). They attract water molecules (in which the distribution of electrical charge also is not uniform). The coating of adhering water keeps the clay particles apart from one another until the river that is transporting clay sediment finally reaches the ocean. Seawater rich in dissolved Na^+ and Cl^- in effect "saturates" the water with electrical charges. At that point, the clay loses its protective armor; the particles clump together and precipitate.

Sedimentary rocks formed wholly of colloidal precipitates are quite rare. The environment of deposition must survive for long periods (col-

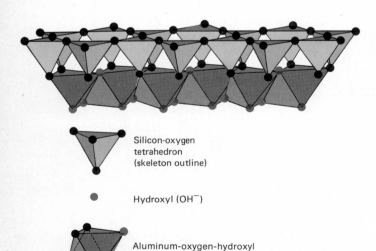

Silicon-oxygen
tetrahedron
(skeleton outline)

Hydroxyl (OH⁻)

Aluminum-oxygen-hydroxyl
octahedron (an eight-sided solid figure)

FIGURE 5-6 Electrical charges are in balance within the sheet structure of kaolinite, a clay mineral, but the charge is not quite uniformly distributed. An excess of negative charge is found on the surface populated by oxygen atoms, and a deficiency of negative charge on the opposite surface, where hydrogen atoms are exposed.

FIGURE 5-7 When matrix rock in this specimen had been partially dissolved away by acid, delicate fossil shells that long ago had been replaced by insoluble chert were exposed. [*Smithsonian Institution photograph.*]

loidal sedimentation is slow). It must be protected from violent storms and from the slightest influx of terrigenous particles. Two to three billion years ago, great accumulations of mingled iron oxides and silicates were precipitated, probably as colloids. The ironstone deposits in Minnesota, Labrador, and elsewhere currently supply most of the iron for the Western bloc of nations. The origin of the ancient ironstone is somewhat a mystery, for no similar sedimentary rock is forming today anywhere in the world. Some experts believe that iron-precipitating bacteria (microbes that oxidize ferrous to ferric iron) were responsible.

Even more of a mystery is the origin of *chert* or *flint*, a dense, hard aggregate of submicroscopic grains of quartz and other minerals of the silica family. Even under the electron microscope, the chert grains often reveal few clues to their origin. Was the chert deposited on the sea floor as microscopic skeletons of one-celled organisms, or was it an atom-by-atom replacement of some previously deposited sediment? Or was it deposited in open pores in a sediment by groundwater, long after the host rock had been uplifted above sea level? In different localities each of these possibilities is supported by the evidence. For example, beautifully preserved fossil shells composed of chert are known (Figure 5-7).

TABLE 5-2

Mass Balance between Rivers and the Ocean

Substance brought by rivers	Number of times the ocean would have been filled during the past 100 million years (the latest 2 percent of geologic time)*
Water	2400
Cl^-	1
Na^+	1.4
Mg^{2+}	7
SO_4^{2-}	10
K^+	15
Ca^{2+}	81
HCO_3^-	1000
SiO_2	5000

* "All the rivers run into the sea, yet the sea is never full; into the place from whence the rivers come, thither they return again," (Ecclesiastes 1 : 7). Today, we realize that water and *all other substances* are recycled.
SOURCE: Robert M. Garrels and Fred T. Mackenzie, *Evolution of Sedimentary Rocks*, W. W. Norton & Company, Inc., New York, 1971.

Since the shell material was originally calcium carbonate or calcium phosphate, this evidence argues that the chert formed by replacement. We are only beginning to understand this common but perplexing substance.

Evaporites Potassium and sodium ions in solution are another product of the weathering of granite; they eventually enter the ocean. If the ocean had retained all these dissolved ions, long ago it would have become a pool of concentrated brine (Table 5-2). Why hasn't this happened? One possible way to remove K^+ and Mg^{2+} ions is by a slow absorption into clay particles resting on the ocean floor. Other deep-ocean reactions are known or postulated, but numerous details are yet to be worked out.

About 3 percent of all sedimentary rocks consist of salt left behind by evaporation of seawater. Though originating from the sea, these *evaporite* deposits of soluble salts have been lifted above sea level in Israel, New Mexico, Germany, and numerous other areas of the world. Even in desert terrains, salt is highly vulnerable to being washed right back where it started. Measurements of the dissolved load of rivers suggest that evaporites are eroded about 10 times more readily than other kinds of rock. Evaporites are recycled rocks indeed! Even so, the amount of salt stored at present in the form of evaporite sediments is about double the quantity of salt dissolved in the ocean.

Carbonate sediments Probably as much as 90 percent of the world's *carbonate* sedimentary rocks have been formed by living organisms in a number of ingenious ways. Corals, snails, clams, etc., build

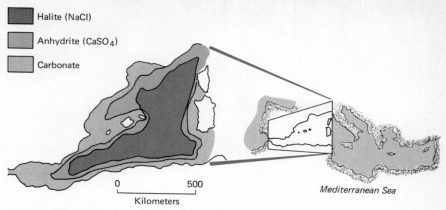

Halite (NaCl)

Anhydrite (CaSO$_4$)

Carbonate

0   500
Kilometers

Mediterranean Sea

FIGURE 5-8 Drilling of the western Mediterranean Sea floor has revealed an astonishing series of concentric deposits of evaporite sediments. This pattern of least-soluble minerals on the rim, to most-soluble in the center, is obtained if an isolated body of salt water grows smaller as it deposits salt from an ever more concentrated residual brine. About 6 million years ago, the Mediterranean Sea, blocked off from the Atlantic Ocean, very nearly dried up. Brine pools lay exposed to the hot sun many thousands of meters below sea level at the bottom of the parched Mediterranean basin. [*After K. J. Hsü and others, "Late Miocene Desiccation of the Mediterranean,"* Nature, *vol. 242, no. 5395, 1973.*]

skeletons or shells of calcium carbonate. Single-celled floating organisms, *Foraminifera*, construct fragile shells of $CaCO_3$ (Figure 5-9). Various species of red algae secrete a calcareous coating over their cell walls.

In addition, the process of photosynthesis indirectly causes $CaCO_3$ to precipitate. Suppose that some carbon dioxide has dissolved in water, forming carbonic acid:

$$H_2O \; + \; CO_2 \; \rightleftharpoons \; H_2CO_3 \qquad\qquad (5\text{-}4)$$

Water   Carbon   Carbonic
dioxide   acid

Carbonic acid in turn can react with calcium carbonate:

$$CaCO_3 \; + \; H_2CO_3 \; \rightleftharpoons \; Ca^{2+} \; + \; 2HCO_3^- \qquad (5\text{-}5)$$

Calcium   Carbonic   Calcium   Bicarbonate
carbonate acid (dis-   ion (dis-   ion (dis-
(solid)   solved in   solved in   solved in
water)   water)   water)

A principle of chemistry states that if a chemical equilibrium is disturbed, an adjustment takes place to counteract the effects of the disturbance. When dissolved carbon dioxide is removed from the water by growing algae, carbonic acid promptly dissociates, driving reaction (5-4) to the left (in the direction attempting to restore the "lost" CO_2).

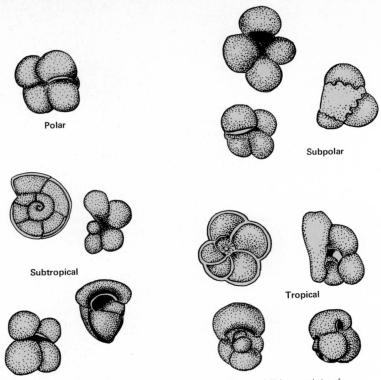

Polar

Subpolar

Subtropical

Tropical

FIGURE 5-9 The more common species of Foraminifera exhibit a variety of forms of calcium carbonate shells. Some of these organisms are cold-water species; others thrive only in warmer waters. Abundances of temperature-sensitive species in buried oceanic sediment reflect the warm and cold intervals of past climates. Magnification: 40 times.

But this would diminish the amount of H_2CO_3 in reaction (5-5), thereby swinging *it* to the left, and causing solid $CaCO_3$ to precipitate. This might happen, for example, in a shallow tropical sea. On the other hand, groundwater charged with carbonic acid would drive reaction (5-5) to the right, causing $CaCO_3$ to enter solution. This water could dissolve out caverns and open fissures as it seeps through the rock.

Actually, the chemistry of calcium carbonate in seawater is rather more complicated than these simple equations would suggest. Other ionic species are present that interact with $CaCO_3$ by roundabout means, so that in our computations, "everything must be corrected for everything else." It turns out that near the surface, the ocean is oversaturated with dissolved calcium carbonate, but at greater depth it is slightly undersaturated. Where the ocean is deeper than about 3 kilometers, calcium carbonate in the bottom sediments is redissolving. Although a gentle rain of shells of dead foraminiferans continually

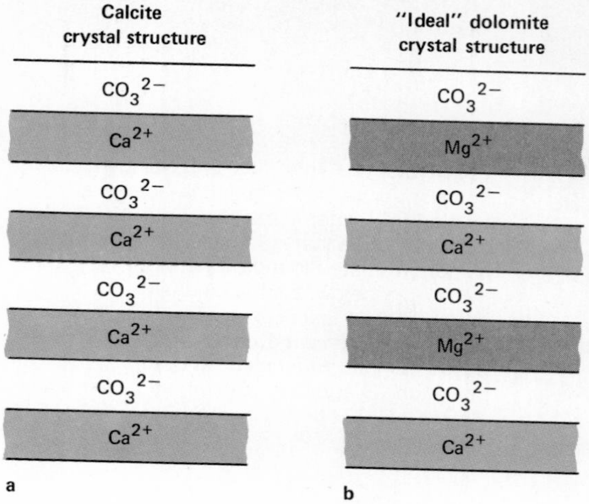

Calcite crystal structure	"Ideal" dolomite crystal structure
$CO_3{}^{2-}$	$CO_3{}^{2-}$
Ca^{2+}	Mg^{2+}
$CO_3{}^{2-}$	$CO_3{}^{2-}$
Ca^{2+}	Ca^{2+}
$CO_3{}^{2-}$	$CO_3{}^{2-}$
Ca^{2+}	Mg^{2+}
$CO_3{}^{2-}$	$CO_3{}^{2-}$
Ca^{2+}	Ca^{2+}
a	b

FIGURE 5-10 Cation layers (seen edge-on) differ in calcite and dolomite. What is the chemical formula of dolomite?

descends to the ocean floor, we do not find carbonate sediments in the very deepest part of the ocean.

Calcite and *dolomite* are the most important carbonate minerals. In the calcite structure (Chapter 3) we have a Ca^{2+} layer, then a $CO_3{}^{2-}$ layer, and so on (Figure 5-10a). "Ideal" dolomite is slightly more complicated: the cation layers are alternately Ca^{2+} and Mg^{2+} (Figure 5-10b). A rock consisting of calcite is called *limestone*. If the mineral is dolomite, the rock itself is also dolomite.

Living organisms do not secrete dolomite. We know that this material forms by a partial replacement of Ca in a carbonate by magnesium ions, but how and when does this take place? Laboratory simulations have succeeded in creating a disordered dolomite at room temperature, but never the ideal dolomite described earlier. Apparently the ordered atomic arrangement is but slowly perfected over hundreds or thousands of years. Dolomite is forming today in only a few environments where the rock is steeped in abnormally salty water (Figure 5-11). We are uncertain of the origin of the enormous volumes of dolomite found in the more ancient rock record.

Limestone is forming today in both shallow and deep water (Figure 5-12). More than a quarter of the earth's surface, in the ocean basins, is covered with a fine calcareous ooze (soft, water-saturated sediment) made of the skeletons of foraminiferans and other microorganisms. Toward land the ooze gradually gives way to other sediments, not because the floating organisms are less abundant there, but because the carbonates are overwhelmed by much clastic sediment shed off the continents.

Organic reefs made up of coral, oysters, calcareous algae, etc., are

FIGURE 5-11. This "flat" in the Bahama Islands lies just above the normal high-tide level, shown by standing water in the background. A broken dolomitic crust that is forming at the surface is somehow related to a drawing up of very saline water from below, but the means of dolomitization is unknown. [*From E. A. Shinn and others, "Recent Supratidal Dolomite from Andros Island, Bahamas," in* Dolomitization and Limestone Diagenesis, *Society of Economic Paleontologists and Mineralogists, Special Publication 13, 1965.*]

FIGURE 5-12 Deep ocean basins are carpeted with an ooze consisting of skeletons of foraminiferans and other floating organisms. Reef complexes border the continents. The Great Barrier Reef (largest in the world), off northeastern Australia, is more than 1600 kilometers long. [*After John Rodgers, "The Distribution of Marine Carbonate Sediments: A Review," in* Regional Aspects of Carbonate Deposition, *Society of Economic Paleontologists and Mineralogists, Special Publication 5, 1957.*]

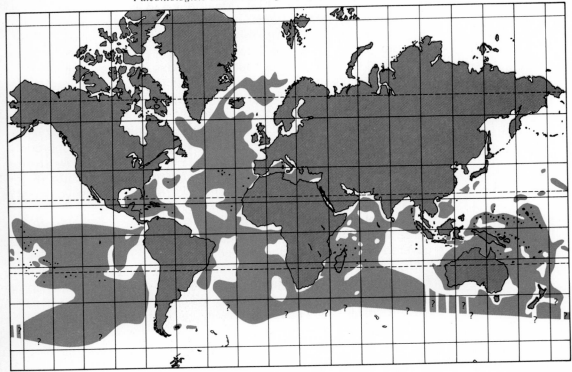

found in shallow tropical waters. With time, these deposits can grow to spectacular size. The Bahama Islands, near Florida, have built up to a thickness of at least 4.5 kilometers. Other areas are no more than a barely submerged shelf populated here and there by small "patch reefs."

Carbonate reefs are accumulating in only a few places, but at certain times in the past, limestone sheets were deposited in shallow seas that flooded the continents (Figure 5-13). Some areas have been submerged on at least 20 separate occasions! At present, there happen to be no such "epicontinental seas" anywhere in the world. Geologists must use some educated imagination in deciding what these ancient environments of limestone deposition must have looked like.

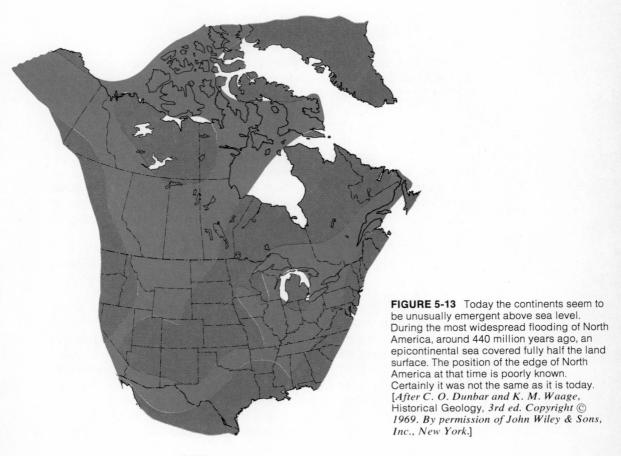

FIGURE 5-13 Today the continents seem to be unusually emergent above sea level. During the most widespread flooding of North America, around 440 million years ago, an epicontinental sea covered fully half the land surface. The position of the edge of North America at that time is poorly known. Certainly it was not the same as it is today. [*After C. O. Dunbar and K. M. Waage,* Historical Geology, *3rd ed. Copyright © 1969. By permission of John Wiley & Sons, Inc., New York.*]

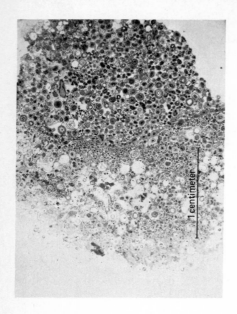

1 centimeter

FIGURE 5-14 Named *oolite* after the Latin word for "egg," this texture of a carbonate sedimentary rock from the Bahamas probably originated by precipitation around small nuclei as they were tossed about by vigorous currents. [*From N. D. Newell and J. K. Rigby, "Geological Studies on the Great Bahama Bank," in* Regional Aspects of Carbonate Deposition, *Society of Economic Paleontologists and Mineralogists, Special Publication 5, 1957.*]

Although carbonate sediments may be rather pure, chemically simple rocks, their *textures* show much fascinating detail. Texture refers to the sizes and shapes of the mineral grains, and to how they are assembled together. Solidified carbonate mud consisting of submicroscopic grains likely precipitated in the midst of the water, later to be wafted quietly to the sea bottom. Other limestones are a mass of broken fossils jumbled together with pieces of carbonate mud that had been torn up by storms. These fragments may be cemented by a "glue" of calcite deposited by groundwater. This texture could have originated on a beach or shoal area exposed to violent waves. Still other limestones are composed of tiny spheres that resemble fish eggs (Figure 5-14).

METAMORPHIC ROCKS

The final class of rocks is in some respects the most complex of all. *Metamorphic* comes from a Greek word meaning "to transform." Since all rocks have been transformed at least slightly, how shall we define metamorphism? For example, is the conversion of limestone to dolomite a type of metamorphism? How about the breakdown of granite into clay minerals and other weathering products? Although these changes no doubt are as thoroughgoing as any that are known to geologists, they do not qualify as metamorphism. The factors essential to metamorphism are *high temperature and pressure*. Pressure can affect a rock

FIGURE 5-15 Metamorphism of this conglomerate in California has flattened the pebbles, but they still can be clearly identified. [*From L. E. Weiss*, The Minor Structures of Deformed Rocks, *Springer-Verlag New York, Inc., New York, 1972.*]

uniformly in all directions, but often it is accompanied by application of *shear stress*—force that is directed into the rock along certain orientations only. Most metamorphic rocks and intrusive igneous rocks originate very deep. They are exposed to view only after prolonged erosion has stripped away the overburden.

Metamorphic transformation may be anything from mere baking or hardening of the rock to a reorganization so profound that every mineral has recrystallized or chemically reacted with other minerals. If metamorphism is very hot, some of the rock may even melt. Igneous bodies are commonly emplaced into metamorphic terrains, but the metamorphic rocks themselves were recrystallized *in the solid state*. For instance, pebbles in a metamorphosed conglomerate may be flattened or stretched into long spindlelike rods (Figure 5-15). If they had actually melted, they could no longer be recognized.

Role of Water in Metamorphism

Early experiments showed that most minerals react so sluggishly that the reaction would not finish even after millions of years at high temperature. However, the reactions were found to be enormously speeded up by the addition of just a single drop of water to the capsule containing the reacting crystals. It is clear now that water is the great promoter of metamorphism. This water eventually escapes from the rock, or is bound chemically into hydrous minerals such as biotite and hornblende.

Furthermore, *absence* of water is responsible for the continued preservation of many metamorphic rocks. Suppose that a quartz sandstone with a small amount of kaolinite (a hydrous clay mineral) became more and more deeply buried beneath a deforming mountain range. As temperature and pressure rise, the quartz and kaolinite react with each other. A series of other assemblages of metamorphic minerals appear, each representing equilibrium in a different environment (Table 5-3). In the meantime, water is driven out of the rock. What happens when temperature and pressure finally begin to decline as the mountains overhead are eroded away? Is the metamorphism simply reversed, step by step? Some rocks do show a faint hint of this reversal, but the great majority do not. The missing ingredient is water, without which the speed of metamorphic reactions decreases almost to nothing. If these changes could be so easily "undone," very few metamorphic rocks would be visible at the surface.

Laboratory simulations help us to discover the conditions of an ancient metamorphism. Take, for example, the metamorphic minerals andalusite, sillimanite, and kyanite (Table 5-3), which have the same chemical formula but slightly different crystal structures. Figure 5-16 summarizes the "stability fields"—the combination of temperature and pressure at which each of these minerals is stable. Intensified metamorphism will cause one mineral to recrystallize into another. If the change were a rise in pressure (vertical arrow), it is possible for andalusite to become sillimanite, which in turn converts to kyanite at very high pressure. If there were a temperature increase at constant pressure (horizontal arrow), then kyanite would recrystallize into sillimanite, which is just the reverse of the previous transformation. A geologist must consider not only the laboratory findings but also other kinds of evidence in his interpretations.

TABLE 5-3
A Simple Metamorphic Series

	Equilibrium assemblage	Chemical formula of silicate* (quartz: SiO_2, always present)	Atomic structure	Remarks
	Kaolinite + quartz	Kaolinite: $4SiO_2 \cdot 2\,Al_2O_3 \cdot 4H_2O$	Sheet	Starting material (surface temperature and pressure)
	Pyrophyllite + quartz	Pyrophyllite: $4SiO_2 \cdot Al_2O_3 \cdot H_2O$	Sheet	Decreasing H_2O content
	Andalusite + quartz	Andalusite: $4SiO_2 \cdot 4Al_2O_3$	Complex	Anhydrous (no H_2O)
	Sillimanite + quartz	Sillimanite: $4SiO_2 \cdot 4Al_2O_3$	Complex	Anhydrous
	Kyanite + quartz	Kyanite: $4SiO_2 \cdot 4Al_2O_3$	Complex	Anhydrous

Increasing metamorphic intensity (shown as downward arrow at left of table)

* Expressed in an unconventional manner to emphasize the changing proportions of silica, aluminum oxide, and water in the mineral that is in equilibrium with quartz.

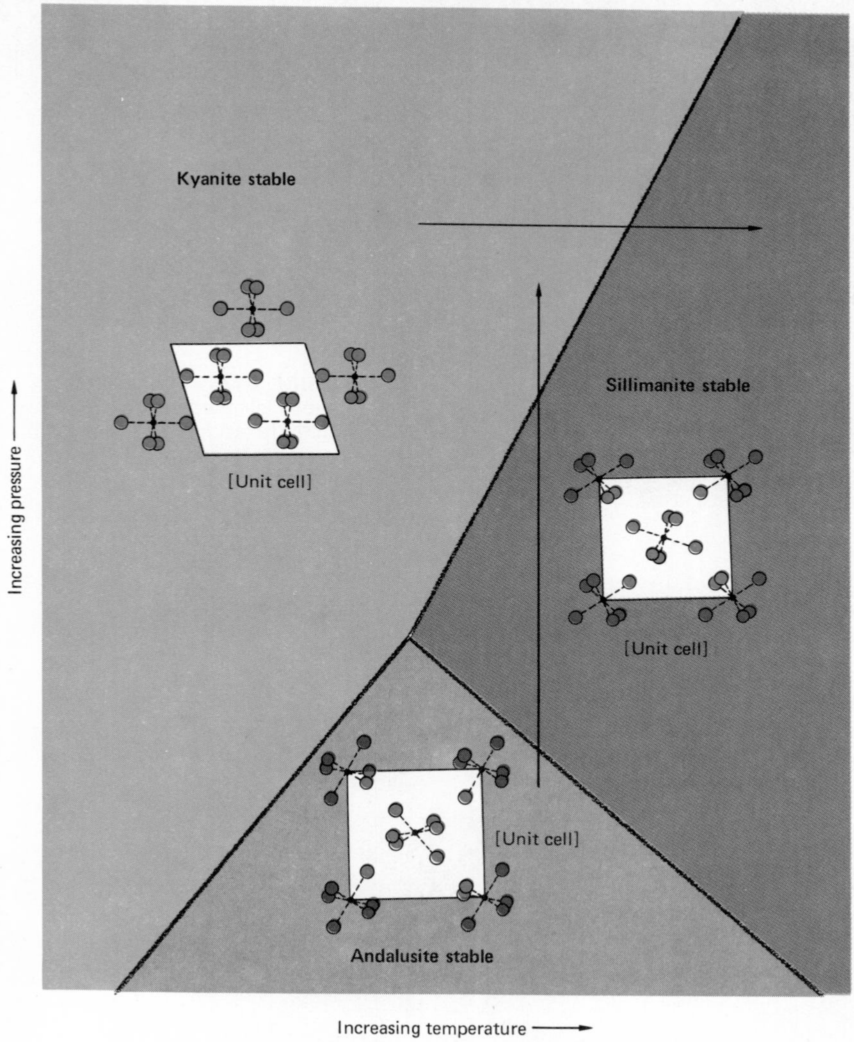

FIGURE 5-16 A stability diagram of andalusite, sillimanite, and kyanite also shows the differing arrangements of oxygen atoms around aluminum atoms in the crystal structures of these minerals. (Not all the oxygen atoms in a unit cell are shown.)

Types of Metamorphism

Temperature, pressure, and shear stress can operate in various combinations, with different results. The earth's crust (and probably the mantle and core too) has always been in a state of unrest. It has been warped upward into plateaus, depressed below sea level, mashed and crumpled into folds (Figure 5-17), torn asunder by faults. Metamorphism is often an important part of this scene.

Major faults are not simple ruptures; they are zones as much as several kilometers wide, and up to hundreds or even thousands of kilometers long (Figure 5-18). Rocks caught up in such zones are crushed

FIGURE 5-17 An outcrop of folded metamorphic rock in central Texas. A rod 1 meter long gives the scale of the photograph. [*Courtesy William Workman.*]

FIGURE 5-18 An ancient fault system can be traced for hundreds of kilometers from Sudan across Uganda into Kenya. Mount Elgon, a colossal volcano on Uganda's eastern border, has erupted through this belt of intensely crushed rock. The White Nile abruptly turns to follow this somewhat more easily eroded zone for some distance before resuming its northward course.

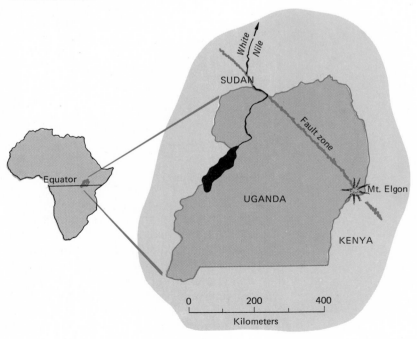

1 millimeter

1 millimeter

a

b

FIGURE 5-19 These photomicrographs show a progressive shear metamorphism of a granitic rock. Unaffected igneous rock (*a*) has been sheared out into zones of fine-grained material (*b*) in which the crystals are ground too small for visual identification.

and smeared out by an almost completely mechanical style of metamorphism. The ultrafine-grained metamorphic rock may be "cold-welded" into a flinty, durable mass (Figure 5-19).

Another metamorphic style is seen along the margins of some igneous intrusions. If the magma was emplaced into relatively cold host rock, a large temperature difference would have once existed at the boundary (Figure 5-20). Heat and fluids (mainly H_2O) flowing outward from the intrusion can *contact*-metamorphose the host rock. Important ore deposits are found in the "halo" of contact-metamorphic rock surrounding some intrusions.

Although a contact-metamorphosed rock may have recrystallized into a mosaic of tiny crystals, the original form of sedimentary layering or fossils may be delicately preserved. This is the opposite result from shear metamorphism, in which the minerals may be unchanged but the original structures of the rock are obliterated.

A third type, by far the most common, is *regional* metamorphism. Shear stress, high pressure, and high temperature are all likely to be important ingredients here. Regional metamorphism is going on wherever rocks formed near the surface find themselves (after a time) at great depth, especially where mountain building is in progress. Intense dislocation—folding and faulting of the rocks—generally accompanies the metamorphism in these mountain belts. An affected area might be as large as the entire Atlantic seaboard of the United States and Canada, where mountains were repeatedly uplifted during the past 600 million years. Today, regional metamorphism is probably active beneath deforming mountain ranges in New Zealand, southern California, and northern India.

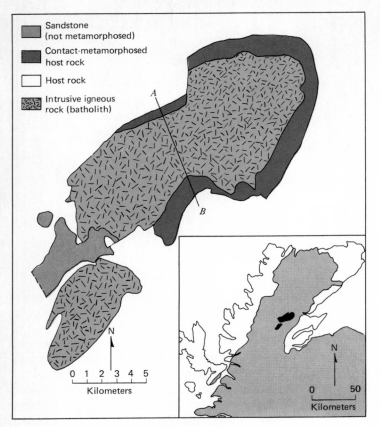

Cross section along line AB

FIGURE 5-20 A granite batholith in Scotland (see inset map) reveals a complex history. When emplaced, the magma contact-metamorphosed the sedimentary host rock along its margins. More than 100 million years later, the entire shaded area (inset map), including the granite and its metamorphic fringe, was metamorphosed on a regional scale. A rock can be metamorphosed more than once. [*After L. E. Long, "Rb-Sr Chronology of the Carn Chuinneag Intrusion, Ross-shire, Scotland," Journal of Geophysical Research, vol. 69, p. 1590, 1964. Copyright by American Geophysical Union.*]

Common Metamorphic Rocks

We have seen that igneous rocks, such as granite and basalt, can be recycled by weathering, erosion, and transportation of the weathering products. A grand sorting takes place as dissolved ions reach the sea, and solid particles are segregated into various types of clastic sediment. Some of the sedimentary rocks are complex mixtures of different minerals, but others, such as sandstone and limestone, may be made up almost entirely of one mineral species. Metamorphism of one of these latter types can do little except to change the texture. Rounded

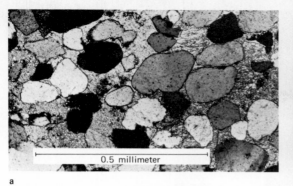

a

b

FIGURE 5-21 Rounded quartz grains, bonded together by a precipitated cement in sandstone (*a*), become recrystallized into an interlocking mass of crystals in quartzite (*b*). Sandstone feels gritty because it fractures *around* the grains which project above the broken surface. Smooth quartzite surfaces have fractured *across* the grains. Quartz grains appear white, or various shades of gray, depending on their orientation, which determines how effectively polarized light can be transmitted through them. [(*a*) *Courtesy Earle McBride.*]

grains in a quartz sandstone are subtly refashioned into a mosaic of interlocking crystals in *quartzite,* the metamorphosed equivalent of sandstone (Figure 5-21). Metamorphism can change limestone into a more coarsely grained mass of interlocking crystals that typify *marble.* Quartzite is the strongest, most resistant to attack, of all the common rocks, but marble is no more resistant than limestone to being dissolved by carbonated groundwater.

Metamorphism of shale (or mudstone) is more complicated. At the outset, shear stress causes the minute clay particles to rotate into a par-

FIGURE 5-22 Quartzite ridges in the Republic of South Africa stand sharply above a plain that is underlain by less resistant rock. [*Courtesy Daniel Barker.*]

125 METAMORPHIC ROCKS

FIGURE 5-23 This fossil trilobite (an extinct marine animal) in North Wales became weirdly deformed as it and the enclosing rock were compressed. [*From L. E. Weiss*, The Minor Structures of Deformed Rocks, *Springer-Verlag New York, Inc., New York, 1972.*]

allel alignment within the rock. Tiny parallel wisps of muscovite develop; they also help the rock to cleave into the thin, flat-sided plates characteristic of *slate*. Deformed fossils that have been shortened along the direction of applied pressure and elongated in other directions can be identified in slate (Figure 5-23). Fossils have been completely destroyed in most other metamorphic rocks.

Beyond the slate stage (especially if water is still present), the chemical effects of metamorphism become evident. Two other rock types deserve mention. *Schist* is a coarse-grained rock containing abundant mica flakes arranged in a subparallel fashion. Breakdown and reaction of the original minerals may have led to the appearance of garnet (an "isolated tetrahedra" silicate) and other metamorphic minerals. In *gneiss* (nice), layers rich in dark ferromagnesian minerals alternate with quartz-feldspar layers (Figure 5-24). With increasing meta-

FIGURE 5-24 Layers in gneiss from New York have been deformed into numerous small folds. Although the gneiss was probably once a sediment or a volcanic rock, these layers are not necessarily original depositional bedding. Gneissic layering is a feature developed in the rock by metamorphism. [*Judy Camps.*]

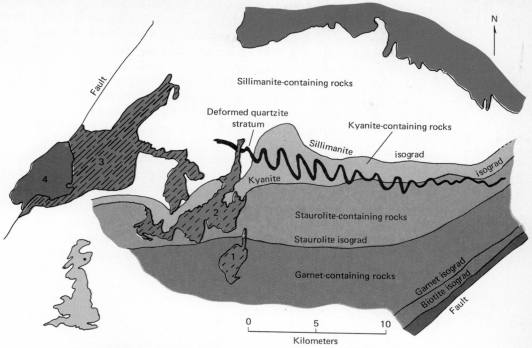

FIGURE 5-25 Barrow's map of an area in Scotland (see inset map) shows several igneous bodies, and isograds crossing a region of metamorphosed sedimentary rock. [*After G. Barrow, "On an Intrusion of Muscovite-Biotite Gneiss in the Southeastern Highlands of Scotland, and Its Accompanying Metamorphism*, Quarterly Journal of the Geological Society of London, *vol. 49, 1893.*]

morphic intensity, it becomes more difficult to tell what the original rock was. Schist and gneiss are descriptive of rocks that could have been derived from any of a great variety of other rocks. Under the most extreme metamorphic conditions, all the hydrous minerals disappear. Some terrains are monotonous expanses of a simple feldspar-pyroxene rock.

Ancient Regional Metamorphism: A Case Study

One of the first systematic studies of metamorphism was reported in 1893 in a classic paper by the British geologist George Barrow. The small area in Scotland that he examined provides an ideal case example of regional metamorphism (Figure 5-25). Barrow found that metamorphism of the rock is hardly perceptible in a narrow strip next to a major fault. A short distance to the northwest, tiny crystals of biotite appear in the rock, and at still greater distances, metamorphic garnet, then staurolite (a complex iron aluminum silicate similar to kyanite), then kyanite, and finally sillimanite appear. (All the rocks contain quartz, feldspar, and other minerals.) Barrow considered a line marking the first appearance of a diagnostic mineral to reflect uniform conditions of metamorphism. The lines of equal grade (or intensity) have thus been termed *metamorphic isograds*.

It could be argued that the isograds simply indicate variations in the original rock. Barrow neatly countered this criticism by pointing to a deformed layer of impure quartzite that surely must have been a rather uniform sandstone initially. Today, the isograds cut *across* the quartzite (Figure 5-25).

What is meant by metamorphic "intensity"? Work in the field and laboratory has shown that metamorphic grade depends in a complicated way upon pressure, temperature, chemical composition, and amount of H_2O contained in pore spaces in the rock. In Barrow's locality, kyanite gives way to sillimanite toward the north. Therefore, the pressure must have been uniformly high, and temperature was the chief variable, corresponding to changes shown by the horizontal arrow in Figure 5-16.

FURTHER QUESTIONS

1   Refer to page 102. Weathering causes a rock to disintegrate. What is the surface area of a cube 1 centimeter on edge? Suppose that weathering broke down this cube to smaller cubes, each 10^{-6} (one-millionth) centimeter on edge. What is the surface area of one of these? How many cubelets could be made out of the original cube? What is the surface area of the material after weathering?

2   As a rule of thumb, the speed of chemical reactions doubles for every 10°C rise in temperature. Which predominates in arctic regions: chemical or mechanical weathering? Which type of weathering is most important in the tropics?

3   Review the chemical formula and physical properties of magnetite (Chapter 3). What happens to magnetite during weathering? Is it attacked chemically, or does it tend to remain as sand-sized particles, like quartz? Why?

4   Refer to a map of southeast Asia. Can you suggest why rivers in this part of the world carry such a heavy load of sediment?

5   What are some of the uncertainties in estimating the correct shape of the abundance distribution curve in Figure 5-4?

6   Why does calcium carbonate dissolve in the very deep ocean? The high pressure and low temperature (about 3°C) at depth in-

crease the solubility of CO_2 in water. Refer to Equations (5-4) and (5-5). Trace through the chain of reactions that result if the concentration of CO_2 increases. Which would tend to have thicker, more robust shells: arctic or tropical foraminiferans? Why?

7 Does the flooding of a continent by a shallow sea require a drastic adjustment of the land elevation? Refer to a topographic map of North America. Where would the shoreline be if the land sank or if the sea level rose 200 meters?

8 Refer to Figure 5-14. How can the spherical shapes and concentric layering in the oolite grains be created? Can you suggest the environment in which oolite is deposited? Would you expect large fossil-shell fragments to be preserved in this environment?

9 Refer to Figure 5-15. In what direction was the applied force the greatest during metamorphism? In what direction was the applied force weaker?

10 Refer to Figure 5-16. Which of the three metamorphic minerals would you expect to be most dense? Least dense? Which has the largest unit-cell volume?

11 Refer to a geologic map of North America. Where are metamorphic rocks located? Are these regions mountainous today? If not, why not?

12 Refer to Figure 5-21. Why does sandstone (unmetamorphosed) break around sand grains, whereas quartzite (metamorphosed) breaks across the grains?

13 Refer to Figure 5-25. The igneous rocks also show the effects of metamorphism. Rock 1 is nearly unaffected, but toward the northwest, deformation increases to the point that rock 3 is sheared and mashed almost beyond recognition. And yet, igneous rock 4 is totally *un*metamorphosed. Can you suggest why? Does evidence seen on the map support your hypothesis?

READINGS

Listed in order of technical difficulty.

Hunt, Charles B., 1972: *Geology of Soils: Their Evolution, Classification, and Uses;* W. H. Freeman and Company, San Francisco, 344 pp.

Garrels, Robert M., and Fred T. Mackenzie, 1971: *Evolution of Sedimentary Rocks;* W. W. Norton & Company, Inc., New York, 397 pp.

chapter 6

"... Milestones on the Eternal Path of Time"

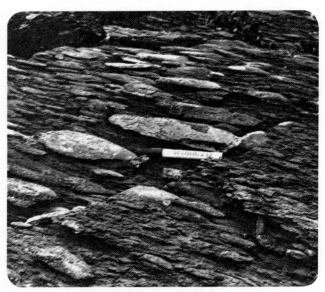

The oldest rock on earth, in southwestern
Greenland. [*Courtesy David Bridgwater.
Copyright 1973 by Grønlands Geologiske
Undersøgelse.*]

In preceding chapters we have seen that the *configuration* of the earth (its composition and the arrangement of its parts) is constantly re-shaped by various geological *processes*. A "memory" of some of these earlier geologic events is preserved in the rocks. Although the record is complex, fragmentary, and distorted, it can offer many tantalizing clues to the *history* of the earth. Ideas of configuration, process, and historical change are tied together in almost every sort of geologic study. The earth has witnessed the passage of unimaginably long stretches of time. Geologists have had to devise means to estimate bygone time and, above all, to become accustomed in their thinking to the special "flavor" of a sweeping view of world history and of man's brief part in it. So distinctive is the idea of seemingly limitless duration that it has come to be known, even in ordinary speech, as geologic time.

In this chapter, we shall review the history of thought about antiquity (a history of historical thinking, if you like). We shall also investigate some geologic "clocks" and denote some of the important events of earth history.

MAN DISCOVERS TIME

As individuals, we shall always be prisoners of our own limitations. Our view of reality extends throughout only a tiny part of the environment, at one point in history. Not many people today are aware of the painful controversies through which civilized man has slowly come to recognize his place in the stream of time. What were some of the major turning points in this long and devious intellectual pathway?

Early Speculations

There were, first of all, cogent reasons why the remote past was once thought actually to be irrelevant. Consider the situation of the classical empires of the Middle East: people knew communal life and

131

FIGURE 6-2 "Hutton's unconformity" on the coast of the North Sea (inset map) is seen in a photograph (*b*) and as a drawing (*a*) in which the unconformity (a buried surface of erosion) and the relationships of the older and younger strata are emphasized. [(*b*) *Crown Copyright Geological Survey photograph. Reproduced by permission of the Controller, Her Britannic Majesty's Stationery Office.*]

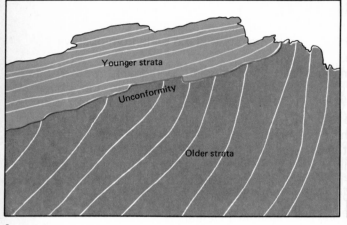

a

b

1 The near vertical beds of sedimentary rock (slightly metamorphosed) must have been deposited as horizontal strata. (Hutton accepted the long-recognized principle of "original horizontality" of stratified rocks.)

2 The beds were buckled by deforming forces.

3 A lengthy interval of stability followed, during which some of the contorted strata were eroded away.

4 Again the area was depressed below sea level, and more sedimentary beds deposited (horizontally) across the beveled, upturned edges of older rock.

5 Once again the region was uplifted, but gently tilted this time. Then came erosion, which has continued until the present day.

The buried erosion surface, or *unconformity*, represents a period during which a portion of the geologic record (embodied in the older sedimentary rock) was destroyed. As Hutton put it:

This earth, like the body of an animal, is wasted at the same time as it is repaired. . . . It is thus destroyed in one part, but it is renewed in another.

In Hutton's time several heated disputes gripped the new discipline of *geognosy* (later renamed *geology*). One of these was a clash between proponents of *catastrophism* and those who advocated what

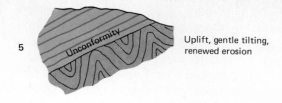

5   Uplift, gentle tilting, renewed erosion

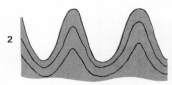

4   Submergence, deposition of "younger strata"

3   Beveling by erosion

2   Folding deformation of older strata

FIGURE 6-3 According to the law of superposition, older rocks (unless disturbed) lie beneath younger rocks. Therefore, cross sections of this sort that display a series of events are normally sequenced from bottom to top.

1   Deposition of "older strata"

was later termed *uniformitarianism*. The catastrophists argued that the earth was shaped by cataclysmic events—sweeping destructions and supernatural interventions on a worldwide scale. (Noah's flood was supposedly the most recent of these.) Hutton came down solidly on the side of the uniformitarian viewpoint, the notion overwhelmingly vindicated by later findings. He aptly summarized the uniformitarian idea by saying that "the past history of our globe must be explained by what can be seen to be happening now." He did not deny catastrophic events (earthquakes and volcanic eruptions were well known), but emphasized rather the impressive results of commonplace events when accumulated over long periods. A more appropriate term, *actualism*, came into use in some circles to describe the uniform operation of "actual causes" in changing the configuration of the earth.

137   MAN DISCOVERS TIME

Hutton perceived that the relationships seen in Figure 6-2 must have required countless ages to develop. Wisely he did not attempt, from the scanty evidence available, to place exact dates on geologic events. Instead, he adopted a noncommital viewpoint expressed in an often quoted statement concerning geologic time: ". . . that we find no vestige of a beginning,—no prospect of an end." Theologically, Hutton was a deist;* his refusal to pinpoint an age of the earth brought upon him much undeservedly harsh condemnation. His philosophy was said to encourage atheism, infidelity, and immorality. To this Hutton replied that he really had not "deposed the Almighty Creator of the Universe from His Office" as accused by his critics, but only acknowledged that "In nature, we find no deficiency in respect of time." Hutton's contribution to the modern concept did not provide a precise method to determine the timing of past events. Nevertheless, his cautious insistence upon seeking from nature itself the answers to its riddles, and the spacious perspective of time that he advocated, laid the foundation for all geologic studies that were to follow.

Hourglass methods Even before Hutton's day, interest began to shift to more quantitative means of estimating the passage of time. Prominent among these were various "hourglass" methods based upon the accumulation of some geologic substance in a reservoir. In 1715 the British astronomer Edmund Halley published a paper entitled "A short account of the cause of the Saltness of the Ocean, and of the several Lakes that emit no Rivers; with a Proposal, by help thereof, to discover the Age of the World." He argued that the salinity of the ocean, measured at intervals, would be seen to increase because of input by rivers. The increase observed during one time interval should have been about the same in previous intervals, and so on back to the time when the ocean contained no salt. Halley complained that the Greek and Latin authors had not recorded the ocean's salinity, for he was certain that a noticeable enrichment could be detected over 2000 years, which seemed to him a very long time. Unfortunately, the time scale he envisioned was far too short. In retrospect, we can state that in all of recorded human history, the amount of salt brought to the ocean is less than 0.01 percent of the amount already there.

In 1899 another widely publicized estimate by the Irish geologist J. Joly used the salinity data in a different manner. Instead of considering the reservoir (the ocean) at two widely separated instants of time, Joly focused on the rate of salt addition by rivers at only a single "instant," one year. (The ages of the earth and of the ocean were presumed to be similar.) The hourglass concept is plainly evident in his simple formula describing the accumulation of salt (Figure 6-4):

* A person who believes that God, after creating and setting the universe in motion, leaves it strictly alone thereafter.

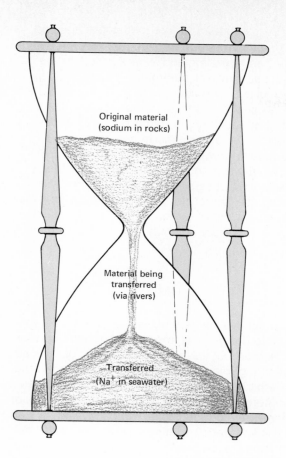

Original material
(sodium in rocks)

Material being
transferred
(via rivers)

Transferred
(Na$^+$ in seawater)

FIGURE 6-4 Joly's hourglass singles out two aspects of a dynamic situation. The total salt content of the ocean (specifically, its dissolved Na$^+$) corresponds to the pile of sand at the base of the hourglass. The rate of salt addition is symbolized by the density of sand grains still in mid-air. Sand in the top compartment (continents waiting to be eroded) is immaterial to Joly's formulation.

$$\frac{\text{Total salt content of ocean}}{\text{Rate of salt addition}} = \frac{\text{grams}}{\text{grams/year}}$$

$$= \text{years (time of accumulation)} \quad (6\text{-}1)$$

After various corrections for leaching of evaporite deposits, salt initially present in the ocean, etc., he obtained

$$\frac{1.24 \times 10^{17} \text{ grams}}{1.39 \times 10^{9} \text{ grams/year}} = \text{about 90 million years} \quad (6\text{-}2)$$

Joly, a brilliant scientist, recognized some 14 possible disturbing factors in his computations, but even so, his age calculation was off the mark by a factor of 50 times! The fatal flaw was his assumption that the ocean simply accumulates salt. Actually, it stores the dissolved salt only temporarily, eventually returning it to the solid earth in the form of evaporite deposits, or by other means. How would this recycling process invalidate Equations (6-1) and (6-2)? The Joly calculation is in fact a special application of the information contained in Table 5-2. It so happens that the "residence time," or average duration a sodium or chloride ion would spend in the ocean before being precipitated, is around 90 million years.

Another version of the hourglass method was applied to the accumulation of an observed thickness of sedimentary rock at a given rate of deposition. Estimates of this type suffered from the same defects as the salinity hourglass. The sedimentary record is incomplete owing to erosion or to periods of nondeposition, to name but two serious difficulties. In summary, the accumulation methods are indeed a sort of clock, but it is subject to resetting, variations in speed, and even to reversals in its movement. What was needed was a more reliable timepiece not susceptible to so many possibilities of disturbance. Shortly before the turn of the century, the search for this new, hoped-for clock was fulfilled in Becquerel's discovery of radioactivity. The radioactivity clock was to revolutionize and solidly quantify our vision of time in earth history.

USING RADIOACTIVITY TO TELL TIME

As noted in Chapter 1, radioactive, or parent, isotopes are unstable; they decay spontaneously into daughter isotopes. Within two decades, the early investigators had traced out the essential features of radioactive decay through a series of ingenious experiments. They learned that radioactivity (and likewise, atomic stability) is a property of the nucleus, and that no normally encountered extremes of temperature, pressure, or variation of the atoms chemically bonded to a radioactive atom can change the probability that it will decay within a given period of time. The decay probability is immutable because radioactivity involves the immensely powerful nuclear force, which far exceeds any other force that might be brought to bear on an atom. This favorable circumstance provides us, at last, with the makings of a superior clock.

Soon the discovery was made that one cannot predict with certainty whether an individual radioactive atom will decay within the next second, next day, or next billion years. But accurate statements *can* be made about a very large population of atoms. A convenient description of radioactive decay rates is through the notion of *half-life*: the length of time required for one-half of an initial number of radioactive atoms to decay. Surely the very term "half-life" is rather odd! Why not speak of "whole-life" or simply "lifetime"? Let us explore how the half-life concept can be used to tell the ages of rocks.

Isotopic Ages

Suppose we start with a large number of radioactive atoms at an arbitrary time which we shall call "time zero" (Figure 6-5). Obviously, decay will have reduced the size of the remaining population at any

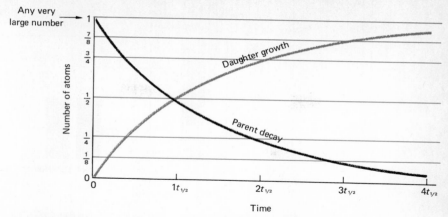

FIGURE 6-5 A parent decay curve and a daughter growth curve describe the smoothly varying, time-dependent abundances of parent and daughter isotopes. The symbol $t_{1/2}$ stands for "half-life."

later time. After one half-life has gone by, half the original atoms are still left. We can apply the same reasoning to halving the remaining atoms during the *next* half-life, and so on. The fraction of radioactive atoms therefore goes like this: (after one half-life) one-half the beginning number still remaining; (after two half-lives), one-fourth; (after three half-lives), one-eighth; then one-sixteenth, one thirty-second, etc. The remaining fraction approaches zero but never quite reaches it. Thus "whole-life" is meaningless because, in principle, some of the atoms survive forever.*

Radioactive atoms are not annihilated, but simply change status from parent to daughter. Accordingly, the number of daughter atoms, initially zero, increases at the same rate that the parent atoms disappear (Figure 6-5). The sum of parent plus daughter remains constant—only the relative distribution between the two species changes with time. Thus the changing ratio (atoms of daughter isotope)/(atoms of parent isotope) becomes the "hand" of the radioactivity clock (Figure 6-6). *Isotopic ages* are calculated on the basis of this ratio, which can be measured in the laboratory.

Each parent isotope has a unique half-life. Some radioactive species are exceedingly unstable; half-lives as short as one-billionth of a second have been measured precisely. At the other extreme are isotopes that are so nearly stable that millions or billions of years are required for half the radioactivity to disappear. (How would you express the half-life of a truly stable isotope?) Since the less accurate hourglass

* What happens when only one radioactive atom remains? (It eventually decays.) Behavior of small numbers must be described by specially modified statistical formulas. In practice, the number contained in a rock specimen is always large. One-millionth of a gram of matter, a speck too small to see without magnification, contains about a million billion atoms!

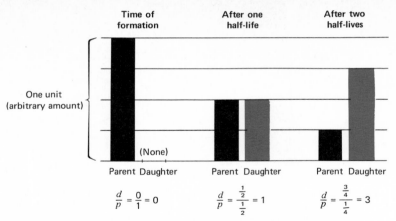

$$\frac{d}{p} = \frac{0}{1} = 0 \qquad \frac{d}{p} = \frac{\frac{1}{2}}{\frac{1}{2}} = 1 \qquad \frac{d}{p} = \frac{\frac{3}{4}}{\frac{1}{4}} = 3$$

FIGURE 6-6 The height of each bar is proportional to the number of atoms of parent or daughter. The ratios pictured here apply only to the situation after a whole number of half-lives has elapsed. A simple formula has been derived to express the daughter/parent (d/p) ratio at any arbitrary time.

methods have already suggested that rocks may be very ancient, it is apparent that a geologist must work with long-lived isotopes. Suppose he foolishly tried to measure the age of a billion-year-old rock by use of a radioactive isotope whose half-life is a mere 10 million years. After a billion years had gone by, an infinitesimal 10^{-30} of the original parent would remain—an amount far too small for accurate measurement.

Nature has endowed us with five isotopes whose half-lives are of suitable values (Table 6-1). These radioactive species were part of the earth from the time of its beginning. Happily, the chemical elements they represent are distributed through a wide variety of rocks. Potassium is a major constituent in feldspar and mica, and is a less abundant but still easily measurable element in hornblende. Rubidium, uranium, and thorium are found in a number of minerals but usually only in parts-per-million amounts.

TABLE 6-1
Parent-Daughter Pairs Used for Isotopic Age Determination

Parent	Daughter	Half-life (billion years)
^{87}Rb*	^{87}Sr	50
^{40}K	^{40}Ar + ^{40}Ca†	1.3
^{238}U	^{206}Pb	4.5
^{235}U	^{207}Pb	0.70
^{232}Th	^{208}Pb	14

* An age based, for example, upon the ratio (daughter ^{87}Sr)/(parent ^{87}Rb) is called a rubidium-strontium (or Rb-Sr) age.
† Potassium 40 decays both to argon 40 and to calcium 40. In practice, only the ^{40}K$\rightarrow$ ^{40}Ar branch is used for age determination.

Mass Spectrometers

Specialized, highly sophisticated (and expensive) instruments had to be developed before isotopic age measurements were to become routinely available in the early 1950s. Most elements are mixtures of isotopes—they consist of a *spectrum* of masses. The analyzing instrument, a mass spectrometer, must be able to sort out and record the relative abundance of each isotope. Why must the isotopes be thus separated? Take the element strontium as an example; although strontium has four stable isotopes, only ^{87}Sr is a decay product of radioactivity.

Suppose we wish to examine the abundances of strontium isotopes in a rock specimen (Figure 6-7). Purified strontium, chemically separated from other elements in the rock, is coated onto metal ribbons

FIGURE 6-7 This mass spectrometer is designed to analyze solid materials. The flight paths of ions traveling along the bent tube (forked white line) are deflected where the tube passes between the poles of an electromagnet (M) that weighs about 1200 kilograms. The mass of this magnet is about a billion billion billion times greater than the mass of the ions it deflects!

which are inserted into the machine at location S. After the machine has been pumped to a high vacuum, the ribbon assembly is heated so as to evaporate part of the sample in the form of positive ions. A high-voltage ion "gun" fires the ion particles between the poles of a powerful electromagnet (M). The magnetic field bends the flight paths of the ions. Light ions are deflected more than heavier ions. Beyond the magnet, the ion beams fly apart in slightly different directions depending upon the masses of the isotopes. By varying the magnetic field strength, the operator can aim first one, then another of the isotope beams into a collector (C) for an intensity measurement. Other types of mass spectrometer can handle gas samples such as argon 40, which is a daughter of the decay of potassium 40 (Table 6-1).

The ponderous appearance of a mass spectrometer is deceptive, for an analysis customarily requires only a few millionths to a few billionths of a gram of material! A typical age determination costs about one man-week of effort and (in commercial value) hundreds of dollars. In spite of the high cost, information about the dimension of time in geology is so valuable that already perhaps 100,000 isotopic age measurements have been completed in laboratories around the world.

Interpretation of Isotopic Ages

We must bear in mind that in the laboratory a *ratio* of isotope abundances is measured. We cannot directly sense the *age* of a rock, which is actually a calculated value based upon this ratio. What event does an isotopic age signify? How do these isotopes participate in geologic processes as "labeled" or as "tracer" atoms? Each of the three classes of rock (igneous, sedimentary, metamorphic) illustrates special problems in the application of isotopic age methods.

Igneous rocks   What if a rock, when it formed, inherited some daughter isotopes generated by earlier radioactive decay? In this case, the rock would start with the indication of an age already "built in." Let us see how this possible objection to the isotopic age technique applies to the igneous rocks. A straightforward example is an isotopic measurement by the potassium-argon (K-Ar) method on a volcanic rock. When molten lava reaches the earth's surface, the argon 40 produced by radioactive decay before the time of eruption is quickly released to the atmosphere. Various minerals that crystallize in the lava may contain ^{40}K, but no daughter ^{40}Ar. Once the temperature drops below about 500°C in the case of hornblende, or 200°C in the case of biotite, the subsequently produced argon (even though a gas) is trapped in the crystals where it continues to accumulate indefinitely.

A potassium-argon age of the lava therefore refers to the time when the clock was "reset"—the time since the material had cooled sufficiently to retain argon once again. Even the simple interpretation of igneous rock ages must be qualified, though. Cooling of a granite batholith may be so slow (beneath thousands of meters of insulating overburden) that argon continues to escape the rock through tiny cracks for millions of years after the granite has become completely solid.

The inherited decay products of rubidium 87, uranium 235 and uranium 238, and thorium 232 are not so mobile as argon, and do indeed remain in the magma, eventually to be frozen into the solid igneous rock. Some very clever procedures have been invented to correct for the initial quantity of these daughter isotopes. Properly applied, the Rb-Sr, U-Pb, and Th-Pb methods are also indicators of the time of crystallization.

Sedimentary rocks The age of a sediment is conventionally taken to be the time since it was deposited. Suppose that a granite were eroded, and that pieces of it were deposited as a conglomerate. What would an isotopic age measurement on a granite boulder in the conglomerate signify? We would obtain interesting and useful information about the granite source rock, but not the age of sedimentary deposition. (Carrying a clock from one room to another is not the same as resetting the clock.) For this reason, most sedimentary rocks cannot be dated directly. Resetting an isotopic clock requires a physical separation of parent from daughter isotopes, which is a more radical process than the mere transportation of rock or mineral grains.

Metamorphic rocks Metamorphic rocks are even more difficult, for they record at least *two* kinds of age: the time since the rock first formed, and the time since it was recrystallized by metamorphism (Figure 6-8). (For that matter, some rocks have been remetamorphosed again and again.) Only in a few localities has the history of metamorphism been resolved in any detail.

A case study of rocks near Baltimore, Maryland, illustrates how different isotopic clocks on a variety of rocks and minerals can be used to unravel a complicated history of metamorphism (Table 6-2). Here, an older gneiss is unconformably overlain by younger marble and schist. Igneous rocks are injected into both the lower and upper series of metamorphic rocks. Uranium-lead ages on zircon* from the older gneiss are about 1.1 billion years. The rather inert zircon was practically unaffected by later metamorphisms. Muscovite "books" and large feldspar crystals from igneous intrusions record an age of 430 million years by the rubidium-strontium method. No doubt, metamorphism accompanied the igneous activity. Regional heating continued until about 320 million years ago, as indicated by potassium-argon and

---

* A zirconium silicate mineral that contains a fair concentration of uranium.

TABLE 6-2

Isotopic Ages from Gneisses and Igneous Rocks near Baltimore, Maryland

Geologic event	Mineral	Isotopic age (billion years)				
		$^{238}U{-}^{206}Pb$	$^{235}U{-}^{207}Pb$	$^{232}Th{-}^{208}Pb$	$^{87}Rb{-}^{87}Sr$	$^{40}K{-}^{40}Ar$
End of latest metamorphism	Fine-grained biotite, feldspar, hornblende, etc.				0.31 (10 samples)	0.33 (17 samples)
Emplacement of intrusive igneous rocks (probably accompanied by metamorphism)	Coarse-grained muscovite, feldspar				0.43 (8 samples)	
Deposition of upper series	(No secure information—probably 0.6–0.7 billion years ago)					
Metamorphism of lower series	Zircon	1.00	1.05	1.02		
		(average of data from 2 samples)				
Deposition of lower series	(Unknown—probably shortly before 1.1 billion years ago)					

Most recent event ← ... → Most ancient event

SOURCES: G. R. Tilton and others, "Ages of minerals from the Baltimore Gneiss near Baltimore, Maryland," *Geological Society of America Bulletin*, vol. 69, pp. 1469–1474, 1958; and from G. W. Wetherill and others, "Age Measurements in the Maryland Piedmont," *Journal of Geophysical Research*, vol. 71, pp. 2139–2155, 1966.

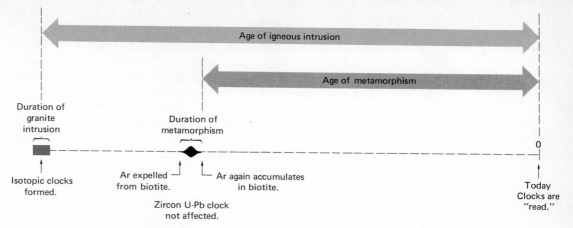

FIGURE 6-8 A time line shows the history of a metamorphosed granite. The isotopic ages refer to the time interval from a past event (such as metamorphism) to the present day, *not* to the duration of that event. In this example, the K-Ar clock in biotite was reset by metamorphic heating, whereas the U-Pb clock in zircon crystals was unaffected by metamorphism. Therefore the U-Pb clock gives the age of intrusion of the granite, and the K-Ar clock gives the age of metamorphism.

rubidium-strontium ages on fine-grained minerals from the ancient gneiss. Isotopic clocks in these small crystals were continually "reset" until heating ceased, whereas large crystals of biotite and feldspar, which were less affected by the gentle metamorphism, record an older age.

SOME MAGIC NUMBERS

Although something is going on somewhere in the earth all the time, the isotopic ages show that geologic history has been punctuated by a few especially momentous episodes of activity. Here are some "magic numbers": the ages of these profound events.

Ages of the Earth

Ironically, the age of the greatest event of all, the origin of the earth, is the most difficult to estimate. How are we to select the sample to be analyzed? Shall we take a mixture of representative pieces of the whole earth? Impossible! What is the difference between the age of the earth and the age of a rock? Is the oldest rock as old as the earth? Rocks can be destroyed and reconstituted; in fact, on the average each atom in the upper part of the crust has, at one time or other, been part of five different rocks.

Moreover, does the "age of the earth" mean the time when its atoms were created, the time when it condensed from a dust cloud, or the time of its differentiation into crust, mantle, and core? The first two of these events presumably involved all the members of the solar system. Perhaps we can attack the problem by assuming that meteorites and the earth shared a common earliest history. Since most meteorites have not been affected by the numerous agents of change that are familiar on earth, these visitors from space may provide just the clues we need.

Age of creation of the elements   Astonishingly, it is actually possible to estimate the time interval between the creation of the chemical elements and the final stages of the creation of the solar system. Since the earth and meteorites were not yet in existence during this time, it is necessary to use the radioactivity clock in a somewhat novel manner.

In Chapter 1 we learned that heavy elements, synthesized in supernova explosions, were mixed into the dust cloud that condensed into the solar system. One of the many products of a supernova is iodine 129, a radioactive isotope admirably suited to our purpose:

$$\text{Iodine 129 } (^{129}\text{I}) \xrightarrow[\text{16.4 million years}]{\text{half-life} =} \text{Xenon 129 } (^{129}\text{Xe})$$

Parent   Stable daughter

No iodine 129 remains today. Compared to the remoteness of the creation event, the half-life of this isotope is so short that not one atom would survive even if all the matter in the visible universe had been iodine 129 initially. However, early in the history of the solar system a considerable amount of this now extinct radioactivity was still around. While the primordial cloud shrank, then segregated into sun, planets, and other objects, the ^{129}I decayed into ^{129}Xe, which simply mixed into all the other gas. Finally, when meteorites condensed, some of the remaining iodine was incorporated into them, but the xenon to which it eventually decayed was trapped in the solid object.

The longer the interval before meteorites took form, the less iodine 129 yet remained to be incorporated into them, and the less the amount of the decay product (xenon 129) that would be found in meteorites today. Of course, an exact "decay interval" cannot be defined if more than one supernova explosion contributed debris to the solar system. The fact that *any* extinct radioactivity was trapped in meteorites shows that the interval must not have been longer than, say, something between 60 and 200 million years. An even more significant finding is that all the meteorites so far studied in detail have formed within an interval of less than 15 million years, and that the common stony variety formed (long ago) within 1 or 2 million years. Compared to the immensity of geologic time, the final stage of accumulation of the solar system was an instantaneous event.

Age of accumulation and differentiation Conventional isotopic age methods, based upon the decay of potassium, rubidium, and uranium isotopes, can be applied to meteorites. In general, the ages point to the time when the meteorites differentiated into stony and iron types. Reliably determined meteorite ages show a pronounced cluster of values of about 4.6 billion years, but no ages higher than this. (Some meteorite ages are shorter, suggesting that the meteorite experienced a later event such as a violent collision in space or passage close to the sun.) The significance of this magic number to earth history is somewhat obscure. Several lines of evidence strongly indicate that the earth is united with all the other solar objects into a single chemical system. If so, the earth accumulated 4.6 billion years ago, then differentiated immediately.

The oldest rock The most ancient known rocks, in southwestern Greenland, are about 3.75 billion years old. Relict areas nearly as old as this are preserved in the Kola Peninsula of the U.S.S.R. (bordered by the Arctic Ocean), in Minnesota, and in localities scattered over southern and central Africa. Other very ancient regions are being discovered from time to time, but it now seems certain that no very large areas have escaped detection.

There is not a single known rock that formed during the first 800 million years of earth history! What did the earth look like then? How much of the continents is of great antiquity, but overprinted so intensely by metamorphism that the more primitive record was obliterated? Were there continents at all, or only a few volcanic islands sprinkled in the midst of a shallow but universal ocean? Was the face of the earth shattered constantly by the bombardment of meteorites, as the moon was? We have no substantial evidence on this subject.

Age of life and of the ocean Sedimentary rock is not likely to survive for several billions of years without being eroded or metamorphosed. By good fortune, the oldest sediments known, in the Republic of South Africa, contain fossils of single-celled organisms resembling modern rod-shaped bacteria. These rocks, which have been dated by Rb-Sr at about 3.4 billion years, prove by their sedimentary character and fossil content that water was present. Life may have appeared much earlier, possibly almost as soon as the surface of the earth was cool enough to permit water to accumulate.

OROGENY

The term *orogeny** (o-RODJ-e-ny) is a convenient shorthand for the phenomena active in the formation of mountain chains. Generally these

---

* From the Greek meaning "origin of mountains."

processes include faulting and folding of the rocks, igneous activity, and at depth beneath the growing mountain range, metamorphism. Even though extensive regions of the continents are rather low plains today, the geologic record clearly shows that almost every part of the world's land masses has experienced orogeny at some time.

In contrast to orogeny is another style of upheaval in which the earth's crust may warp upward or downward on an extremely broad scale. The movements of *epeirogeny* (EP-i-RODJ-e-ny) are dominantly vertical, involving little or no folding of the rocks. Plateaus and broad basins are examples of land forms that may be produced by epeirogenic movements.

North America: A case study In order to understand the interplay of orogeny and epeirogeny in better perspective, let us consider the structure of North America, a familiar continent that is also one of the best known geologically (Figure 6-9). Fringing North America along its eastern and western sides are mountain chains that have taken form through a complex series of orogenies within the past 500 million years (the latest 11 percent of geologic time). A lowland stretches some 5000 kilometers from the Arctic Ocean south to the Gulf of Mexico through the central part of the continent.

This interior lowland in turn can be divided into three subunits. An immense tract, called a *shield*, includes the eastern two-thirds of Canada and a small part of the northernmost United States (Figure 6-9). Numerous characteristics of the Canadian Shield attest to its involvement in orogeny. Gneissic layering stands vertically in large areas of high-grade metamorphic rocks, as though crumpling forces had deformed the rock into an accordionlike mass of tight folds (Figure 6-10). Rocks in the shield have been torn by faults and invaded by dikes and other intrusive igneous bodies ranging up to batholithic size. Mineralized zones containing gold, copper, nickel, and other valuable metals are found in some of these once deep-seated rocks.

And yet, the land surface of the shield is only tens of meters to 100 or so meters above sea level. After the last orogeny subsided, the region must have stabilized into a rigid, nearly unyielding block that was reduced by erosion to a landscape of low relief over the following millions of years (Figure 6-11). Here and there, small patches of younger, little disturbed marine sedimentary rock lie unconformably upon the eroded rocks of the Canadian Shield. The shield region was therefore not *perfectly* stable—it was epeirogenically depressed slightly beneath sea level and reelevated, perhaps several times. But never again was it caught up in extreme mountain-forming movements.

A second parcel of lowland borders the shield on the west in Canada, expanding greatly in the United States to include the Great

FIGURE 6-9 This highly generalized map of North American physiographic provinces shows the Canadian Shield, Great Plains, and Coastal Plain. Areas of relatively undisturbed sedimentary rocks lie unconformably on the Canadian Shield. Small squares indicate the locations of Figures 6-10 and 6-11 (and the area from which the data in Table 6-2 were obtained). A cross section drawn along line *ABC* is seen in Figure 6-12.

Plains states (Figure 6-9). Almost everywhere, this area is overlain by a thin veneer (typically about a kilometer thick) of younger sedimentary strata. In North Dakota, Illinois, Michigan, and the Oklahoma-Texas panhandle, local epeirogenic movements have depressed basins that are now filled to the brim with sedimentary rock. Some 5600 boreholes in the United States have pierced all the way through the sedimentary blanket. Samples recovered from these deep wells demonstrate that a floor of shield-type rocks lies beneath the sediments; an erosional unconformity separates the two groups of rock. Geologists refer whim-

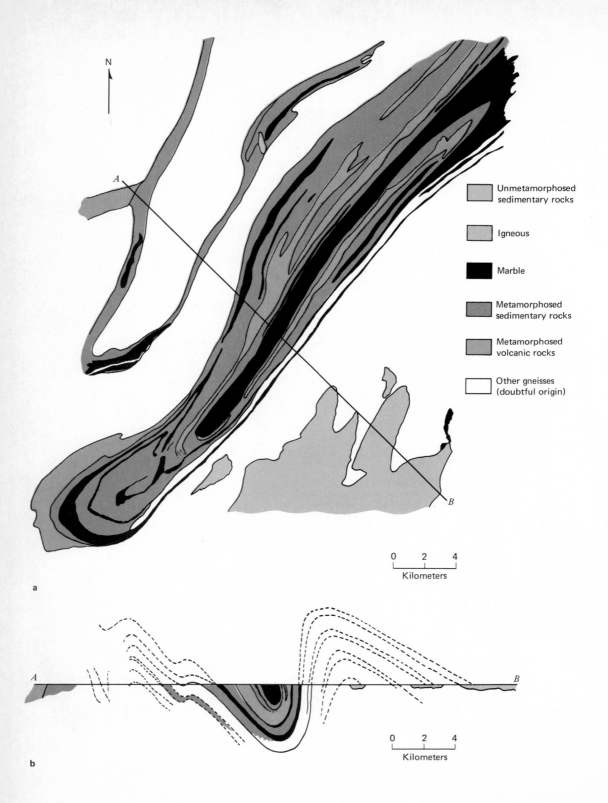

	Unmetamorphosed sedimentary rocks
	Igneous
	Marble
	Metamorphosed sedimentary rocks
	Metamorphosed volcanic rocks
	Other gneisses (doubtful origin)

N

A

B

0   2   4
Kilometers

a

A   B

0   2   4
Kilometers

b

FIGURE 6-10 (*a*) A geologic map of a small portion of the shield in Ontario shows a mass of layered metamorphic rocks that have been tightly pinched into a canoe-shaped downfold. The keel of the fold runs northeast through the middle of the rock outcrop pattern. (*b*) In cross section along line *AB*, the surface of the shield is seen to be essentially flat. Once-horizontal layers now stand nearly vertical, and in the core of the downfold they are even slightly overturned. [*After J. W. Ambrose and C. A. Burns, "Structures in the Clare River Syncline: A Demonstration of Granitization," in* The Grenville Problem, *Royal Society of Canada, Special Publication 1, 1956.*]

FIGURE 6-11 Much of the Canadian Shield is bare rock, scoured clean by an ice sheet and now, after the melting of the ice, dotted with countless thousands of lakes. This scene in Manitoba illustrates the typical low relief of some tens of meters.

sically to these metamorphic and igneous underpinnings of the continents as the "basement." Local high spots have been planed off by erosion in South Dakota, Missouri, and central Texas. Small patches of basement show at the surface in these areas. A shield, on the other hand, is a *very large* area of basement exposed at the surface by erosion.

Thirdly, the Coastal Plain, especially along the Gulf of Mexico, is comprised of great thicknesses of even younger sediment borne by rivers to the continental margin. Basement rocks here, if present at all, lie far beneath the penetration of the deepest drill.

On the grandest scale, the surface of the basement in central North America is a somewhat irregular plane inclined gently downward to the

153 OROGENY

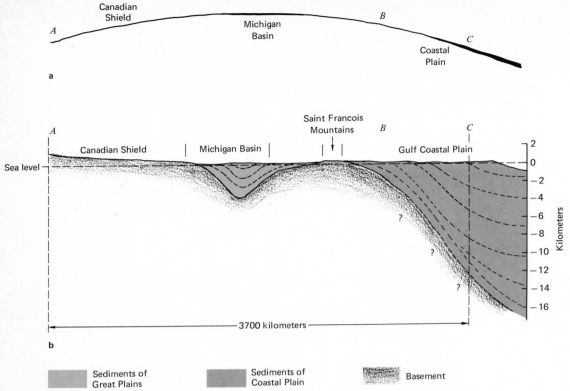

FIGURE 6-12 (*a*) A cross section along line *ABC* (Figure 6-9) indicates the earth's curvature and the sediments of the Michigan Basin (an epeirogenic sag in the basement) and the Coastal Plain in true scale. Surface irregularities and the thickness of the sedimentary "skin" are insignificant on the scale of the whole earth. (*b*) Most cross sections are drawn with the earth's curvature removed and, to convey more information, with a large exaggeration of the vertical scale (a factor of 80 in this drawing). Here, the cross section has been deliberately drawn through the Michigan Basin and the Saint Francois Mountains, Missouri (a "high" where basement is exposed). A cross section oriented in practically any other direction through interior North America would show the top of the basement to be a rather monotonous flat surface.

south (Figure 6-12). At its highest point—the Canadian Shield—erosion has already stripped off the surficial rocks. Erosion is presently nibbling away at the sedimentary cover fringing the shield, as well as reducing the elevation of the landscape generally. The weathering products, transported to the Gulf by the Mississippi River system, are building new continent as the shore is pushed farther into the Gulf of Mexico. North America is a cannibal; it grows in one place at the expense of its own substance elsewhere.

154 "... MILESTONES ON THE ETERNAL PATH OF TIME"

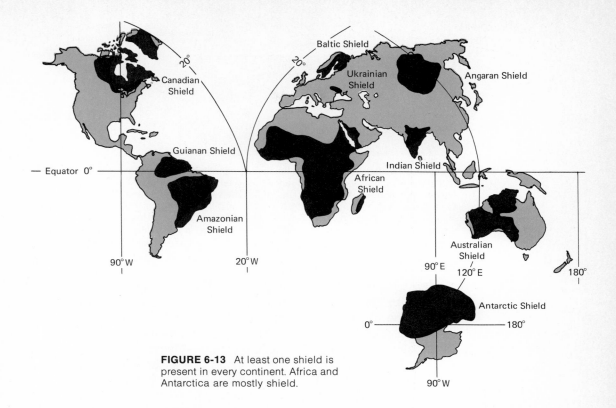

FIGURE 6-13 At least one shield is present in every continent. Africa and Antarctica are mostly shield.

Isotopic Age Patterns

Ever since geologists came to accept Hutton's concept of an actively destroyed and regenerated earth, they have recognized that the shield rocks must be almost inconceivably ancient. So much history is obviously impressed upon the distorted visage of these rocks—they *look* old. Many people thought that shields must be remnants of the original crust, survivors from the time of primordial fire. Hutton never accepted the idea that original crust could be found, though we may suspect that his conclusion was based more on wishful thinking than upon independent data.

For many years the geology of the basement was widely ignored as being impossibly difficult to unravel. Outcrops of basement complex were symbolized on geologic maps in a monotonous brown color, even though the same maps resolved the form and field relations of all the more surficial rocks in elaborate detail. Then, shortly before the turn of the century, the field geologists summoned the courage, it seemed, to tackle harder problems. They found that they could work out an involved geologic history even though the basement rocks are structurally disturbed and metamorphically recrystallized.

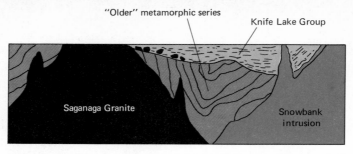

FIGURE 6-14 To clarify the essential field relationships, the more intricate faults and folds have been edited out of this idealized cross section of basement in northern Minnesota. The "older" metamorphic series and Knife Lake Group consist of iron-bearing beds, volcanic flows, conglomerate, shale, etc., all metamorphosed. Cobbles of Saganaga Granite lie locally at the base of the Knife Lake Group. [*After C. O. Dunbar,* Historical Geology. *Copyright © 1949. By permission of John Wiley & Sons, Inc., New York.*]

An idealized cross section of an area in northern Minnesota, mapped by some of the pioneers of basement geology, illustrates several of the principles of interpretation (Figure 6-14). What is the sequence of origin of the four rock bodies, as deduced from the law of superposition and the law of crosscutting relationships? What geologic events took place between the intrusion of the Saganaga Granite and the later emplacement of the Snowbank intrusion? How much time would be required for these intervening events to take place? A hundred million years? Two billion years? Isotopic ages have revolutionized our understanding of the ancient orogenies. The ages have upheld the interpretations of the field geologists, but sometimes in an odd and unexpected way. For example, both these igneous intrusions in Minnesota were found to be about 2.7 billion years old. They formed within a time interval too brief to be resolved by the isotopic ages!

Hundreds of localities in the North American basement (both the shield and buried basement) have been sampled for age determination. A plot shows that the continent is organized into a sort of crazy-quilt patchwork pattern of ages (Figure 6-15). Continuous areas of millions of square kilometers are characterized by uniform isotopic ages (within a plus-or-minus of 100 million years).* Analyses completed thus far have been mostly by the K-Ar and Rb-Sr methods on biotite. Since biotite is so easily recrystallized, these ages indicate when a terrain cooled off following its *latest metamorphism.* These metamorphic ages seem def-

---

* Compared to, say, 2.5 billion years, a variation of 100 million years is only a rather minor ±4 percent. The isotopic ages vary partially because of the ever present experimental error, but in part the variation shows that orogeny may have been in progress for tens of millions of years.

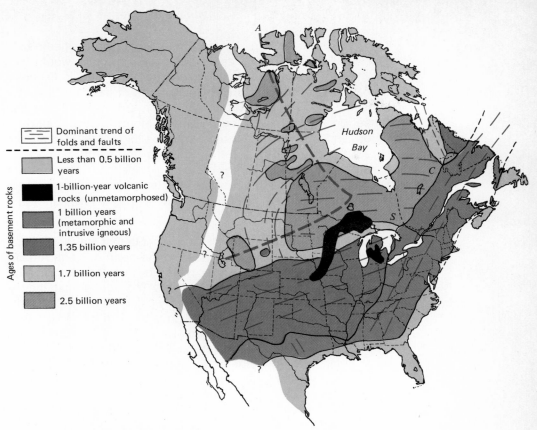

FIGURE 6-15 Some of the apparently younger areas in this map of North American isotopic age provinces are actually more ancient rock that has been caught up in renewed orogeny a second, or even a third, time. Trends of deformation of older orogenic belts are abruptly cut off and transected by trend directions of younger belts. [*Data compiled from W. R. Muehlberger and others, "Basement Rocks in Continental Interior of United States,"* Bulletin of the American Association of Petroleum Geologists, *vol. 51, 1967*]

initely to signify times of orogeny, but not necessarily the original age of a rock. For example, several small "islands" of older rock are strung out along trends *A* and *B* (Figure 6-15) away from the main 2.5-billion-year-old area centered below Hudson Bay. Probably this ancient orogeny affected a much larger area (perhaps half the continent), but much of the region was strongly overprinted by renewed metamorphism 1.7 billion years ago. Isotopic age techniques designed to "peer through" an obscuring metamorphism to detect the primary age of a rock are laborious and difficult. We shall not have a detailed picture of the multiple overlays of orogeny in North America for at least one or two more decades.

Suppose that an ancient basement terrain were reheated at a much later time. Since the heat can diffuse through rocks, would we not expect the metamorphism to taper off gradually and indistinctly at the margins of the affected area? Some parts of the Canadian Shield do show that old isotopic clocks have been only partially set back by a gentle, pervasive rewarming. However, along line *SC*, between the 1-billion- and 2.5-billion-year-old provinces (Figure 6-15), the transition is incredibly abrupt. It is more narrow than the width of the line on the scale of the drawing. Field evidence supports the indications of the isotopic ages. In some places the metamorphic grade can be seen to change dramatically over a distance of just a few hundred meters.

Do continents grow larger? You will note that the most ancient ages are preserved in a block of territory in the heart of North America (Figure 6-15). Successively younger belts rim this core, and finally, mountains of the modern orogenic episode (so young that they are still rugged) trend along the eastern and western seaboards. Could North America have grown larger by a series of additions that welded new crust onto old shield blocks? Perhaps the growth of continents would signify that the crust is still differentiating from the mantle. How much of the continent is simply older rocks, reworked over and over? A decisive answer must wait until the primary ages of these complex terrains can be established. Concentric age patterns were not found in the other continents. Their most ancient rocks are just as likely to be located on a coast as in the interior.

Ages of orogeny Is mountain building going on all the time, or does it take place in worldwide episodes? It is difficult to tell, for the older geologic record is poorly preserved. Ancient orogenies are easily overlooked. Moreover, since orogenies last tens to hundreds of millions of years, some overlap in timing is expected. Even with these qualifications, a decided clustering of isotopic ages suggests that orogeny affected several continents at about the same times (Figure 6-16). In the Northern Hemisphere, the active periods were centered on 2.5, 1.7, and 1.0 billion years ago. Barest traces of older orogenies show through in a few places. Similar ages are found in the Southern Hemisphere, except that a conspicuous 600-million-year-old orogeny is seen in place of the 1-billion-year event. In the more recent geologic record, several orogenies can be finely resolved. The isotopic ages show that orogenic pulses in different parts of the world were not strictly simultaneous. Periods of multiple, closely spaced pulses were set apart by other

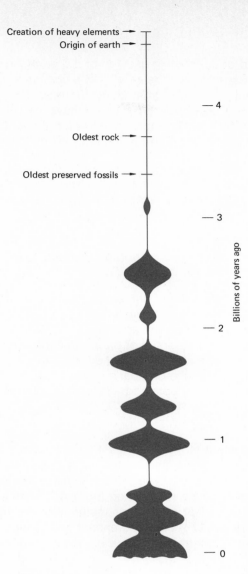

Creation of heavy elements →
Origin of earth →

— 4

Oldest rock →

Oldest preserved fossils →

— 3

Billions of years ago

— 2

— 1

— 0

FIGURE 6-16 A time line for earth history shows some of the most prominent events. The solid figure corresponds to the distribution of isotopic age values from orogenic belts. Orogeny is very active today.

periods of relative quiet lasting for hundreds of millions of years. These intervals were distinguished by deep erosion, gentle epeirogenic warping of the crust, and the advance and retreat of shallow seas across the beveled continental surfaces.

The ages of orogeny join our list of magic numbers of worldwide significance. They are perhaps the most astonishing results of all, and we shall make good use of them in a following chapter.

1 Instead of calculating the age of the earth, Joly actually calculated an average "residence time" that a sodium ion spends in the ocean before it is removed from the water mass. Refer to Table 5-2. If data had been available, Joly could have calculated "ages," or residence times, of other ions in seawater. Would these ages have been longer or shorter than his value of 90 million years for the sodium ion?

2 Refer to Figure 6-5. What is the *sum* of the quantities of daughter and parent isotopes after one half-life has gone by? After two half-lives? After three half-lives? After four half-lives?

3 Refer to Table 6-1, for the half-life of ^{238}U. What fraction of the ^{238}U that was originally incorporated into the earth has already decayed away?

4 Zircon crystals from ancient basement rock were analyzed isotopically, and found to contain 0.0285 units of daughter ^{207}Pb and 0.0019 units of ^{235}U. The amount of ^{235}U *originally* present in the mineral equals the quantity of parent isotope remaining today plus the quantity that has already turned into daughter. What is this sum? What fraction of the original amount of parent is present today? This fraction would be obtained after how many half-lives have gone by? One half-life of ^{235}U equals 700 million years (Table 6-1). How old is the zircon?

5 Deposition of terrigenous clastic sediments (Chapter 5) cannot be dated by isotopic age methods. Is this unfortunate circumstance true of all sediment types? For example, would you expect the isotopic clock to be reset when the evaporite mineral sylvite (KCl) is deposited? What method or methods might be suitable to date a sylvite deposit? Would you expect any of the isotopic clocks to be suitable to determine the age of limestone or dolomite?

6 How would the near passage of a meteorite to the sun affect the apparent age that would be measured by the K-Ar technique? Why is this age changed by the event?

READINGS

* Eicher, Don L., 1968: *Geologic Time*, Prentice-Hall, Inc., Englewood Cliffs, N.J., 149 pp.

* Faul, Henry, 1966: *Ages of Rocks, Planets, and Stars*, McGraw-Hill Book Company, New York, 109 pp.

 Toulmin, Stephen, and June Goodfield, 1965: *The Discovery of Time*, Harper & Row, Publishers, Incorporated, New York, 280 pp.

* Available in paperback.

chapter 7

Origin and Evolution of Life

"Ammonia! Ammonia!"

Life on earth originated in an atmosphere devoid of free oxygen. Could the chemistry of life on some distant planet require a nonoxidizing atmosphere, like that of the early earth? [*From* The Scientific Endeavor, *Rockefeller Institute Press, National Academy of Sciences, Washington, D.C., 1965.*]

All life is one. In what sense can such a sweeping statement be true? What is its relevance to human society in the midst of a world of other living things? No wonder that such intense controversy has beset Charles Darwin's book *On the Origin of Species* ever since its publication in 1859. Perhaps his conclusions suggest that man is related to other animals a little too closely for comfort!

Darwin was the first to present a carefully documented set of arguments in support of the theory of *organic evolution*—the hypothesis that organisms are alike not only in being alive, but also in being linked together by a common *historical* development. He proposed that modern organisms have evolved, or gradually emerged, from ancestral types to which the modern types may bear only faint resemblance.* He accepted the obvious principle that the present state of the earth is a result of events that happened during all the previous moments, clear back to the time of the earth's beginning. Darwin wedded the historical viewpoint to the observation that each generation of living organisms is slightly different from its parent generation.

It follows that evolution of life *has* to take place. No matter how subtle the changes may be, the fact remains that the complexion of life on earth can never exactly repeat at any two moments. There are simply too many possibilities of variation. Still, it will take more than a set of truisms about historical antecedents and inheritance to explain the origin of the 10 million species of organisms that are estimated to exist. How came this immoderate deluge of ants and apple trees, of bacteria, birds, and human babies? Why did dinosaurs, dodos, and other creatures once flourish, only to vanish later?

And what has geology to do with all this? The answer is threefold. Other hypotheses of the origin of species, advocated before Darwin's day, required only a short time scale for the earth's history. Vistas of

* We have since discovered that on the level of the chemistry of a living cell, resemblances among organisms have remained rather close.

163

time opened by the newly developing science of geology in the mid-nineteenth century profoundly influenced Darwin's thinking. The Darwinian idea of imperceptibly slow change over long periods would not have appeared to be reasonable without supporting evidence from geology. Ideas about the origin of species developed in close parallel with the growing awareness of geologic time noted in Chapter 6. Secondly, a successful theory must account for the living kingdom in both space and time. The biological sciences focus upon the aspects of evolutionary theory that can be observed and tested "here and now." Geology has contributed to this picture through information provided by *fossils*.* These remains of organisms entombed in the rocks are documents of what actually happened as the history of life unfolded. Thirdly, important parts of the sedimentary rock record are of biological origin. Evolutionary theory can help to explain, for example, why limestone (and marble) is rare among the very ancient rocks, but abundant among rocks that were deposited after $CaCO_3$-secreting organisms had evolved.

ORIGIN OF LIFE

There is a curious parallel between the living and nonliving states of matter. We saw that silicon-oxygen tetrahedra in silicate minerals may be joined together into chains, sheets, and more complex structures. In a similar manner (Table 7-1), the atoms of living material are organized into organic molecules, and the molecules in turn are arrayed in a regular manner in *organelles*—discrete structures within cells.† Like the organs in the body, various organelles joined together in a working relationship can form the simplest possible living organism, a single *cell*. An agglomeration of many cells performing a specific function forms tissue. Generally, multicelled animals and plants are fashioned from a considerable variety of tissues. Aggregates of many individual organisms of the same type may be considered as a population, which may live in an association or community of populations.

Could it be that the linkage of atoms into organic molecules, of molecules into organelles, of organelles and certain other ingredients into cells, of cells into tissues and finally into the whole organism is in fact the most fundamental declaration about all life? A classification based upon the degree of organization is both straightforward and natural, and just might be an important clue to the origin and development of life.

---

* From the Latin word meaning "dug up."
† Chromosomes are an example of an organelle whose function is to hold and transmit hereditary information.

TABLE 7-1
A Hierarchy of the Organization of Life

Level	Example	Scientific discipline*	Comments
Species	Grizzly bear, *Ursus arctos horribilis*		
Population	Grizzly bears in Yellowstone National Park		
Organism	One grizzly bear		
Organ system	Circulatory		
Organ	Heart		
Tissue	Muscle		
Cell	Muscle cell		
Organelle	Chromosome		
Giant molecule	Deoxyribonucleic acid (DNA)		
Small molecule	Nitrogenous base		
Atom	H, C, N, O, P, etc.		
Subatomic particle	Proton, neutron, electron		

Scientific disciplines (spanning levels): Anatomy, Taxonomy, Genetics, Embryology, Ecology, Histology, Cytology, Paleontology, Biochemistry, Chemistry, Particle physics, x-ray crystallography. Comments: Living / Nonliving; "Higher" animals and plants only; Increasing complexity of organization.

* Ecology and taxonomy extend above the chart to include the analysis and interaction of broader groupings of organisms. Boundary lines of the various disciplines are only approximate.

Of course, no one presumes that life extends down to the lowest rungs of the organizational ladder. Even the most complicated organic molecules do not have the ability to reproduce, to respond to stimuli, to ingest food, etc., that we associate with "being alive." At the same time, is the old saying true that "Only life begets life"? The world received a rude surprise when Wöhler accidentally synthesized an organic compound (urea, NH_2CONH_2) in 1828. Up to that time it was firmly believed that only life could beget organic molecules, let alone the more complicated task of begetting another living organism. Today we can artificially make thousands of kinds of molecules that are also produced by living cells. Within the last two decades, progress has greatly accelerated toward closing the gap between the more complex states of organization of "mere matter" and a truly living entity. Perhaps nature followed a similar course as a long period of "chemical evolution" preceded the present era of "biological evolution" that is the theme of Darwin's book.

Ancient and Not-so-ancient Myths

Undoubtedly most people find the ancient mythologies to be fun reading, even if unscientific. Far from being trivial, these venerable stories about gods, monsters, and superhumans are an attempt to grapple with cosmic questions of life, death, and the place of man in the universe on a level that is too deep for ordinary language. Myths are forever unproven and unprovable because they describe events that are supposed to have happened in the distant past, and *only once*. These same qualities of mythology meet our endeavors to understand how life began. Life is unquestionably of cosmic importance (to earthlings, anyway), and the unique circumstances of its origin probably will never be known precisely. The ancient myths once must have seemed plausible enough, though we would regard them as based upon inadequate data. Will some of *our* most cherished beliefs sound mythological to future generations?

Up until the mid-1800s, the origin of life was not considered to be a problem. Although the transmission of life through sexual and asexual reproduction was well known, it was also widely believed that life could occasionally arise through *spontaneous generation*. In the fourth century, Saint Basil the Great, an authority of the Eastern church, wrote:

For if there are creatures which are successively produced by their predecessors, there are others that, even today, we see born from the earth itself. In wet weather she brings forth grasshoppers and an immense number of insects which fly in the air and have no names because they are so small; she also produces mice and frogs. . . . We see mud alone produce eels; they do not proceed from an egg, nor in any other manner; it is the earth alone which gives them birth.

In the seventeenth century, F. Redi disproved the accepted notion that fly maggots can arise spontaneously. He showed that no maggots appeared on meat placed inside jars covered by muslin, whereas they appeared in abundance in similar uncovered jars. A long series of experiments culminated in 1862 in Louis Pasteur's elegant demonstration that decay is possible only if invading microbes can infect the decaying object. His experiments led to the pessimistic conclusion that the origin of life was beyond the scope of science, and further work on the subject was discontinued for some 50 years. However, Pasteur did not prove that spontaneous generation *never* had occurred, but only that it is not a commonplace event today.

If spontaneous generation was not a "first cause," could life on the earth have been seeded from another part of the universe? Calculations suggest that a bacterium, having a mass of a trillionth of a gram, could be propelled from Alpha Centauri (the nearest star) to the earth under the pressure of starlight in 9000 years. Probably the low temperature of interstellar space, the long hibernation while en route, and the meteoric descent of the organism into our atmosphere would not kill a bacterium, but the intense ultraviolet radiation in space almost certainly would. Philosophically, this theory is unsatisfying. It merely removes the problem from the earth, about which we know something, to some other solar system, about which we know nothing.

A Modern Scientific Myth

Much of the modern life mythology (still an appropriate term for these studies) we owe to the work of a biochemist, A. I. Oparin (U.S.S.R.) and the mathematical geneticist J. B. S. Haldane (Great Britain). They view life as originating in three stages as the original simple compounds were organized into more and more complex aggregates. These stages (1) brought the simplest substances together into organic molecules, which (2) became further associated into giant organic molecules. These in turn (3) formed the organized structures in organelles and cells. Let us examine the evidence for these stages in more detail.

Stage I: From atoms to simple molecules Is life probable? An origin by finely graduated steps will seem more acceptable if we can establish a high probability for each step. Figure 1-8 reminds us that hydrogen, carbon, nitrogen, oxygen, phosphorus, and sulfur, the main chemical ingredients in living cells, are also among the most abundant elements in the universe and in our solar system.

Meteorites were analyzed in the search for clues to the precursors of life. A brief flurry of excitement in the early 1960s followed the alleged discovery of fully developed microfossils in these visitors from space. Further search did not always confirm the findings, even when different pieces of the same meteorite were examined. Now it appears that the so-called organized particles are either contamination or peculiar external forms of mineral grains. Many meteorites do contain a rich variety of organic molecules, but according to present consensus, they are not of biological origin. Probably the earth contained a similar assortment of compounds during its prebiological stage.

Complex organic molecules in living cells are made up of a few primary "building blocks." Among these are 5 *nitrogenous bases* and a group of 20 *amino acid* molecules (Figure 7-1). Long chains of nitrogenous bases, linked by sugar and phosphate, comprise the DNA* molecule, the information storehouse of the cell. Amino acids may be joined into giant molecules, the proteins. Protein molecules, typically representing the linkage of around 500 amino acid units, perform a variety of the ongoing functions of the cell. In addition, a cell contains fat and carbohydrate. These 29 building blocks (20 amino acids, 5 bases, sugar, phosphate, fat, carbohydrate) do not quite account for all of cell chemistry, but they make an excellent start. They are "letters" of a living alphabet that has only a few more characters than the 26 letters of the English alphabet.

* Short for *deoxyribonucleic acid.*

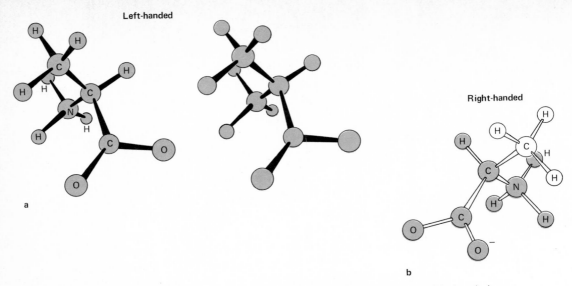

FIGURE 7-1 Amino acids can exist in both left-handed and right-handed forms, but only the left-handed type is synthesized by living cells. (*a*) In this perspective view of alanine, the second simplest amino acid, the connecting rods represent chemical bonds. Relax your eyes as you stare straight ahead at the paired figures of alanine. The visual images will fuse together, permitting you to see the molecule stereoscopically, in three dimensions. (*b*) The shaded atoms are common to all 20 amino acids. Unshaded atoms (CH_3 in the case of alanine) are different in each amino acid.

We have already seen that the ocean and the atmosphere are volcanically derived; thus water, carbon dioxide, and nitrogen were initially present, but no free oxygen. Some authorities insist that the early atmosphere also contained ammonia and methane. In 1953, the American biochemist Stanley Miller performed a pioneering experiment in which he cycled a mixture of water, methane, ammonia, and hydrogen through a sparking chamber that simulated lightning discharge (Figure 7-2). After several days of sparking, the solution (which had turned yellow) contained amino acids. Biochemists have since used ultraviolet radiation, heat, light, and radioactivity as energy sources, all successfully. Most of the 29 building blocks have been synthesized in this primitive and inefficient, but realistic, manner. The requirements are so simple that if there were any standing water, if the sun shone, and if there were thunderstorms in those early days, the completion of stage I was guaranteed. Practitioners of life mythology fondly refer to the end product as the "primordial soup." Some have even suggested that the ocean was covered by an "oil slick" of condensed organic compounds.

A major unsolved problem of stage I is illustrated by our chosen

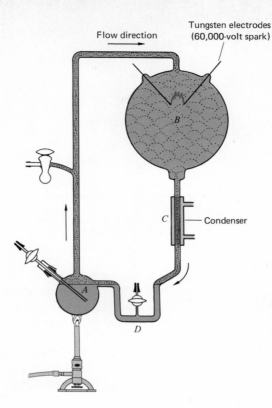

Flow direction

Tungsten electrodes
(60,000-volt spark)

B

C — Condenser

A

D

FIGURE 7-2 This simple apparatus was used in Miller's classic experiment. Water boiled in flask *A* promoted the circulation of the primordial brew. Products formed by the spark in chamber *B* were condensed at *C* by contact with a jacket through which cool tap water was kept circulating. A U-tube at *D* prevented backflow of vapor.

example of the amino acid called alanine (Figure 7-1). Although amino acids exist in equivalent right-handed and left-handed forms, it happens that in living things they are almost exclusively the left-handed sort. In one sense, the situation is inevitable because a cell containing both forms would not function, and an organism whose chemistry is right-handed could not mate with, nor even eat and digest, a left-handed organism. And yet, either variety of amino acid is equally able to participate in chemical reactions. Could it be that the very first living organism just happened to have a left-handed chemistry, and that all its descendants followed suit?

Stage II: From simple molecules to giant molecules The organic soup provided by stage I had to become even more highly organized before further reactions were likely. A remarkable photograph (Figure 7-3) emphasizes the power of these substances to condense into microbe-like droplets. A preparation of linked amino acids was boiled in seawater to simulate the action of an underwater volcano. Microspheres that condensed from this brew were roughly of bacterial size. They were composed of a double-walled membrane (like bacteria) and could

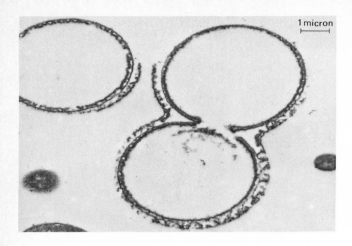

FIGURE 7-3 An electron microscope picture of stained microspheres shows one unit in the act of dividing. The double wall shown here is considerably thicker than the outer double membrane of a bacterium. A micron is one-millionth of a meter. [*Courtesy S. W. Fox.*]

even be made to form chains and to divide in half! We must not suppose from such experiments that the artificial creation of life is just around the corner, but the possibilities are certainly intriguing.

One of the chief unanswered questions of stage II is how the *genetic code* came to be embodied in molecular form. Information needed by a cell in order to carry out its functions is contained in the millions to billions of atoms in the DNA molecule. X-ray studies and arguments based on chemistry led the biochemists J. D. Watson and F. H. C. Crick to propose that DNA is a double helix twisted, like a spiral staircase, around a central space (Figure 7-4). DNA is now known to be composed of bases, phosphate, and sugar. The latter two components are attached together along a strand, and the bases are stretched, like the treads of the staircase, between the strands.

This model of the structure of DNA provided great insight into the means by which it operates. First, its long, threadlike shape makes it easy for other constituents of the cell to get next to DNA in order to "sense" it. A fundamental unit of the DNA molecule, the *gene*, is understood to be a section of about 1500 nucleotide pairs (see Figure 7-4) that codes for the production of a chain of amino acids in protein. Genes are hereditary factors; they determine the bodily construction, chemistry, involuntary functions, and many of the behavior patterns of organisms.

Secondly, DNA is a *double* strand. When a cell divides, the DNA strands unwind (Figure 7-5). Each strand can then serve as a pattern or gauge for the formation of a new, complementary strand. In this manner the genetic information encoded in the DNA is transmitted throughout the growing organism, and from generation to generation.

At present, stage II in our proposed origin of life is mostly conjecture, and in fact it would seem impossible to construct such compli-

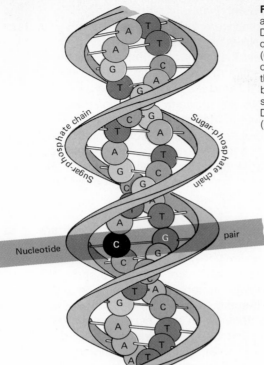

FIGURE 7-4 Double twisted chains of sugar and phosphate (left) form the "backbone" of DNA. Four nitrogenous bases—thymine (T), cytosine (C), adenine (A), and guanine (G)—are attached to the backbone by chemical bonds. Adenine always pairs with thymine, and guanine with cytosine. Two bases, together with their accompanying sugar and phosphate, form a segment of the DNA molecule known as a nucleotide pair (shaded bar).

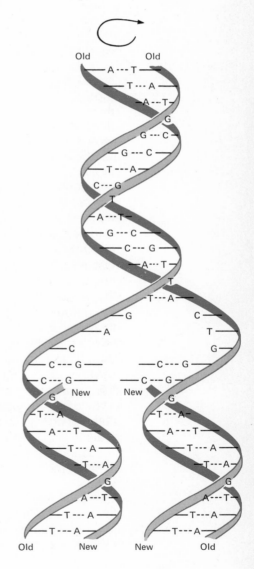

FIGURE 7-5 An unwinding double strand of DNA serves as a template, or gauge, for the formation of two new DNA molecules.

cated molecules as DNA by purely natural means. A simple calculation suggests that if numerous atoms were agitated together, they would not organize by chance into a DNA molecule in all of geologic time! But is this chancy assumption correct? Is every configuration of atoms just as likely to form as any other configuration? Sodium and chloride ions do not form a random heap, but instead they organize into a crystal structure of rock salt. Perhaps DNA and protein, though far more complex, may be "preferred" configurations that can be assembled by nonliving means. The x-ray crystallographer J. D. Bernal commented, "This argument is equivalent to saying that in the origin of life nature is always playing with loaded dice—sometimes very heavily loaded." This tendency of matter to form complex molecules has been called "biochemical predestination" by other authors.

Stage III: From giant molecules to cells The Oparin-Haldane third stage of assembly lies far beyond the frontiers of experimental work, which has only just entered the realm of stage II. Already stage III was well underway when the earliest fossil record began to appear. Among the first organisms were the lowly bacteria and blue-green algae. They are about as simple as a self-sufficient cell can be, which is none too simple, considering that a bacterium can synthesize some 3000 to 6000 compounds at a rate of about 1 million reactions per second! Cells of bacteria and blue-green algae contain just a single molecule of DNA, and they lack well-defined internal structures, such as a nucleus, chromosomes, and internal membranes. If abundance is a badge of success then these creatures are highly successful, as they are by far the most numerous organisms on earth and possess the most ancient lineage. Organisms similar to rod-shaped bacteria are found in the 3.3-billion-year-old Fig Tree Chert, a distinctive group of sedimentary beds in South Africa (Figure 7-6).

Many organisms, such as algae and green plants, can manufacture their own food from the simple inorganic compounds H_2O, N_2, and CO_2. Animals and fungi, on the other hand, must obtain the more complex sugars, carbohydrates, amino acids, etc., as primary food. Which type of organism came first in the development of life? Unquestionably, life today depends upon organisms that employ photosynthesis as a source of energy to create food out of the simple molecules. Were it not for algae and green plants at the base of the food chain, life would rapidly cannibalize itself out of existence. But photosynthesis is an advanced process dependent upon the rather complicated chlorophyll molecule (Figure 7-7). It is likely that the very first organisms simply "ate" the remaining organic soup as food. According to this version of life mythology, they quickly multiplied and used up the food supply, thereby bringing on an early crisis in the history of life. If photosynthetic organisms had already appeared, life could continue without interrup-

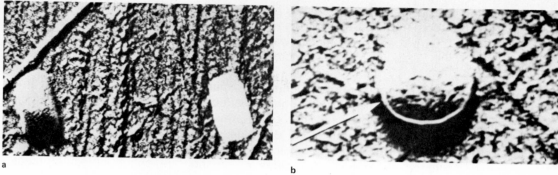

FIGURE 7-6 (*a*) During the preparation of a specimen of Fig Tree Chert, a fossil bacterium (white area) was torn loose from its original embedded position (at the left). Vertical marks are polishing scratches. (*b*) In another bacterium, cut transversely, details of the cell wall are preserved. [*From E. S. Barghoorn and J. W. Schopf, "Microorganisms Three Billion Years Old from the Precambrian of South Africa," Science, vol. 152, pp. 758–763, 1966. Copyright 1966 by the American Association for the Advancement of Science.*]

FIGURE 7-7 The rather complicated chlorophyll *a* molecule has been synthesized in the laboratory, but how plants put chlorophyll together is not known.

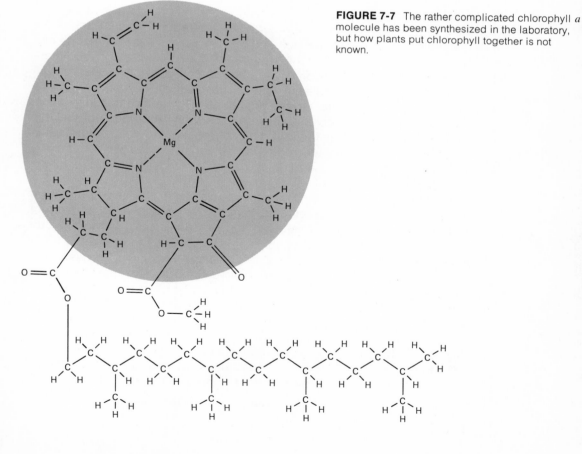

tion. Otherwise, there may have been several "population crashes," and perhaps even several disappearances and new origins before a self-perpetuating community was established.

Study of organic molecules in sedimentary rocks may someday give us a detailed account of these ancient organisms. In the meantime, the organopaleobiogeochemists* must learn how to distinguish true chemical fossils from contamination later introduced into the rock. Another problem is to recognize previous complex molecules from their breakdown products. For example, the side chain and the part of the chlorophyll structure enclosed in the circle (Figure 7-7) have been found separately in sediments, but not the entire molecule intact. Already it is possible to detect the former presence of blue-green algae in sediments by identifying its unique chemical "signature."

Moreover, the first organisms were anaerobic: they lived in an environment devoid of free oxygen. In fact, they probably had none of the elaborate biochemical defenses against an oxidizing atmosphere possessed by the aerobic organisms. Not only does the latter type function in an oxygen-rich environment; many representatives, including all the animals, *require* the presence of free oxygen. When did aerobic organisms emerge? Isotopic ages indicate that the great sedimentary ironstone deposits, noted in Chapter 5, were formed in many parts of the world between about 3 billion and 2 billion years ago, but that no significant iron deposition of this type has occurred since then. The American paleontologist Preston Cloud speculates that photosynthetic organisms first began to release free O_2 (the waste product) during this interval. This oxygen immediately combined with ferrous iron in solution, precipitating it as insoluble ferric oxides. Only after much of the iron had already been precipitated did oxygen begin to accumulate in the atmosphere. Sedimentary "redbeds," brilliantly stained with ferric oxide compounds, first appeared in the geologic record around 2 billion years ago.

The earliest photosynthesizers had to be in a position to receive sunlight and yet to be protected from the lethal ultraviolet radiation of the sun. A water layer 10 meters or so thick would have provided a sufficient ultraviolet shield. Possibly the organisms arose in shallow lakes or seas, or, what is more likely, they perhaps first gained command in the open ocean. In the latter environment, the thermocline (an abrupt transition at about 200 meters between surface water and denser water beneath) could have provided a "floor" beneath the organisms' habitat.

Once free oxygen (O_2) appeared in the air, a minute trace of ozone (O_3) was also established. Ozone is a highly effective absorber of the ultraviolet radiation that is so disruptive of DNA and protein molecules. When the oxygen content reached about 1 percent of its present level, two more dramatic improvements became feasible. A few centimeters

---

* These researchers must familiarize themselves with several overlapping disciplines of science.

of water sufficed to complete the necessary shielding, thus opening up the vast regions of shallow water fringing the continents to an invasion by organisms that swim in shallow water, grovel through the mud, or live solidly attached to the bottom. In addition, a phenomenon first noted by Pasteur began to affect the manner in which an organism respires (utilizes food energy). We assume that the first organisms derived energy by fermentation, a simple but inefficient process that goes on in the absence of free oxygen. The increasing oxygen level actually suppressed fermentation while opening the possibility of aerobic respiration, which is about 30 times more efficient than fermentation. Finally, when the oxygen level reached 10 percent of today's value, about 400 million years ago, life could move out on the land without excessive danger of sunburn.

The earliest fossils   Isolated bacterial cells have been revealed by the electron microscope only recently, and only in a few localities. Another type of ancient fossil that is common and well studied occurs as thin layers that are arranged in biscuitlike form or in simple or branching columns (Figure 7-9). Similar structures can be seen forming today, for example in Shark Bay, Australia, where frequently wetted mud flats are encrusted by a dense mat of algae. Actually, the outline of fine sediment trapped in the algal mat is preserved, not the form of algal cells as such. Geologists have interpreted the multitude of forms of the fossil algal colonies in different ways. The North American school claims that differences in water level between low and high tide, the swiftness of the currents, etc., are responsible for the differing shapes.

FIGURE 7-8 A time line summarizes some of the hypothesized major events in the origin of life.

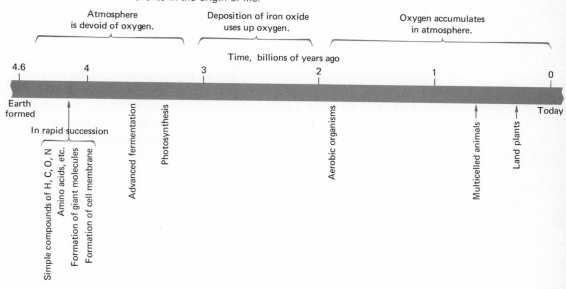

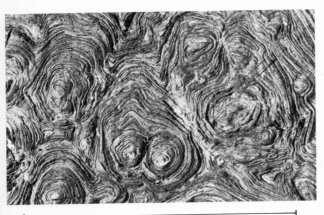

0.5 meter

FIGURE 7-9 (*a*) Highly saline Shark Bay, Australia, is one of the few places in the world today where algal colonies can gain a foothold. (These biscuit-shaped structures are solid fine sediment, bound by algal filaments and cemented by calcium carbonate.) Once abundant, the algal colonies nearly vanished when grazing marine animals (snails, etc.) appeared, about 500 million years ago. (*b*) These ancient, fossil algal colonies in Canada have been planed off by an advancing glacier. After the ice disappeared, the finely layered internal structures were exposed. [*Courtesy P. F. Hoffman.*]

Soviet geologists, while recognizing the factor of environment, insist that in addition there was a gradual change of these structures with the passage of geologic time. They believe that on this basis they can identify the age of their ancient sedimentary rocks to within 200 million years. Although this "clock" may seem incredibly crude, it would be helpful if we could use these fossils to divide the sedimentary rocks deposited during the middle 3 billion years of geologic time into, say, 5 to 10 units. (Most sedimentary rocks, you recall, cannot be dated by isotopic age methods.)

TOWARD A THEORY OF ORGANIC EVOLUTION

With the appearance of the living cell came a new era in the evolution of life. We have noted the pioneering efforts of Charles Darwin, who thrust upon the world these radically new biological principles. Ever

MR. BERGH TO THE RESCUE

THE DEFRAUDED GORILLA: "That Man wants to claim my pedigree.
He says he is one of my Descendants."
MR. BERGH: "Now, MR. DARWIN, how could you insult him so."

FIGURE 7-10 To this day, the theory of evolution has been a controversial subject, especially with those who feel, as this cartoonist did, that it threatens the uniqueness of human beings.

since the first publication of *On the Origin of Species* (which completely sold out on the day it was released), it would be difficult to find a work that has been so eagerly embraced by some people and so maligned by others (Figure 7-10). That is not to say that Darwin was seeking a quarrel. (He once confided in a letter that to assert that species are not immutable is "like confessing a murder." His childhood upbringing, his professional training, and his cautious and sensitive personality all restrained him from the task of presenting the theory that was to be the grand watershed of paleontology and biology. In fact, Darwin delayed publication for almost two decades while he continued to gather supporting data. His ideas might never have seen print had not A. R. Wallace, another eminent but more impatient biologist, reached the same conclusions and was about to publish them. (The two men finally announced their views jointly but Darwin received most of the credit, as he well deserved.) Another influential figure was Charles Lyell, the greatest geologist of the day, who continually urged his friend Darwin to commit the evolutionary ideas to print.

Why was opposition to evolutionary thinking so strong? The reasons were partly theological, partly philosophical, and partly the result of the lack of critical evidence. The image of creation held by the church was essentially borrowed from the Greeks, who were accustomed to seeing a sculptor engrave the minutest details upon a statue of some complicated object, such as a human figure. It was only natural to assume that the Divine Sculptor had instantly created a full-fledged biological world in which living things thereafter "bred true" to their respective kinds. Any fool knows that dogs give birth to puppies, not to kittens! To this observation the theory of evolution replies, "It's a matter of scale. On the short time scale, emphatically yes; on a vastly extended time scale, not quite." The science of geology provided the critical evidence for this extended time scale.

From a different quarter, philosophers to whom Plato, Aristotle, Kant, and Hegel were sacred text reproached Darwin not because he consciously invented a new school of philosophy (which would have been acceptable), but because he simply ignored them. Darwin was concerned, not with the standard philosophical categories, but with physical processes. He was the father of "population thinking"—an analysis of statistical variations in very large aggregates of organisms. The uncertainty that is attached to any statistical statement ran contrary to the idea of direct cause and effect so important to philosophical systems of the nineteenth century.

Classification of Organisms

Having reached the highest levels of organization shown in Table 7-1, how should we classify life beyond this point? The best route may not be obvious, as witnessed by the zoologists of the seventeenth and eighteenth centuries who debated whether the profusion of individual organisms might be just that and nothing more. If "Nature recognizes only the individual," then, since every individual is unique, no classification is possible.

This noncommital viewpoint was soon found not to be very fruitful, however. We could classify an organism in at least three useful ways. One classification based upon habitat would distinguish terrestrial from aquatic organisms. The aquatic type could be further divided into attached forms, swimming forms, floating forms, etc. Another possible classification based upon trophic level takes regard of the "food pyramid." The broad base of the pyramid is made up of the vast numbers of primary producers (photosynthetic organisms) on which a higher level, herbivores, depend for food. These in turn become food for an even smaller number of carnivores at the apex.

A classification that we shall consider here in more detail is based upon inheritance; its basis is evolutionary, and therefore historical. A great conceptual advance came when the English naturalist John Ray described a *species* as a group of organisms that can interbreed among themselves. Obviously *some* individuals are alike in this regard, and although the notion does present difficulties (especially to the paleontologist), Ray's concept of species is generally accepted today. Just as matter is made up of fundamental particles, so the species is a "fundamental particle" of evolution. Two species may be set apart by many factors, but most commonly the barriers are genetic. That is, the DNA or perhaps the chromosomal structures between different species are not compatible.

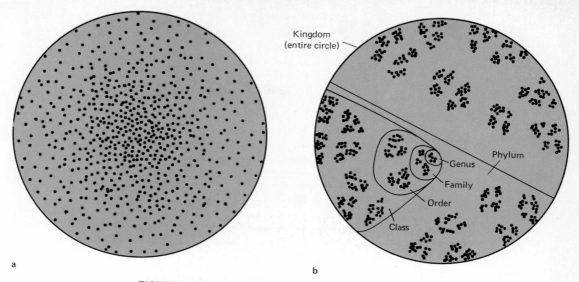

FIGURE 7-11 In each of these shotgun patterns, a dot represents a species.

Soon other naturalists began systematically to compare species to see if further likenesses are apparent. Most notable of these men was the Swedish naturalist Carl Linné, who attempted the herculean task of classifying the entire living kingdom. Even here, the recognition of natural species did not guarantee that higher degrees of classification would turn out to be useful. Suppose we were to diagram species as dots on a plane (Figure 7-11). The distance between any two dots indicates how dissimilar the species are. Our basis of comparison could be strictly biochemical: for example, differences in the sequence of amino acids in blood-serum albumin (a protein molecule) from the two species. Another approach could be to judge similarities through "multivariate analysis," a comparison of the degree to which many characteristics (appearance, structures, biochemistry, etc.) are held in common (Figure 7-12). No matter what criteria we choose, the broad picture of classification that we obtain is much the same. (Linné applied the multivariate analysis method in a rather intuitive way to the comparison of physical characters.)

Two distinct possibilities are presented by our diagrams (Figure 7-11). If the shotgun pattern (a) should be found, we might conclude that species are no more than an "unstructured multitude." Even if they sprang from a single source, they are all different and not obviously related. Happily, the shotgun pattern that actually emerged (b) abounds with possibilities for further interpretation. One is that each pellet was carefully aimed from a separate gun, but a more reasonable idea starts with a large number of pellets shot from one gun. Clumps of pellets stuck together for a time as they approached the target. Large clusters

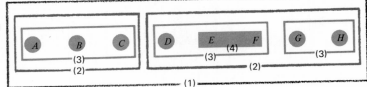

Step 1 Populations described	Population	A	B	C	D	E	F	G	H
								1	1
						2	2		
	Distribution of comparable structures numbered 1 to 6				3	3	3		
		4	4	4					
					5	5	5	5	5
		6	6	6	6	6	6	6	6

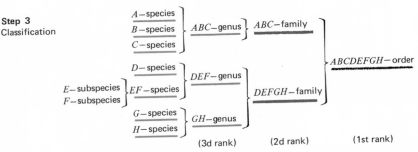

Step 2
Groups ranked
according
to degree of
similarity

Step 3
Classification

A—species
B—species
C—species
} ABC—genus } ABC—family
} ABCDEFGH—order

D—species
E—subspecies } EF—species
F—subspecies
} DEF—genus
G—species
H—species } GH—genus
} DEFGH—family

(3d rank)   (2d rank)   (1st rank)

FIGURE 7-12 Multivariate analysis is used to relate groups of species into
larger assemblages. Natural populations are first carefully described
(step 1), then assigned to larger groups of populations depending on
degrees of similarity (step 2). The groups are designated as genus, family,
etc., and baptized with formal names (step 3).

separated early, but fragments continued to break off right up to the
moment of impact. In like manner, living things can be arranged into a
hierarchy of fairly distinct clusters within ever larger clusters. Linné
marked off ascending groups known today as species, genera, families,* orders, classes, phyla,* and kingdoms (Figure 7-11*b*). Could it
be that species, genera, and the higher degrees of organization appeared and split apart throughout the entire time interval since life
originated?

It is interesting to note that this reasoning failed to convince Linné,
a person of nonevolutionary persuasion who preceded Darwin by more
than a century. For most of his life Linné clung to the belief that species
(or at the very most, genera) were fixed by separate divine creations.

* These categories have been added since Linné's day.

Could it be that he lacked the vital sense of geologic time so important to the evolutionary interpretation? Figure 7-11 is a diversity map as seen in a single "time plane"—that of today. We shall presently see this map again with an added dimension.

PROCESSES OF EVOLUTION

What makes evolution go? By now, evolutionary processes have become quite well understood, although they are still being intensively studied by numerous paleontologists and biologists. Let us examine the chief mechanisms of evolution, some of which were first outlined more than a century ago in Darwin's *Origin*.

Natural Selection

One of the germinal ideas of Darwin's scheme was proposed by Thomas Malthus (an English clergyman) in *First Essay on Population*, surely one of the most powerful and somber books ever written. Malthus brought forward a premise that came to be known as his "dismal theorem." He saw how rapidly organisms can reproduce themselves. Until restrained by external forces, the number of individuals continues to increase according to a *geometrical* progression, in a sequence such as 1, 2, 4, 8, 16, 32, etc. Malthus further considered the abode of these species—a finite earth that cannot provide support for unlimited multiplication of any organism. He noted that even with man's best efforts, our food supply can be improved only according to an *arithmetic* progression such as 1, 2, 3, 4, 5, etc. (The food supply of other organisms tends to remain stable, or at least it is beyond their ability to manipulate.) For both man and beast the end result is the same. Inevitably the supply of hungry creatures must outrun the supply of food necessary to sustain them, and the population will be held in check through starvation or predation of the young, the diseased, and the aged.

Malthus went on to make a number of social and political recommendations that need not concern us. Darwin too was unconcerned with these, but recognized that Malthus's principle held an enormous potential for explaining the course of evolution. Darwin summed up the situation thus:

As many more individuals of each species are born than can possibly survive; and as, consequently, there is a frequently recurring struggle for existence, it follows that any being, if it vary however slightly in any manner profitable to itself, under the complex and sometimes varying conditions of life, will have a better chance of surviving, and thus be *naturally selected.* . . . Any selected variety will tend to propagate its [own kind in] new and modified form.

In Darwin's view, nature is a strict schoolmaster that continually rewards the fittest individuals with prolonged survival while "failing" others with premature death.

We are not speaking here of ruthless cruelty, of Tennyson's poetic phrase, "Nature red in tooth and claw." After all, the cruelty of biological death ultimately befalls *every* individual. A more useful understanding of natural selection is furnished through the ideas of *habitat* and *niche*. Habitat refers to the "address" of an organism—for example, burrowed in marine mud, or attached to the underside of a rotting log, or roaming about freely. An organism's niche includes not only its address but its "profession." Niche refers to behavior patterns—migration, mating habits, how the creature obtains food, and how it may become food for some other creature. Many species can occupy the same habitat indefinitely , but not the identical niche, because the best-suited species will win the struggle for existence. The Nevada desert has probably harbored numbers of coyotes and rabbits in a rather tense but stable state of balance for millions of years. Both occupy the same habitat but different niches.

Hence the most stern competition arises not among different species, but among individuals of the *same* species. In a given area they are all striving for the same means of support; they occupy the same niche. Suppose that an individual somehow inherits a modified body structure or behavior pattern that slightly enhances its ability to cope with the environment. This individual will more likely survive to the age of reproductive maturity, and so pass along its superior genes to the next generation. Suppose that the modified gene bestows only the tiniest advantage: for example, making it possible for 0.01 percent more of the "favored" individuals to reproduce than do the individuals of the "original" type. A mathematical analysis shows that the new gene will become established throughout an interbreeding population (a *gene pool*) in an instant of time, geologically speaking. And so, the survival of the fittest is usually not a direct confrontation of tooth and claw, nor a test of endurance in hard times, though both of these forms of competition may be important on occasion. In the end, the earth is inherited by those individuals whose forebears outreproduced all the others.

Recombination

Natural selection cannot operate unless individuals in a population differ from one another. Geneticists have discovered that these variations are caused by extraordinarily subtle and complex factors. Both heredity and environment ("nature" and "nurture") are important. Most genes do not code for inherited characters in a forthright, one-gene-equals-one-trait manner. Many genes confer only the *potential* to react in a particular way to the environment. Moreover, in every organism a large number of genes are not expressed at all; they are latent or *reces-*

sive. The recessive genes are potential in the sense that they will surface again in future generations.

Genetics has made quite clear the essentially *mechanical* and *particulate* nature of inheritance. Genes are segments of DNA molecules, which are composed of nucleotide pairs, which in turn are made up of atoms (particles). Therefore, a characteristic acquired by one generation cannot be passed on to the next unless it is coded for in the DNA in reproductive cells. This observation dispels the once popular but mistaken theory that acquired characteristics can be inherited generally. Did the giraffe's neck gradually become longer because generation after generation of animals stretched their necks to nibble at higher leaves? Does the blacksmith's son have brawny arms because his father wielded a massive hammer? No; we must seek other explanations for these data because no amount of stretching or hammering can alter the molecular structure of DNA.

Sexual reproduction can reshuffle the thousands to millions of genes carried by a species in an astronomically large number of ways. The range of variation of a characteristic (such as adult height or weight) among individuals of an interbreeding population can be pictured by a *frequency distribution curve* (Figure 7-13). (A complete description of a population would call for many such curves.) If a char-

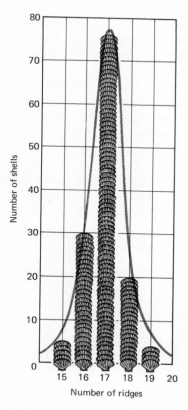

FIGURE 7-13 The number of ridges on each shell in a pailful of *Pecten* (a common scallop) collected on an Atlantic beach produced this result. More than half the individuals had 17 ribs, and 93 percent had 16, 17, or 18 ribs. Only a handful had a number of shell ribs that was above or below this range. [*After R. C. Moore, C. G. Lalicker, and A. G. Fischer,* Invertebrate Fossils, *McGraw-Hill Book Company, New York, 1952.*]

Number of shells (vertical axis: 0, 10, 20, 30, 40, 50, 60, 70, 80)

Number of ridges (horizontal axis: 15, 16, 17, 18, 19, 20)

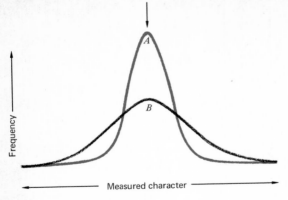

FIGURE 7-14 Frequency distribution curves A and B are of two equal-sized populations (the area contained beneath each curve is the same). In both populations, the most frequently observed value of the measured characteristic is the same (both are centered beneath the arrow). But the range of variability in population A is small (sharply peaked curve), whereas in population B, the range of variation is large (low, flattened curve).

acteristic is critical, if it must be "exactly so" for the creature to survive, the limits of variability will tend to be narrow (Figure 7-14, curve A). On the other hand, if the characteristic is only incidental to survival, a very wide range of variability can be tolerated without harm (curve B).

Shuffling of genetic material (*recombination*) is accomplished by several means. One of these is simply the fertilization of an egg by a sperm, thereby bringing together genetic contributions from the two parents. Another process, called *crossing over*, takes place while eggs or sperm are being made. In this situation, cells containing paired chromosomes are transformed into germ cells (sperm or eggs), each containing half the normal chromosome number. During crossing over, the chromosomes may be broken and relinked in an endless variety of new combinations.

Through natural selection, the extreme ends of the frequency distribution curve get "clipped off." Individuals who are not adapted well enough to their niche are eliminated from the race for survival. If the environment remains unchanged for long periods, natural selection will keep the peak of the frequency curve centered under the arrow (Figure 7-14). The fossil record suggests that certain environments have been remarkably persistent. Modern brachiopods of the genus *Lingula* (a burrower in mud flats) look just like fossil *Lingula* that are more than 300 million years old (Figure 7-15). The evolution of organisms

FIGURE 7-15 Shell a is a modern specimen of *Lingula*. Shell b is an ancient fossil. Since this organism appears to have evolved little through geologic time, it is often referred to as a living fossil. [*Judy Camps*.]

a

b

that dwell in the rather stable marine environments has long been a favorite subject for paleontologists. Marine sedimentary rocks are abundant, and the step-by-tiny-step changes can be precisely followed upward through a succession of fossil-containing strata.

Speciation

Not many environments, though, have persisted in any one spot for geologically long intervals. Climate may have changed; water in a restricted basin perhaps became more saline; more dangerous predators may have evolved; etc. A species that was once well adapted may find life rather difficult under the new conditions. In fact, the inability of a "brittle" species to evolve rapidly enough to keep pace with a changing earth is probably the chief cause of extinction.

The response of a species to changing circumstances brings us to the heart of evolution. Suppose that individuals on the far left side of the frequency distribution curve (Figure 7-14) are discriminated against. One result could be to shift the peak of the distribution slowly to the right as time passes (Figure 7-16). If populations are somehow isolated into different environments that continue to change, a parent stock may split up into new species; *speciation* has taken place (Figure 7-17).

How much time is required for speciation? The process is so slow by human standards, and so many factors are working, that we must look to those rare localities where nature has performed an isolation experiment for us. Oceanic volcanic islands provide an ideal situation because they are far from land, they originate as a fiery sterile mass, and they are geologically recent.

The Galápagos Islands, located a thousand kilometers west of Ecuador, became an important datum point in evolutionary theory when they were visited in 1835 by (you guessed it) Charles Darwin. There, the

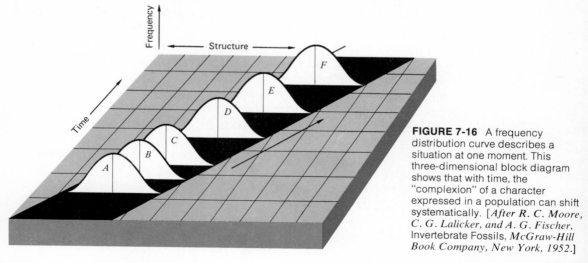

FIGURE 7-16 A frequency distribution curve describes a situation at one moment. This three-dimensional block diagram shows that with time, the "complexion" of a character expressed in a population can shift systematically. [*After R. C. Moore, C. G. Lalicker, and A. G. Fischer,* Invertebrate Fossils, *McGraw-Hill Book Company, New York, 1952.*]

185

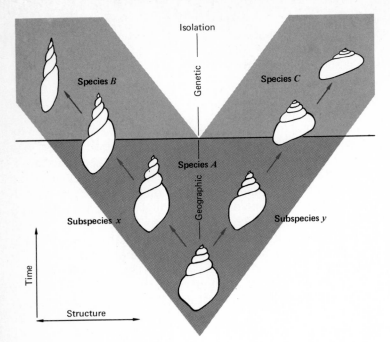

FIGURE 7-17 Geography is also an important factor in evolution. Here a snail population (species *A*) is distributed in different areas in which the niches vary slightly. Evolution leads to the establishment of subspecies *x* and *y* that are geographically isolated, but which *can* interbreed if brought together. With time, the genetic divergence creates new species *B* and *C* that *cannot* interbreed even if the geographic barrier were to be removed. [*After R. C. Moore, C. G. Lalicker, and A. G. Fischer,* Invertebrate Fossils, *McGraw-Hill Book Company, New York, 1952.*]

presence of some 23 species of finch, perfectly adjusted by size and type of beak to life in different niches, deeply impressed Darwin (Figure 7-18). They caused him to wonder why, if these species arose by separate divine creations, the creative power was so prodigally expended just in these islands, and why the finches so closely resemble one another. And why do animals in the Galápagos (including Darwin's finches) resemble the fauna of South America? Why are animals in the Cape Verde Islands (also visited by Darwin) completely different, resembling instead the fauna in nearby Africa? Darwin concluded: "One might really fancy that from an original paucity of birds in this archipelago, one species had been taken and modified for different ends" (or as we say today, adapted to different niches). The Galápagos finches must have speciated in less than 1.5 million years, the age of the oldest volcanic rock in these islands dated by the K-Ar method.

Speciation can occur even more rapidly if the organism has a short reproductive cycle. Hawaii is famous as uniquely harboring about 250 species (one-fourth of all that are known) of genus *Drosophila*, the common fruit fly. Genetic comparisons suggest that these species were founded by the chance colonization of Hawaii by a single pregnant female! Speciation took place after the island emerged above water less than 1 million years ago.

Consideration of the time factor raises some interesting questions about evolution. In most cases, it is a simple matter of observation to

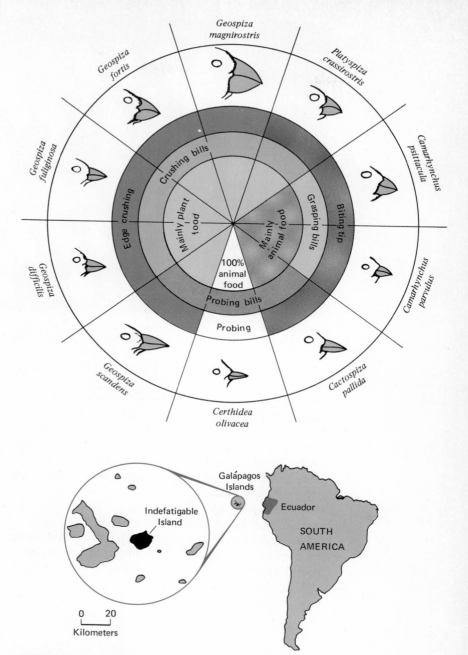

FIGURE 7-18 Ten species of Darwin's finches from Indefatigable Island (see map) can live in noncompetitive harmony today because they occupy different niches. [*After R. I. Bowman, "Morphological Differentiation and Adaptation in the Galápagos Finches," University of California Publications in Zoology, vol. 58, 1961.*]

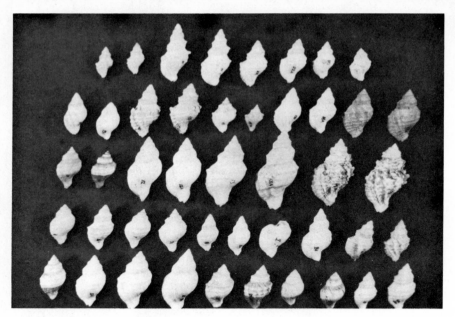

FIGURE 7-19 These considerably varied snail specimens, all of the same species, were collected on beaches from northern California to Washington. [*Courtesy Richard Miller.*]

distinguish today's species. Only populations that are rapidly undergoing speciation would be likely to give ambiguous answers. But as we peer back into time, the idea of species becomes more inexact. Are populations *A* and *F* (Figure 7-16) the same species? Was the DNA of our great-great-grandparents, our forebears a hundred generations back or a million generations ago, compatible enough with our DNA to have made interbreeding possible? Since we have no time machine, a direct test will never be possible. Paleontologists generally must assume that if fossil organisms *look* different, they probably did not interbreed either. That may be an often mistaken interpretation, but it is the best that can be offered. Some modern populations are highly variable (Figure 7-19). No doubt many fossil assemblages are thought to include several species when in fact there is only one (Figure 7-20).

Mutation

So far, we have seen that natural selection is able only to remove something that was already present. It can help us understand extinctions, but how can something genuinely *new* be created? The answer to this question is found at the molecular level of inheritance. What if the DNA molecule, when replicating, forms an inexact copy? Of course the

FIGURE 7-20 Numerous specimens of this fossil ammonite (a marine animal related to the pearly nautilus) were studied in Germany. Adults in the population were present in distinct sizes, as though two species were present. Smaller specimens were equal in size to the early stages of larger ones, and frequently both were about equal in number. For several reasons, it is more likely that the smaller specimens were the males (♂) and the larger specimens were females (♀) of the *same* species. (In most modern invertebrate animals, females are larger than males.) [*Courtesy U. Lehmann.*]

imperfect self-copy, or *mutation*, is only a modified configuration of atoms, but that new arrangement *does* code for an altered sequence of amino acids in protein (Figure 7-21). Usually the change in structure, body chemistry, or behavior resulting from mutation is harmful, even lethal to the recipient. In the unhurried pace of nature the unfortunate mutant is selected against, together with countless other "normal" individuals that fail to live to maturity. On rare occasions, a mutation may give an individual a trait with definite survival advantage in times of change. At such times the mutant would be selected *for*, not against. We might say that natural selection is like a player in a game of cards. Recombination is like shuffling the cards, and mutation is the wild card in the deck.

An interesting idea that needs to be tested further is that viruses play a major role in evolution. When virus particles infect a cell, the viral DNA becomes attached to the DNA of the host cell. Whole viruses are known to be transmitted within germ cells from generation to generation. Viruses can infect any living thing from bacteria to man. They can

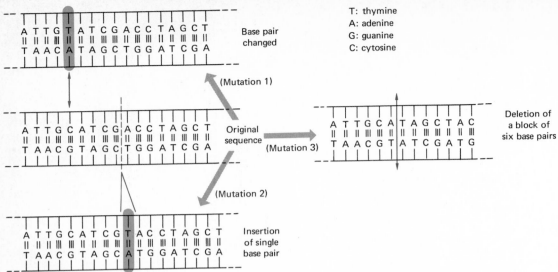

FIGURE 7-21 The original sequence of nucleotide pairs in part of a DNA molecule is shown in the center (with strands untwisted). Mutations can occur by at least three different means, as shown in the accompanying figures. [*After J. D. Watson*, Molecular Biology of the Gene, *W. A. Benjamin, Inc., New York, 1970.*]

carry genes from one host to the next. In principle, every living organism may have the potential to receive or donate big pieces of genetic information to any other organism. If so, all life is indeed one!

Phylogeny

Evolution is four-dimensional—it encompasses both time and the conventional three dimensions of space. The working out of evolutionary processes in space and time is nicely crystallized into the notion of *phylogeny*, a description of the line of origin and descent of species. A phylogenetic interpretation explains some apparently very awkward features of the classification of organisms. For example, why are birds, bats, and pterodactyls (extinct flying reptiles) considered to be closely related, whereas insects that also possess wings are distantly related (Figure 7-22)? Why does a porpoise look much like a shark (Figure 7-23)? In each of these cases, the wings or fishlike shapes are similar, but does this indicate that the similar organisms are closely related? They may be, or they may not. Possibly the resemblances mean only that the organisms are adapted to life in similar habitats.

Natural selection (acting through genetic recombination and mutation) can so modify a body structure through geologic time that the line of descent would be completely obscured but for the evidence from

a

b

c

d

FIGURE 7-22 Only four major groups of organisms have ever developed the power of sustained flight. These pictures illustrate how they acquired their prowess in different ways. In the bat (*a*), a mammal, the skin is stretched between the leg and all fingers of the hand. Feathers cover the wrist and forearm of the bird (*b*). In the pterodactyl (*c*), a reptile, the skin is stretched between the leg and the fourth finger of the hand. (*d*) An exterior skeleton covers the insect's back. [*American Museum of Natural History photographs.*]

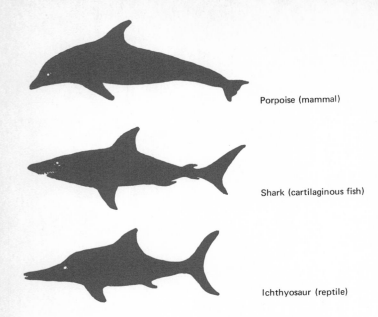

Porpoise (mammal)

Shark (cartilaginous fish)

Ichthyosaur (reptile)

FIGURE 7-23 Body outlines of a porpoise, a shark, and an ichthyosaur (extinct aquatic reptile) show adaptation to similar niches.

fossils. When phylogeny is taken into account, it turns out that a porpoise is more closely related to a hummingbird than to a shark! Wings of birds, mammals, and reptiles (Figure 7-22) are *homologous* organs; they developed through modification of a primitive five-digit, external appendage present in the ancient ancestors of these animals. Other organs that are homologous to wings are human arms and hands, the front legs of cattle, front flippers of seals, etc. Insect wings, on the other hand, are merely *analogous* to the wings of a vertebrate. They perform the same function, but they appeared through modification of an external skeleton rather than from a bony, flesh-covered appendage.

A species diversity map, repictured in the context of geologic time, summarizes the major features of a typical phylogeny (Figure 7-24). In study after study, the fossil record shows that a generalized ancestral species had suddenly exploited a favorable situation. Its evolution "exploded" into many previously unoccupied niches. As a result of this *adaptive radiation*, the stock species disappeared, not by outright extinction, but by evolutionary modification into new and different groups. (Indeed, the original species would not have been able to compete with its better-adapted descendants even if it had persisted.) The initial development was so rapid that the fossil record of adaptive radiation is commonly incomplete and poorly understood. Once the niches were occupied, a long period of relative stability ensued (shown by the upward-trending parallel phylogenies).

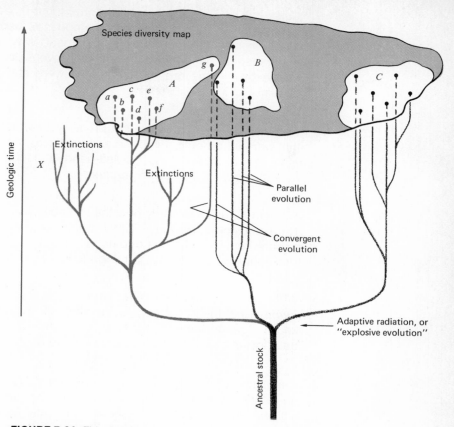

FIGURE 7-24 This three-dimensional model of phylogeny shows the present-day diversity map of species as a two-dimensional plane. (Compare with Figure 7-11.) Ancestors of genus *A* experienced a rather turbulent phylogenetic history. The habitats or niches occupied by one offshoot (genus *X*) became unfavorable, resulting in extinction of the entire genus. Another related genus became extinct except for a single species (*g*) whose niche resembles that of successful genus *B* sufficiently for species *g* to survive. Linné would have classed species *g* incorrectly into genus *B* on the basis of similarity. The fossil record clearly indicates that species *g* belongs historically with genus *A*.

FURTHER QUESTIONS

1   Redi's experiment with the uncovered and muslin-covered jars of meat was a total success. Why did no fly maggots appear in his muslin-covered jars?

2 Scientists who are working on the problem of the origin of life must be careful to keep the organic compounds used in their experiments in a strictly germ-free environment. Why is this so? Why is spontaneous generation an unlikely event today? What would happen while the organic molecules were still in the process of assembling?

3 Can you suggest some natural process by which the amino acids that were incorporated into the earliest living organisms could have been "prearranged" into the left-handed configuration? Can your hypothesis be tested by experiment?

4 Refer to Figure 7-5. Verify how the two new DNA molecules come to be exact copies of the original molecule, no matter which of the separating strands of the original serves as a template for the complementary strand that is forming alongside it.

5 What are some of the chief arguments that have been quoted *against* the theory of organic evolution? In the light of modern knowledge, do they appear to be valid objections? What are some data for which evolutionary theory has no obvious explanation?

6 A mutant gene confers a tiny survival advantage upon organisms in an interbreeding population. Slowly the individuals with the new, superior gene displace the individuals with the original gene that once was the only type present. Each generation, the fraction of the number of individuals with the original type of gene is reduced by 0.01 percent. What fraction of the population still carries the original gene type after 10,000 generations have lived and died? [*Hint:* If we designate the fraction as F, then $F = 0.9999^{10,000}$. Make use of the relationship: $\log F = 10,000 \times \log(0.9999)$.] For most organisms, a new generation is created in a period of minutes to, at most, a few years. Would 10,000 generations seem to be a long time, geologically speaking?

READINGS

Barghoorn, Elso S., 1971: "The Oldest Fossils," *Scientific American*, vol. 224, no. 5, pp. 30–42.

* Cloud, P. E. (ed.), 1970: *Adventures in Earth History*, W. H. Freeman and Company, San Francisco, 992 pp. A compendium of articles, of which pages 361–477 pertain to the origin and evolution of life.

Darwin, C. R., 1964: *On the Origin of Species*, Harvard University Press, 502 pp. A facsimile of Darwin's 1859 edition, with an introduction by Ernst Mayr.

De Beer, Gavin, 1964: *Atlas of Evolution*, Thomas Nelson & Sons, New York, 202 pp.

* Available in paperback.

chapter 8

Life and Time

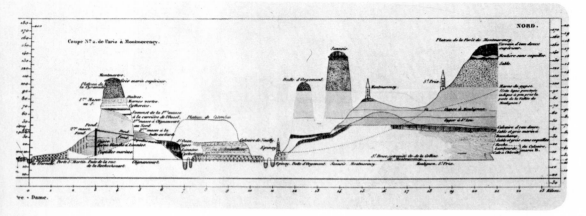

The French naturalist Georges Cuvier made
some of the earliest correlations of sedimentary
strata that have been partially eroded away. This
drawing shows a correlation in the vicinity of
Paris. [*From Georges Cuvier*, Description
Géologique des Environs de Paris: *Edmond
D'Ocagne, Éditeur-Libraire, Paris, 1835.*]

Again and again we have seen the evidences of a dynamic earth. Sediments are deposited; radioactive isotopes decay into daughter atoms; groups of organisms have evolved, flourished, and become extinct. All the while, erosion has been effacing the geologic record of these events, though only in part. The imprint of the competing processes of creation and destruction is left behind in the sedimentary rocks in a fascinating variety of ways. Each of these rocks was deposited under unique circumstances, at a unique point in history. Data from fossils, interpreted in the light of the relationships of the sedimentary strata, have given biostratigraphers a valuable insight into the history of the earth.

KINGDOMS OF THE LIVING AND THE DEAD

Today we recognize that the traditional plant and animal kingdoms do not suffice for an adequate classification of living organisms. Recently, systems of four or five kingdoms have been invented to take into account a division between cell types that is far more profound than the mere distinction between animals and plants. The greatest differences lie at the microscopic level, amongst single-celled organisms. On the one hand are *prokaryotic** organisms, represented by bacteria and blue-green algae (Figure 8-1*a*). These minute cells, 1 to 10 microns across, do not contain distinct organelles (nucleus, mitochondria, etc.). They do not carry on predator-prey relationships, but rather, they take in small molecules directly from the environment. DNA is synthesized constantly during the cell's life. In contrast to this type are *eukaryotic†* organisms; they include protozoans (amoebas, foraminiferans, etc.), certain algae, and all animals and green plants (Figure 8-1*b*). Inside

* From the Greek: "prenuclear."
† "Truly nuclear."

197

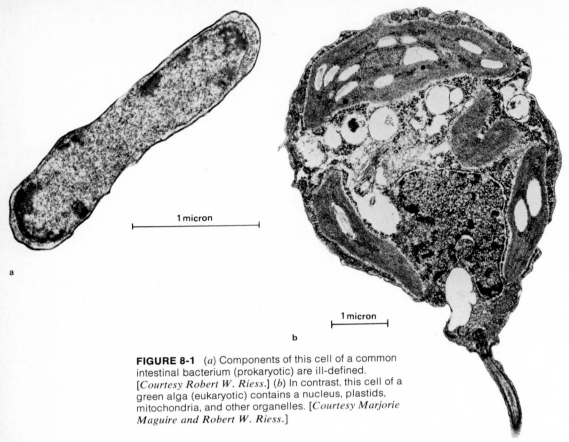

1 micron

1 micron

a

b

FIGURE 8-1 (*a*) Components of this cell of a common intestinal bacterium (prokaryotic) are ill-defined. [*Courtesy Robert W. Riess.*] (*b*) In contrast, this cell of a green alga (eukaryotic) contains a nucleus, plastids, mitochondria, and other organelles. [*Courtesy Marjorie Maguire and Robert W. Riess.*]

eukaryotic cells (which are typically about 500 times larger than the prokaryotic type) are located a variety of organelles enclosed by membranes. Eukaryotic cells can act as predators, even to the point of engulfing whole organisms. DNA is manufactured only during cell division.

Evolutionary origins of eukaryotic cells are a subject of much debate and speculation. Thus far, all researches seem only to confirm the absolutely sharp distinction between prokaryotes and eukaryotes, both among modern organisms and in the fossil record. A theory that has received wide attention asserts that eukaryotes arose by a series of abrupt evolutionary "jumps" (Figure 8-2). Mitochondria, the organelles in which energy-releasing chemical reactions take place, originated when ancestral cells incorporated bacteria. At first, this was no more than a mutually beneficial liaison between the bacterium and its host, but with time and evolutionary modification, the mitochondria became firmly integrated into the life of the cell. In a similar manner, spirochete bacteria supposedly fused with a host to become flagellae (the whip-

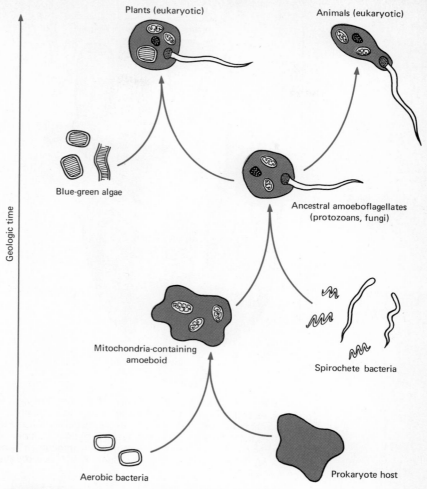

FIGURE 8-2 The most ancient fossils, some 3.3 billion years old, are of prokaryotic organisms. Fossil eukaryotes at least 1.3 billion years old are known. [*After L. Margulis*, Origin of Eukaryotic Cells, *Yale University Press, New Haven, Conn., 1970.*]

like tail present in sperm and in other cell types). Blue-green algae became accommodated as plastids, the chlorophyll-containing organelle present in green plant cells. One piece of evidence for this rather startling theory is that the mitochondria, flagellae, and plastids in a sense still carry on a life of their own. They retain their own DNA, and they divide independently when the host cell divides.

Distinctions among prokaryotes and eukaryotes and between single- and many-celled organisms lead to a convincing fivefold classification of living kingdoms (Figure 8-3). This classification also takes note of the mode of nutrition. Plants manufacture their own food through

199   KINGDOMS OF THE LIVING AND THE DEAD

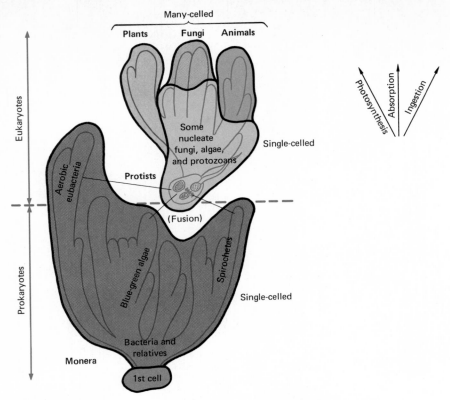

FIGURE 8-3 This classification emphasizes the cleavage between prokaryotes and eukaryotes. Some maverick organisms (such as slime mold) do not fit well into this or any other classification. Viruses are considered not to be alive. Microbes account for three-quarters of the mass and an even higher proportion of the energy flow through all living cells. [*After L. Margulis*, Origin of Eukaryotic Cells, *Yale University Press, New Haven, Conn., 1970.*]

photosynthesis. Animals employ a digestive cavity, whereas fungi simply absorb the food. Differences in mode of nutrition are present, but less marked, among the protists, and even less distinguishable among the members of kingdom monera.

Taxonomists have variously divided the kingdoms into 50 or 60 phyla (see Figure 7-11). Perhaps two or three phyla have become extinct, but by and large the basic structural plans and means of coping with the environment that are adopted by the various phyla have been highly successful, and therefore persistent throughout geologic time. In contrast, untold millions of species have arisen and disappeared.

How Fossils Are Made

Of the five kingdoms, animal remains are the best preserved in the sedimentary record; plant fossils are second, and representatives of the

FIGURE 8-4 This coiled shell of a snail (left) is fully preserved; a cast in which only the external form, not the original substance, is preserved would look much the same. A steinkern (right) provides some evidence of the animal's internal structures. [*Judy Camps.*]

other kingdoms are very poorly preserved. Exceptions are foraminiferans and radiolarians, members of the protist kingdom that build insoluble skeletons of calcium carbonate or silica.

By far the most abundant fossils are the "hard parts" that once served the organism as a supporting stiffener. These may be tiny fragments, such as the spines, or spicules, in a sponge, or the platelets embedded in the body of a starfish. Almost always they simply scatter abroad when the dead animal decomposes. Other animals construct more massive hard parts, such as an external shell. Only rarely is the original, shiny, pearly-luster shell material preserved in its original form. Most often, the shell has become porous and chalky in appearance as the more easily dissolved components were leached away. Many shells have dissolved totally, leaving in the sediment a *mold*, or hollow impression, that preserves the shell's surface irregularities. Sometimes the mold is refilled by fine sediment that later hardens. If the shell had already disappeared by the time of the infilling, the solid *cast* of the animal preserves a direct replica of the outer surface of the shell (Figure 8-4, left). In yet other situations, the flesh rotted and the mud filled in immediately, while shell was still present. When the shell fi-

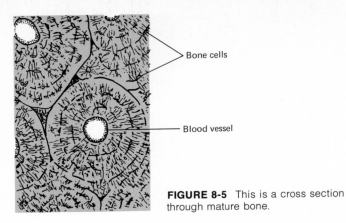

FIGURE 8-5 This is a cross section through mature bone.

FIGURE 8-6 Cave paintings in France dating from the period just before mammoths became extinct consistently pictured a large fleshy hump atop the head. Since this feature is not suggested by the skeleton alone, it was a source of considerable puzzlement to the paleontologists. Later findings of mammoths frozen in entirety have vindicated the authentic realism of these remarkable paintings. [*American Museum of Natural History.*]

nally dissolved, a *steinkern,** or internal cast, was left behind faithfully preserving the shape of the interior of the animal (Figure 8-4, right).

Living bone is a rather porous structure filled with a maze of cavities that contain blood vessels, maturing blood cells, connective tissue, etc. (Figure 8-5). Upon burial, the bone may be *permineralized* by groundwater which slowly deposits insoluble material in the pore spaces. Fossil bone is therefore mostly very dense and solid. The permineralization often preserves delicate features of tissue right down to the cellular (microscopic) size.

A revealing, but much rarer, mode of fossilization is simply the preservation of the whole animal. Best known in this regard are frozen whales and mammoths that were trapped during the most recent glacial cold cycle, some 10,000 to 20,000 years ago (Figure 8-6).† These

---

* German for "stone kernel."
† What business did a woolly mammoth have on top of an ice sheet? Perhaps the beasts were roaming about on partially melted, stagnant glaciers like the modern Malaspina Glacier, Alaska. It is mantled by dirt in which a dense forest has taken root. Except for the hazard posed by occasional deep cracks in the ice, the mammoths enjoyed a normal habitat.

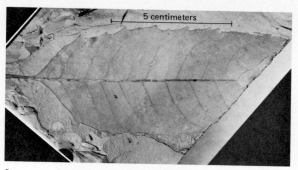

5 centimeters

a

b

FIGURE 8-7 (*a*) Only the outline of the vein system and of the leaf margins remains in a leaf impression. [*Courtesy David Dilcher.*] (*b*) A carbonized film remains in this compression of fern leaves. Often the fossil-bearing sedimentary rock can be split along a bedding plane to reveal a compression on one slab and a counterpart impression on the other slab. [*American Museum of Natural History photograph.*]

remains are considered to be fossils because of their great antiquity, even though they do not fulfill a handy rule of thumb that says, "If it still stinks, it's not yet a fossil." The stench is reported to be unbearable when one of these patriarchs of the Ice Age becomes exposed and starts to thaw.

Another fossil type common especially for plants is a compression. An *impression* is simply an outline of a shape, say of a leaf (Figure 8-7*a*). A *compression* of a leaf is a flattened film of carbon that is a residue of the actual leaf (Figure 8-7*b*). Sometimes compressions of leaves, seeds, and fruit can be dissolved from the rock matrix and restored by soaking (inflated) to something resembling their three-dimensional original appearance.

Petrifaction is another mode of fossilization that is more important in plants than in animals. Here, the original material is precisely replaced, atom by atom, by minerals. The plant fossil, perfectly preserved in its original three-dimensional form, can be reconstructed by means of serial sections (parallel slices cut at closely spaced intervals).

Through the serial section technique, some extinct fossil plants hundreds of millions of years old have become better known than many a species of modern plant.

Animal Fossils

After many decades of work, paleontologists and biologists are still far from completing even the first, descriptive studies of animals (Table 8-1). Every year, around 15,000 species are described for the first time, and according to some estimates, more than 8 million species (mostly insects) await discovery. Therefore a thorough acquaintance with the living kingdoms is a necessary beginning to every paleontologist's repertoire of knowledge. (Turn to Appendix B for descriptions of some of the major animal phyla.) Modern paleontology goes on to interpret the *function* of the organism (its life style in a particular environment) from the *form* of the fossil remains and the enclosing sedimentary matrix.

Certain species in all the phyla make water their permanent abode,

TABLE 8-1
Some Animal Phyla

Phylum	Some representative classes	Symmetry (of adults)	Habitat	Number of extant* species (thousands)
Chordata	Mammals Birds Reptiles Amphibians Fish	Bilateral	Water or land or air†	40
Echinodermata	Crinoids Echinoids (sea urchins) Asteroids (starfish)	Radial or bilateral or none	Water	4+
Arthropoda	Trilobites (extinct) Crustaceans (crabs, etc.) Insects	Bilateral (sometimes modified)	Water or land or air	9000+ (?)
Mollusca	Gastropods (snails) Pelecypods (clams, etc.) Cephalopods (squid, etc.)	Bilateral (sometimes modified)	Water or land	80
Annelida	Oligochetes (earthworms, etc.)	Bilateral	Water or land	7
Brachiopoda	Inarticulates (unhinged valves) Articulates (hinged valves)	Bilateral	Water	0.22
Coelenterata	Scyphozoans (jellyfish) Anthozoans (stony coral)	Radial	Water	9
Porifera	Hyalosponges (siliceous sponges)	Radial or none	Water	10+

* Living today.
† The organism can fly. To our knowledge, no phylum has ever made its *permanent* habitat the air.

but only the arthropods and the chordates have successfully colonized all three possible habitats (water, land, air). Almost all land species can move about, whereas many waterbound animals spend their adult lives fastened securely to the sea bottom. The latter are subject to predators (from which they cannot flee), and to being buried or clogged by sediment. They spread to new territory by releasing countless eggs and sperm which must fuse together, then develop into larvae that settle to the bottom haphazardlv, in hit-or-miss fashion. On the other hand, the sense organs, digestive system, and blood circulation in the attached animals need to be only of the simplest sort. These animals can gather food by stinging unwary passersby with tentacles (which bring the food into the mouth), or by filter-feeding: sieving out food particles by pumping water through a complex set of fringed structures. Therefore the symmetry of many sedentary organisms is cylindrical or *radial*—the creature reaches out equally in all directions for food that must come to it (Figure 8-8). The mobile organisms, active food seekers, mostly have *bilateral* symmetry, and well-defined heads and tails.

Organisms at the bottom of Table 8-1 (sponges, jellyfish) are "simple" or "primitive," whereas birds and mammals, listed at the top, are "advanced" organisms (more highly organized, more efficient users of energy). The concepts of primitive and advanced must not be

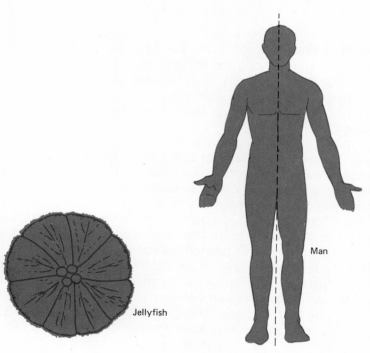

FIGURE 8-8 Radial symmetry is organized about a central point. Bilateral symmetry is based around a central dividing plane.

Man

Jellyfish

Radial symmetry

Bilateral symmetry

Ediacara
Hills

FIGURE 8-9 This fossil is not closely related to any known living organism. Magnification: 4 times. [*Courtesy M. F. Glaessner.*]

pressed too far, however. According to the fossil record, the sponges and coelenterates were among the earliest phyla to appear, but it is clear that *all* the phyla have very illustrious and ancient lineages. In fact, which phylum is descended from which, and how was the transition accomplished? That question is still one of the unsolved mysteries of paleontology.

Windows into the past The chief difficulty is that all the animals were soft-bodied while the critical evolutionary steps were still in progress. By the time animals had acquired skeletons, the various phyla had already developed the fundamental body plans that we see today. A few tantalizing fossil deposits of primitive soft-bodied animals—"windows" into that obscure interval of evolution—show quite a few phyla already sharply distinguished from one another.

Two of these fossil sites deserve special mention. In the Ediacara Hills of South Australia, at a locality discovered in 1947, are found hundreds of fossil impressions well preserved in the sands of an ancient beach (Figure 8-9). Representatives include jellyfish, soft corals, various types of worms, and some animals that do not resemble any living today. The jellyfish fossils at least can be assigned to phylum Coelenterata, but they cannot be tied confidently to any extant families or orders. Searches have confirmed at least 10 localities in Asia, Australia, and Africa where Ediacara-like fossils are preserved.

Another site famous in fossil lore is a small quarry in the Burgess Shale, where it is exposed high up the side of Mount Wapta, British Columbia (Figure 8-10). Fossils in a few thin beds of this black shale afford a glimpse into early animal life found nowhere else in the world. Nearly 150 species of sponges, jellyfish, and annelid worms, etc., are unique to this place (Figure 8-11). In addition, trilobites and other animals that built hard parts are present in abundance. Fossil assemblages in which an *entire* community of organisms is preserved are exceedingly rare and precious.

Recent field studies indicate that the Burgess Shale was deposited in moderate water depth just beyond the foot of a large organic reef. Many of the fossils are buried in the positions they assumed during life. This situation could result if the burials were catastrophic, as, for example, by occasional rushes of muddy water pouring down from the reef above. The sudden burials also kept scavengers away from the soft-bodied corpses.

Plant Fossils

Evolutionarily advanced members of the plant kingdom contain *vascular* tissue—an aggregate of specialized cells that conduct water and nutrients efficiently throughout the plant. Vascular plants have come to dominate the vegetation almost everywhere on land.

FIGURE 8-10 Legend has it that a slab of shale kicked over by a pack horse caught the eye of the paleontologist C. D. Walcott, discoverer of the Burgess Shale fossil site. His party followed the rubble upslope to a highly fossiliferous bed only about a meter thick, in which this quarry was excavated. Walcott collected some 35,000 speciments early in this century. In the 1960s, another extensive collection was made that is now under study. [*Courtesy James Sprinkle.*]

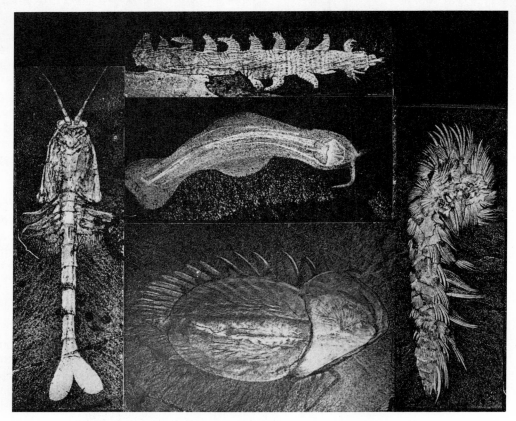

FIGURE 8-11 Soft-bodied and soft-shelled organisms are preserved in the Burgess Shale in such detail that even the internal organs are visible. The fossils are compressions—films of carbon pressed flat, like ink on a page. This assortment includes annelids, arthropods, and problematical animals. [*Smithsonian Institution photograph.*]

Table 8-2 summarizes a plant classification according to the American botanist H. C. Bold. (Refer to Appendix C for descriptions of these plants.) We have already noted that classifications are useful only if they provide insight into the origin or history of change of things. Evolutionary theory gives us exactly this unifying insight in the realm of paleontology. Linné's perspective was limited to living organisms only. More recent classifications are actually much the same as his (at least on the species and genus level), but the fossil record has richly deepened our understanding by tying together the web of life historically.

So it is with the animal kingdom, at least. Plants are much less satisfactory. Paleobotanists have had to be cautious because they are so uncertain of the evolutionary affinities of the various groups. Instances are known in which the underground part of a plant was assigned to

TABLE 8-2
Some Plant Divisions*

Division	Found in the most primitive fossil record?	Seed-bearing?	Distinguishing characteristics	Extant species (thousands)
Psilophyta	Yes	No	Underground and exposed stems only—no leaves or roots	0.003
Microphyllophyta	Yes	No	Stems, roots, and tiny, simple leaflike structures with unbranched veins. Spore-bearing organs on fertile leaves.	1
Arthrophyta	Yes	No	Hollow, jointed stems (both above and below ground) the dominant organ. Tiny, scalelike vascular leaves. Separate spore-bearing appendages.	0.025
Pterophyta	Yes	No	Ferns. Leaves, the dominant organ, are large, complex, with branching veins.	10
Coniferophyta	Yes?	Yes	Woody stems, strongly branched, with treelike or shrublike habit. Leaves are numerous, leathery, scalelike or needlelike.	0.55
Cycadophyta	No	Yes	Stems are stout, fleshy trunks armored with leaf scars. Relatively few, fernlike leaves present at any moment.	0.1
Ginkgophyta	No	Yes	Richly veined leaves of medium size. Woody stems, with numerous branches. Treelike habit.	0.001 (only one species!)
Anthophyta	No	Yes	Exceedingly diverse. Leaves with branching veins are of intermediate size in most species. Stems may be woody or nonwoody (herbaceous). Flowers produced in the reproductive cycle.	300

* A plant division is substantially the same thing as a phylum.

one "form" genus, fossil stems to another genus, leaves to a third, and spores to yet a fourth genus. Then later finds showed all these parts as belonging to the same natural species! Commonly, an entire individual must be present for a positive identification, and failing that, the reproductive organs are the most diagnostic single part. Ironically, these are the least often fossilized.

Origins of plants that appeared comparatively recently can be just as puzzling as the advent of the most ancient species. For example, the makeup of angiosperms (AN-djee-o-sperms) (flowering plants) demonstrates the familiar evolutionary trend toward complex structures. What was the parental stock from which angiosperms evolved? Nobody knows! When they first appeared, some 100 or more million years ago, they were already modern in every important respect. [Some authorities speculate that magnolias and their near kin are the most primitive (most ancient?) angiosperms.]

Possibly the angiosperms became established first in the highlands—an unlikely place for fossils to be preserved. In support of this idea is the rare occurrence of battered and broken logs of "advanced" plants, apparently carried downstream by floods, associated with fossils of more primitive plants. An opposing argument is that angiosperm pollen appeared abruptly, and no earlier than leaves and other plant organs. Since the durable, dust-fine pollen grains easily could have been blown or washed to distant points, we may wonder if angiosperms got started in the high country after all. Then again, the first angiosperms may have borne only the type of sticky pollen that must be transported by insects from plant to plant. Clearly, piling conjecture upon conjecture is getting us nowhere. The origin of angiosperms remains as Darwin put it, "an abominable mystery."

In summary, the more recent plant-fossil record contains sharply distinguished lines of descent. In the more ancient record, certain of these lines appear to converge toward a common ancestry, but the rocks that *should* contain the definitive evidence have no plant fossils at all! Thus details of the emergence of vascular plants, probably from algae, remain obscure. It is a disconcerting state of affairs.

Imperfections of the Fossil Record

Some of the inadequacies of the fossil record have become quite evident by this point. Nearly all information about the nature of past life, except for what the hard parts of an organism can tell us, has been irretrievably lost. In many instances surprisingly little is left. Teeth are about the only remains of sharks whose skeletons otherwise are composed of soft cartilage. Often not even the hard parts are preserved—they are dissolved, slowly but completely decayed, or chewed up and digested by some hardy scavenger. Fossil-bearing sedimentary

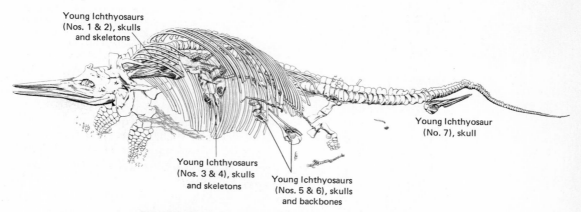

Young Ichthyosaurs
(Nos. 1 & 2), skulls
and skeletons

Young Ichthyosaur
(No. 7), skull

Young Ichthyosaurs
(Nos. 3 & 4), skulls
and skeletons

Young Ichthyosaurs
(Nos. 5 & 6), skulls
and backbones

FIGURE 8-12 The fishlike ichthyosaur, actually a reptile, apparently delivered fully developed offspring not enclosed by a shell, just as certain snakes do today. Several embryos are visible inside the well-preserved skeleton of this female. [*American Museum of Natural History.*]

rocks may be weathered, eroded, and transported to some new site of deposition. In the process, the fossils are reduced to sand- or mud-sized particles. Complete skeletons, such as the fine specimen of an ichthyosaur (Figure 8-12), are rare and lucky finds indeed among land animals, but are much more common among aquatic creatures. Metamorphism usually destroys all traces of fossil remains.

Incompleteness of the fossil record shows up in other, more misleading guises. Organisms may live in one place, but become fossilized somewhere else. If fossils are found still preserved in "living position," we may confidently make further judgments about the ecology of the fauna from the associated species, sediments, etc. At the other extreme, and very common in the sedimentary record, are dense concentrations made entirely of transported fossil fragments (Figure 8-13). These accumulations may attest to the power of waves or currents to heap together similarly sized particles, but the fossils are no more than

FIGURE 8-13 This modern beach is littered with clam shells cast up by storms. Some ancient sedimentary strata are similarly paved with shells. Apparently hurricanes were common long ago, just as they are today. [*Courtesy Daniel Houston.*]

"death communities" bearing little relationship to their occurrence in life.

The trilobites (extinct arthropods, Appendix B) illustrate yet another interesting class of nuisance problem. Arthropods occasionally molt their external skeletons. A paleontologist must learn not to confuse skeletal fragments of juveniles with the remains of other fossilized species. At the same time, these fragments give useful clues to the growth habits of the long-departed trilobites.

It is difficult to estimate the number of species that have ever existed, but the number probably is in the vicinity of 3 *billion*. Only about 300,000 fossil species have been described—an insignificant fraction of the total (Figure 8-14). Some organisms are likely to be fossilized, and some are not. We seem to have a near-complete record of all the brachiopod species, both extinct and living, but practically no fossil insect species are known in comparison with their vast numbers.

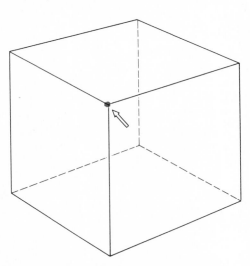

FIGURE 8-14 Only about one species has been discovered as fossils for every 10,000 species that ever existed. The volume of the shaded cube (arrow) relative to the larger cube indicates this minute fraction.

At least a century and a half before radioactivity was put to use to tell geologic time, geologists had recognized that fossils are an index to the age of deposition of the sedimentary rocks. The fossil time scale is the venerable child of geologic research in stratigraphy and paleontology. All types of geologic clocks are subject to misreading, and for similar reasons. The difficulty is that only a *physical quantity* (such as a daughter/parent ratio) can be measured. Geologic time, that elusive fourth dimension, is not sensed directly; it is an *interpretation* based upon a physical configuration. Fossils and sedimentary strata, too, are physical objects from which an interpretation of the passage of time can be made.

For several decades the pioneer stratigraphers, though they were learning much from rocks, were hampered by incorrect ideas about the connection between processes of deposition, organic evolution, and the passage of time. The development of the geologic time scale is a fascinating chapter in the history of science showing how a consistent scheme gradually appeared as invalid assumptions were weeded out.

The Art and Science of Stratigraphy

The guiding principles of stratigraphy gradually developed from consideration of "rocks as they are" to ever deeper levels of abstraction. An obvious starting point is the law of superposition which states that in an undisturbed sequence, older strata lie beneath younger. This simple deduction, fortunately, was taken to be self-evident by the early stratigraphers. It promptly establishes the relative ages of deposition in a local area.

These geologists soon encountered a more difficult problem, illustrated by a cross section (Figure 8-15). Suppose that a distinctive stratum is visible in the faces of various cliffs in a deeply eroded region. Did the stratum once continue across the eroded interval? In short, do the exposures *correlate*: are they equivalent to one another? If the strata are lens-shaped, they may "pinch out" deep underground, or

FIGURE 8-15 Thick sandstone strata erode back as bold cliffs in arid regions, as in parts of the western United States. This cross section shows that an upper, persistent thin shale bed correlates across the entire area. The lower shale unit does not.

[Cross section]

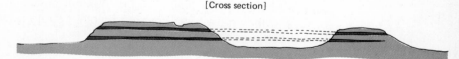

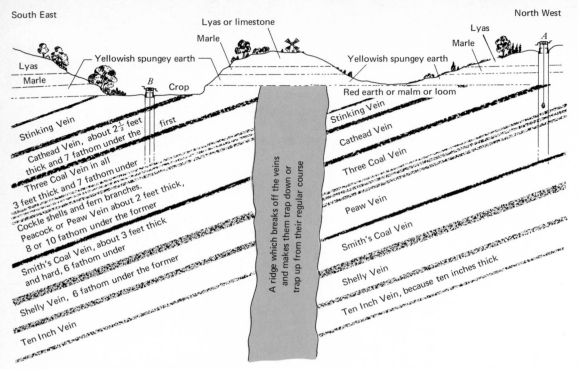

South East

North West

Lyas or limestone

Marle

Lyas

Marle

Lyas

Marle

Yellowish spungey earth

Yellowish spungey earth

A

Lyas

B

Crop

Red earth or malm or loom

Stinking Vein

Stinking Vein

Cathead Vein, about 2½ feet thick and 7 fathom under the first

Cathead Vein

Three Coal Vein in all 3 feet thick and 7 fathom under

Three Coal Vein

Cockle shells and fern branches.

Peacock or Peaw Vein about 2 feet thick, 8 or 10 fathom under the former

Peaw Vein

Smith's Coal Vein, about 3 feet thick, and hard, 6 fathom under

Smith's Coal Vein

Shelly Vein, 6 fathom under the former

Shelly Vein

Ten Inch Vein

Ten Inch Vein, because ten inches thick

A ridge which breaks off the veins and makes them trap down or trap up from their regular course

FIGURE 8-16 Strachey's early stratigraphic correlations made use of some rather quaint descriptive terms, listed here as he formulated them. It seems a shame that modern technical vocabulary, in the interest of precision, has had to give up colorful language. [*After John Strachey*, Philosophical Transactions of the Royal Society of London, *vol. 30, 1719.*]

in a place where erosion has removed the record. Direct correlation of exposures may be impossible in such cases.

A more complex example was worked out in the early 1700s by John Strachey, who was interested in the occurrence of coal in southwestern England (Figure 8-16). His cross section shows a set of dipping beds (actually one side of a large, eroded fold), broken by a fault that has displaced the left-hand block relatively upward by a few meters. Horizontal beds, themselves somewhat eroded, lie unconformably above the coal-bearing strata. Clearly, an accurate correlation from one region to another would be of economic importance in a coal district. A miner needs to know, for instance, if a coal seam is at the bottom of the sequence, or whether deeper layers of coal are likely to be present. Layer-by-layer correlation is not possible. Coal seams look alike, and they have been covered by other strata or vegetation, or they have been eroded away in places.

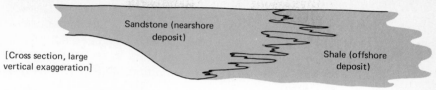

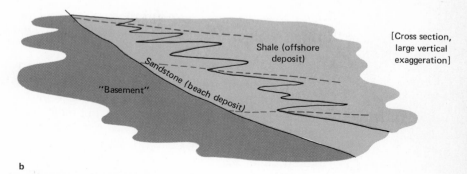

FIGURE 8-17 (*a*) Nearshore sand and offshore shale were deposited simultaneously in this delta. (*b*) Several dashed time lines indicate surfaces of deposition (hence, surfaces of equal age) at selected instants of geologic time.

Correlation of formations   Strachey did what geologists continue to do today. He examined the entire sequence in as many places as possible to ascertain the succession, thicknesses, and composition of the beds. He grouped the beds into what is now called a *formation*: a persistent stratum or a group of strata that are distinctive and are thick enough to be shown on a map. Formations are given local geographic names such as "Rochester Shale." If the strata are a mixture of rock types, they are simply called a formation (e.g., Chinle Formation). Scale is important, as most maps cover a rather large area. Seldom would a stratum less than a meter thick merit the term "formation." On the other hand, strata a kilometer thick would normally be subdivided into several formations. Ultimately, though, the definition and naming of a formation is a stratigrapher's personal judgment.

Recognition of distinctive formations is a big advantage in making correlations, but serious errors of interpretation are possible. For example, suppose that a river had deposited sediments in a delta. Heavy sand grains were dumped immediately at the river mouth, whereas clay particles came to rest in deeper, more quiet water several kilometers offshore (Figure 8-17*a*). We might recognize two formations—a sandstone and a shale—while failing to realize that these rocks were deposited together, in similar environments, at the same moment of geologic

215   GEOLOGIC TIME SCALE

time. Again, suppose that a shoreline had slowly advanced across a low coastal plain (Figure 8-17*b*). The ancient beach deposit left by this event may be a single continuous sandstone formation today, though it took millions of years to form. In regard to time of deposition, it could correlate with many formations that were being deposited somewhere else.

Thus a stratigrapher must bear in mind at least two possible correlations: an interpretation of *environment of deposition* and of *geologic time*. Some of the early controversies (and some modern ones too) arose because not all the factors were recognized.

Correlation of fossil assemblages   Another advance in the art of biostratigraphy came when fossil evidence began to be taken seriously. (Certain of the ancient Greeks and Romans had shrewdly concluded that fossils are remains of organisms, but most people in the eighteenth century regarded them as freaks of nature or tricks of the Devil.) The first person to correlate using fossils was William "Strata" Smith, an untutored canal surveyor. Smith benefited from a highly favorable set of circumstances. Britain's need for canals around the turn of the nineteenth century took him on surveying jobs to many parts of the country. There Smith eagerly examined and described the numerous fossils unearthed by construction activities. In many places he saw that fossil assemblages are different in the lower (older) beds and in overlying (younger) beds. This generalization came to be known as the *principle of faunal succession.* Smith perceived that the assemblages match, in stratum for stratum, over widely separated areas of southern Great Britain. At the urging of friends, he was encouraged to seek money to publish a grand summary of his life work, showing the correlations in map form. His masterpiece appeared in 1815, the first geologic map ever compiled on a large regional scale. It looks strikingly like the most sophisticated modern geologic map of England and Wales (Figure 8-18).

William Smith showed no special curiosity over the *reasons* why fossil assemblages in rocks of different age should be different. (He died before the publication of Darwin's evolutionary explanation of the phenomenon.) A contemporary of Smith was the brilliant French naturalis Georges Cuvier. Cuvier worked out a sedimentary succession in the vicinity of Paris, where he discovered that each group of strata, bearing distinctive marine fossils, is separated from strata above and below by unconformities. Cuvier surmised that a series of catastrophes had wiped out all life. After each extinction, he said, God had repopulated the earth "instantly" with new and different faunas.

An evolutionary interpretation of Cuvier's data is quite different. It would assert that during a "missing" interval of time signified by an unconformity, organisms were gradually evolving into new species in

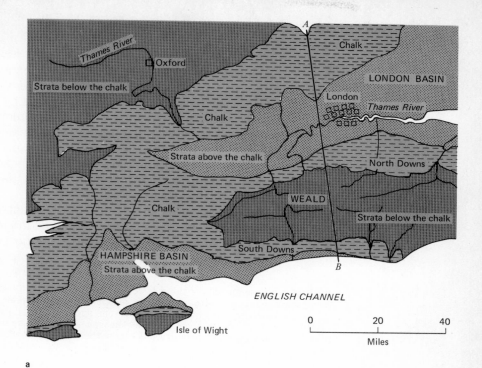

a

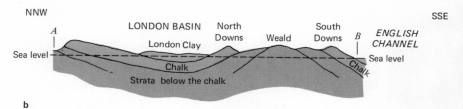

b

FIGURE 8-18 (*a*) In William Smith's early map, sedimentary strata show a looping pattern because they are not horizontal. The rocks were gently warped into giant open folds, and erosion has beveled the surface, removing strata where they formerly projected to the highest elevations. (*b*) In the cross section along line *AB* (see map), the beveling effect of erosion is apparent. Older strata are exposed at the surface in the core of an upfold (the Weald). The London Clay, youngest in the local sequence, is preserved in the keel of a downfold but has been removed everywhere else. A resistant formation, the Chalk, trends across the countryside as a prominent ridge where its upturned edge is being eroded back. [*After William Smith, 1815.*]

some other part of the world. Each time the sea invaded the region of northern France, a considerably modified group of marine organisms came in with it. Extensive collections of fossils have convincingly pieced together these missing links of evolutionary development, thereby vindicating Darwin's explanation over Cuvier's.

The principle of faunal succession, established by Smith and later given a rational basis by Darwin, proved to be an immensely powerful method of establishing the relative ages of sedimentary strata in distantly separated regions. Since the complex network of life is ever changing, the fossil record deposited during one interval of geologic time can never look quite the same as the record laid down during some other interval (Figure 8-19).

Fossils are not foolproof, though. Not all sedimentary rocks contain fossils. Another difficulty is the limited geographic range of most organisms. Millions of years from now, a stratigrapher would be hard

FIGURE 8-19 Sequences of strata (*a, b, c, d*) can be correlated over long distances, even on a worldwide scale, by matching the successions of fossils. Assemblages are correlated, not just one species. (Recall the saying, "There is safety in numbers.") Thicknesses and rock types of the correlated strata may vary from place to place. These characteristics closely depend upon the environment of deposition. On the other hand, appearances and extinctions of cosmopolitan species are indications of the passage of time.

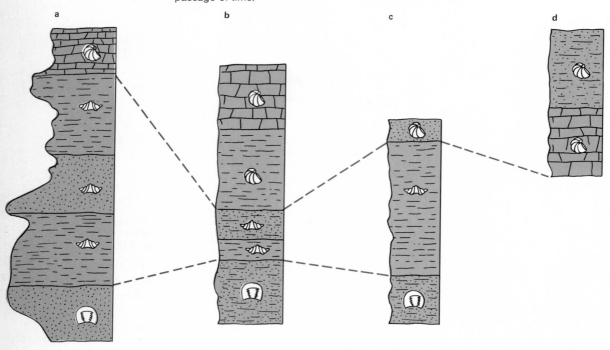

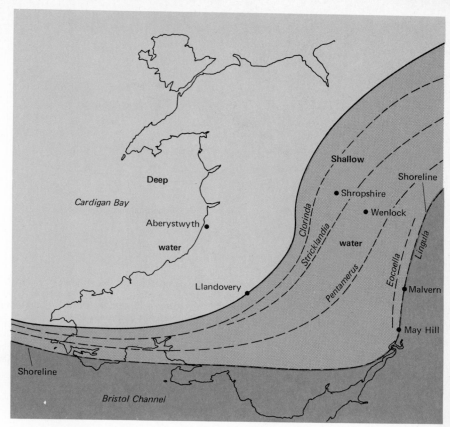

FIGURE 8-20 Brachiopod species from lower Silurian sediments in present-day Wales lived in sharply zoned communities. These attached organisms apparently were highly endemic and water-depth sensitive. [*After A. M. Ziegler, "Silurian Marine Communities and Their Environmental Significance,"* Nature, *vol. 207, no. 4994, pp. 270–272, 1965.*]

pressed to establish on fossil evidence that the North American polar bear, the African lion, and the Antarctic penguin were contemporaries. His best evidence would doubtless be fossils of *Homo sapiens* (you and me) associated with these animals. Some species are *endemic*, or confined to a local area, whereas other species are *cosmopolitan* (found universally,* in all sediment types). The endemic species are better indicators of the nature of the environment of deposition (Figure 8-20). Cosmopolitan species are a more certain index of correlation of the time of deposition (Figure 8-21). Probably the most cosmopolitan of all organisms today are the marine foraminiferans. And yet, out of

* More exactly, marine cosmopolitan species are confined to marine sediments, and land-dwelling cosmopolitan species mostly to terrestrial sediments.

219 GEOLOGIC TIME SCALE

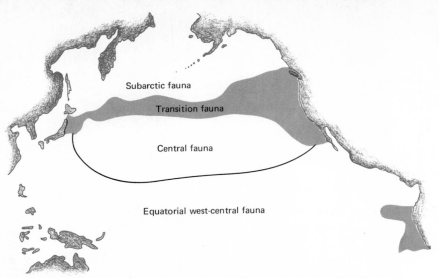

FIGURE 8-21 Assemblages of floating foraminiferans may be found over large areas of the Pacific, but even they are not universal. They are restricted to water masses of characteristic temperature and salinity. Only five species are truly cosmopolitan in marine waters.

thousands of known species, only about 50 are floaters, hence likely to be widely distributed.

Toward a Time Scale

The stage was now set for geologists to construct a time scale. Nowhere, not even on the deep ocean floor, is a continuous sedimentary record preserved. Instead, there are only local regions in which the ages of sedimentary deposition have overlapped. Stratigraphers had to correlate from place to place in piecing together a complete scale (Figure 8-22). In an effort to keep the physical objects strictly separated from the interpretation of time, they designate the fundamental time interval as a *period*; the rocks laid down during a given period they call a *system*. These are given the same identifying name—for example, Devonian Period and Devonian System.

Naturally, no one could have known in advance how many or what fossil species would be useful for correlation, how many periods or systems would be convenient, or what part of a system would be represented in any local sequence of strata. Like the blind beggars describing an elephant, the early workers were compelled to feel their way along. Consequently, the time scale grew in a disorganized manner. Geologic periods were named in random order after ancient Welsh tribes, rock types, mountain ranges, and what-have-you. At least one friendship ended when R. I. Murchison and A. Sedgwick worked

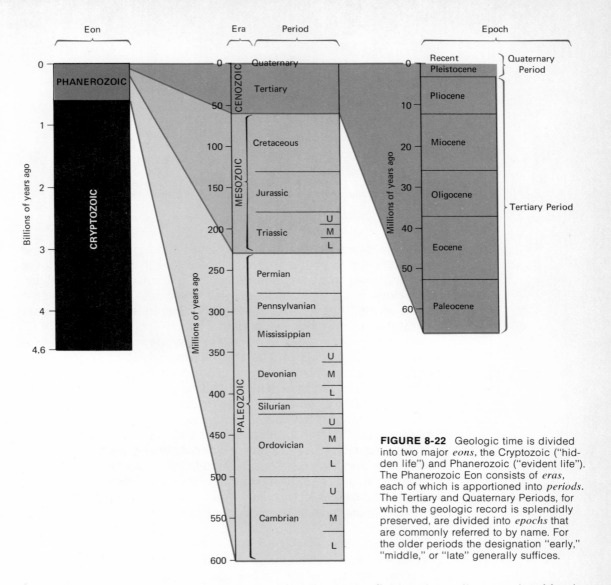

FIGURE 8-22 Geologic time is divided into two major *eons*, the Cryptozoic ("hidden life") and Phanerozoic ("evident life"). The Phanerozoic Eon consists of *eras*, each of which is apportioned into *periods*. The Tertiary and Quaternary Periods, for which the geologic record is splendidly preserved, are divided into *epochs* that are commonly referred to by name. For the older periods the designation "early," "middle," or "late" generally suffices.

their way toward each other in the field, only to discover that Murchison's Silurian rocks were identified by Sedgwick as Cambrian rocks. The argument was finally settled through a redefinition, though not in the lifetimes of Murchison and Sedgwick. (The disputed series of strata were assigned to a third system, the Ordovician.)

Dating the Time Scale

Isotopic age methods have recently made it possible to assign ages, in millions of years, to the geologic periods (Figure 8-22). Vol-

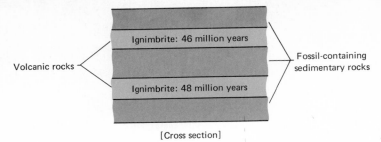

[Cross section]

FIGURE 8-23 A portion of the Eocene Epoch is dated by noting the close association in this cross section of volcanic rocks with sedimentary strata bearing Eocene fossils. What is an approximate age of the middle stratum?

canic strata interbedded amongst the fossil-bearing strata provide the best opportunity to calibrate (set numbers to) the time scale (Figure 8-23). Since volcanic rocks are abundant in the Cretaceous, Tertiary, and Quaternary Systems, these are well keyed with scores of isotopic age "tie points." Ages of the Triassic and Silurian boundaries remain the most poorly known for lack of data.

With the isotopic ages in hand, we may view the entire span of earth history in perspective. The scarceness and rather uninformative nature of Cryptozoic fossils suggest that a classification of time based on these most ancient of organisms will probably never be possible. Abundant organisms with hard parts appeared only after some six-sevenths of geologic time had already gone by. In view of the uncoordinated manner in which the Phanerozoic time units were defined, it is surprising that such a useful small number of periods were established, and that their durations were approximately equal (within a factor of 3). Perhaps we should credit the first stratigraphers with the good sense to apply their definitions to *very thick* sequences of strata—deposits that we now see must have taken a goodly fraction of Phanerozoic time to be laid down.

Kinds of scales Although isotopic ages in no way played a part in the original definition of the geologic time scale, their application has opened up an interesting new dimension. Different kinds of scales have differing properties that can be compared. One use of names is purely as a means of *identification*. People's names illustrate this, the simplest type of scale. A second class of measurement, of which the geologic time scale is an example, is used to establish an *ordering* or *ranking*. It enables us to answer questions that imply both identity and sequence, such as: Which came first, the Jurassic or the Cretaceous Period? A third class of scales, represented by isotopic ages, not only identifies and establishes sequence, but in addition can answer questions that depend upon forming *ratios*. For instance, isotopic ages are needed to answer the question, How much longer was the Cretaceous Period than the Jurassic Period, and what are the ages of the period boundaries?

An Accident of Geography

Early progress in stratigraphy took place in northern Europe and especially in Great Britain. Even the Permian System, first described

FIGURE 8-24 This geologic map of southern Greece is *greatly simplified!*

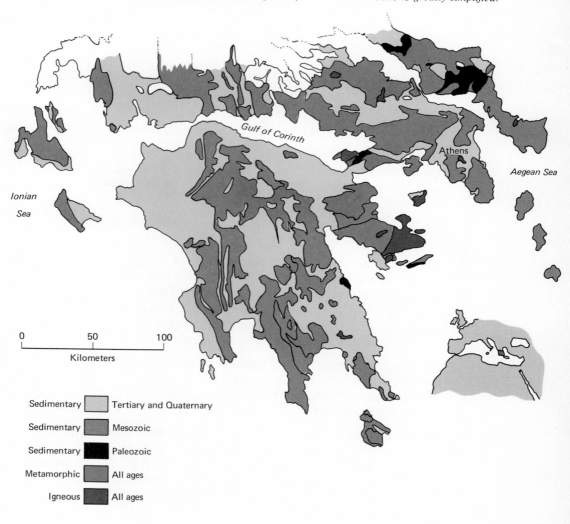

near the Ural Mountains of the U.S.S.R., was formally established by the Englishman R. I. Murchison. Undeniably the British and French were the eminent geologists of the day. (America was, geologically speaking, an unknown wilderness in the mid-1800s.)

Still, why had not the Greeks, with their good start in speculating about fossils, developed a time scale centuries earlier? The reason surely must lie in an accident of geography. A modern geologic map of southern Greece (Figure 8-24) shows that the rocks have been structurally contorted by intense orogenies. The mangled strata of Greece, which are not very well understood even today, were emphatically no place to begin research! Compare the map of Greece with that showing the stately simplicity of southern England, as seen in Figure 8-18.

FURTHER QUESTIONS

1 Refer to Table 8-1, and note the habitat that is occupied by at least some members of every animal phylum. Is this observation in accord with current ideas about where life originated?

2 Why are complete fossil skeletons more commonly found in marine sediments than in terrestrial sediments?

3 Refer to Figure 8-12. Can you state a connection between the live birth of the young of ichthyosaurs and the habitat of this extinct reptile?

4 Refer to Figure 8-17b. What are two reasons why the sea might slowly encroach upon the land?

5 A skeptic claims that successions of similar fossils in isolated groups of strata have no time significance except locally. (For instance, Devonian fossils may be 400 million years old in one place, and 10 million years old somewhere else.) What observations or measurements might be used to persuade him that the principle of faunal succession is valid within narrow limits?

6 Geologic periods were defined on the basis of faunal succession. Would the transition, say from the Jurassic to the Cretaceous Period, necessarily have been accomplished at the same time everywhere in the world? Why, or why not?

7 A series of volcanic layers in west Texas were dated by the K-Ar isotopic age method. One such stratum gave an age of 35 million years, but a stratum above it gave an age of 37 million years. What are some possible explanations of this discrepancy? Is the law of superposition or the K-Ar method untrustworthy?

8 Another way to calibrate the geologic time scale is to date cobbles in a conglomerate that contains fossils diagnostic of a given geologic period. What do the isotopic ages signify? Does this field relationship offer a precise means to establish the age of deposition of the sediment? Why, or why not?

9 Beneath each square meter of nearshore sediments is contained roughly 1 kilogram of nutrients available to bottom-feeding organisms. Offshore, the nutrient supply steadily decreases to about 10,000 times less in the mid-ocean (0.1 gram per square meter). A similar pattern probably obtained in times past. How can a paleontologist use this information to identify ancient sedimentary environments? How would the population density of fossils, and the diversity of species, have been affected?

READINGS

* Berry, William B. N., 1968: *Growth of a Prehistoric Time Scale,* W. H. Freeman and Company, San Francisco, 158 pp.

Fenton, C. L., and M. A. Fenton, 1958: *The Fossil Book,* Doubleday & Company, Inc., Garden City, N.Y., 482 pp.

MacFall, R. P., and J. C. Wollin, 1972: *Fossils for Amateurs,* Van Nostrand Reinhold Company, New York, 341 pp.

Margulis, Lynn, 1971: "Symbiosis and Evolution," *Scientific American,* vol. 225, no. 2, pp. 48–57.

* Available in paperback.

chapter 9

A Bestiary of Vertebrates

"Very well then, hands up all those who propose to become birds."

[© *PUNCH, LONDON*]

During the Middle Ages, when people had rather different explanations from those they accept today for many natural happenings, the *bestiary* was a favorite literary device. A bestiary is a collection of fables in which a description of an animal's habits is followed by an interpretation of their moral significance (Figure 9-1). We have inherited many of these delightful allegories in our common language—"slothfulness," "eager beaver," stupid ass," etc. This chapter is a modern version of the time-honored old bestiary; it tells a story about the geologic history of the vertebrates (animals with backbones). The fossil record, of course, does not admonish us morally. Instead, we shall give the bestiary a scientific twist by showing how vertebrates provide the most splendid confirmation to be found anywhere of the evolutionary principles outlined in the preceding chapters. So let us catch a glimpse of the procession of vertebrate life through time, pausing here and there along the way to examine how some of these long extinct creatures illustrate special lessons in evolution.

Why do the vertebrates furnish the best evolutionary case histo-

FIGURE 9-1 People in the fifteenth century were also concerned with the origin of vertebrates. One famous bestiary explained the annual appearance of migratory geese which were never observed to mate (their arctic breeding grounds were then unknown) in this way: The goose embryos matured within the shell-like fruit of a certain tree growing near the seashore. In another version, the embryos grew like barnacles on floating logs. Somehow the two themes were later merged into the Barnacle Goose Tree. [*After J. D. Bernal*, The Origin of Life, *The World Publishing Company, Cleveland, 1967.*]

a

b

ries? One reason is that they are well studied, billions of dollars having been spent for medical research on vertebrate animals. In addition, we humans, ourselves vertebrates, are just naturally curious about our biological pedigree. Still another factor is that dinosaurs and other spectacular fossils, so well known to every schoolboy, have long fired the public imagination. Museums have continued to place high priority upon sponsoring the prolonged and tedious job of preparing vertebrate fossils for display (Figure 9-2).

However, the most important reason for our success in piecing together vertebrate history is that these creatures are astonishingly similar, notwithstanding the bizarre shapes and sizes of some of them. Differences among vertebrates are minor compared with differences between vertebrates and other animal phyla. Consequently the fossil record, fragmentary though it is, rather conclusively documents the relatively modest evolutionary changes in the vertebrate lineage.

c

d

FIGURE 9-2 (*a*) An unusually complete specimen of *Dimetrodon*, a grotesque reptile with a huge, spiny "sail" on its back, was discovered in Texas. (*b*) Still embedded in matrix, a similar specimen was swathed in protective burlap and plaster for the trip to the preparatory lab at the museum. (*c*) The reassembled skeleton is displayed in lifelike position. The animal's general form guides the preparator who must "dub in" missing parts (occasionally representing most of the skeleton). (*d*) Only the bones shown in black were recovered from the left side of a mosasaur (extinct aquatic reptile) now on display in the Texas Memorial Museum, Austin. Reconstructing an entire skeleton from so few parts may seem like a monstrous joke. It is not. The preparator is aided by a keen knowledge of comparative anatomy, by other fossils of the same species, and by the mirror-image bones in the animal's other half. [(*a*), (*b*), *and* (*c*) *from A. S. Romer*, Man and the Vertebrates, *University of Chicago Press, Chicago, 1941*; (*d*) *after W. Langston, Jr., "The Onion Creek Mosasaur,"* Texas Memorial Museum Notes No. 10, *1966.*]

What *do* vertebrates have in common? As far as the soft parts are concerned, they have well-developed nervous systems including brains (however primitive). Except for a few animals adapted to specialized niches, all vertebrates have nostrils, eyes, and ears. The cluster of delicate sense organs in the head serves as a guidance system for these active animals; they are "always going somewhere." Blood circulates through a network of vessels of broadly similar plan in each organism. All vertebrates are covered by naked skin or by secondary structures originating in the skin (scales, horny plates, feathers, etc.). All possess gills or gill derivatives; fish and amphibians actually use gills for breathing (during at least part of the life cycle), and the other vertebrates retain the gill structure temporarily during the development of the embryo. Vertebrate embryos are so similar that distinguishing a fish from a bird or mammal is difficult during the early stages (Figure 9-3).

Much valuable information about soft parts can be obtained from openings in the bone that accommodate nerves or blood vessels, and from muscle attachments on the skeleton. In a very few localities, fossils are found in which a compression of the body outline and even of soft internal organs is preserved. But as we have already noted, teeth and bones are usually all the fossil legacy we receive. In respect to hard structures, too, we see a remarkable similarity in the general features of

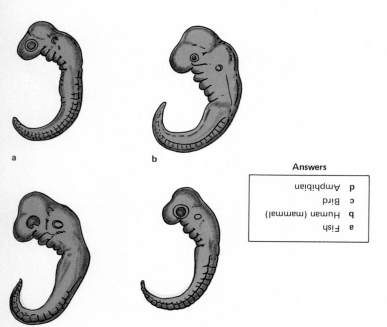

a

b

c

d

FIGURE 9-3 Vertebrate embryos are look-alikes. Which embryo is destined to become a fish? A frog? A fowl? A furry animal? (Turn the page over.) [*After E. O. Dodson*, Evolution: Process and Product, *Van Nostrand Reinhold Company, New York, 1960.*]

Answers

a	Fish
b	Human (mammal)
c	Bird
d	Amphibian

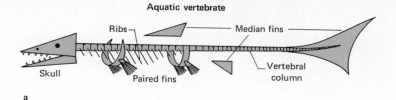

Aquatic vertebrate

Ribs — Median fins

Skull

Paired fins

Vertebral column

a

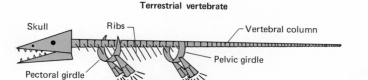

Terrestrial vertebrate

Skull

Ribs — Vertebral column

Pectoral girdle

Pelvic girdle

Paired limbs

b

FIGURE 9-4 (*a*) Aquatic vertebrates typically have median (centrally located) fins, and paired fins or flippers. (*b*) Land-living vertebrates have shed the structures homologous with median fins. [*After E. H. Colbert, Evolution of the Vertebrates, 2d ed. Copyright © 1969. By permission of John Wiley & Sons, Inc., New York.*]

all vertebrate skeletons (Figure 9-4). A segmented backbone, comprising the long axis of the animal, develops in the embryo around a preexisting, flexible rodlike structure called a *notochord*. A few primitive animals possess a notochord, but a backbone never materializes. In view of these similarities, the vertebrates are assigned as the predominant subphylum within the more inclusive phylum Chordata.

The hallmark of vertebrates is thus a bilaterally symmetrical skeleton of bone and cartilage, featuring a backbone and ribs, a tail, a skull, and paired appendages (fins, legs, wings, etc.). Beyond this, the variation of the skeletal form among living and fossil vertebrates would seem to be almost limitless—or is it really? The biologist D'Arcy Thompson pointed out that if a square grid pattern is laid, for example, over the outline of a human skull, the skull shape of other primates or even of a dog is obtained by a systematic transformation of the grid (Figure 9-5). From this we see that not only are the proportions of the human braincase larger and the jaw smaller than in a chimpanzee, but that *every* bone shape and size is varied in a coordinated manner between the two animals. And so, apparently great differences among vertebrates may result from minor changes in skeletal growth factors related in turn to mutations in just a few controlling genes. These comparisons, though not sufficient to establish lines of descent, are at least strongly suggestive of evolutionary relationships.

The vertebrate subphylum may be divided into eight classes (Table 9-1). The distinctions among "fish," amphibians, reptiles, birds, and mammals are fairly decisive, but the further separation of the fish group into four classes may come as a surprise. Closer examination, though, confirms that fish are exceedingly diverse. An intelligent codfish would tartly inform us that the difference between a slimy hagfish and himself is greater than the difference between a frog and a man!

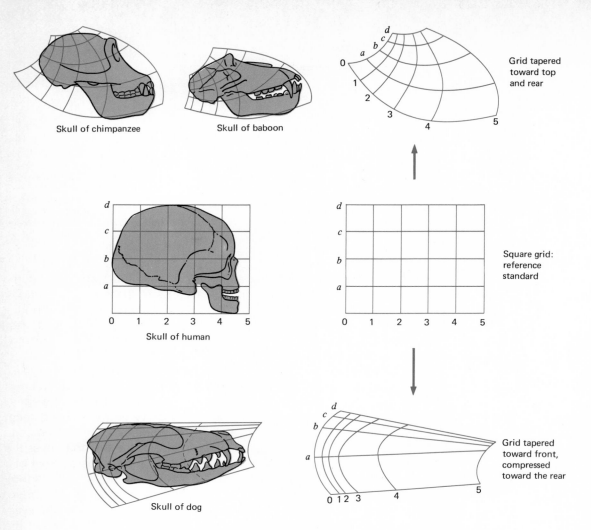

Grid tapered
toward top
and rear

Square grid:
reference
standard

Grid tapered
toward front,
compressed
toward the rear

Skull of chimpanzee

Skull of baboon

Skull of human

Skull of dog

FIGURE 9-5 The skeleton shape is invaluable in the classification of vertebrates. A human skull overlain by a square reference grid (center) transforms into the skull of a chimpanzee or baboon (top) or dog (bottom) by a distortion of the grid. A human is shaped more like a chimpanzee, less like a baboon, and even less like a dog. Moreover, human and chimpanzee skulls correspond bone for bone and tooth for tooth. The correspondence is close but not quite exact between humans and dogs. No amount of stretching and squeezing can make a coordinate pattern for a beetle (arthropod) or cuttlefish (mollusk) fit over that of a vertebrate. [*After D'Arcy W. Thompson,* Growth and Form, *Cambridge University Press, New York, 1942.*]

TABLE 9-1
The Vertebrate Subphylum

Class	Representative subgroups*	Distinguishing characters	Time of origin	Number of extant species (thousands)
Agnaths	Cyclostomes (example: lamprey) Ostracoderms	Jawless. Cartilaginous internal skeletons. No paired fins. Asymmetrical tail. Third eye. External fertilization. Simple heart in this and other fishes. External bony covering in ostracoderms. Ears used dominantly for balance in all fish classes.	Early Paleozoic	0.045
Placoderms	Arthrodires	A catchall class. Jaws and paired fins. Asymmetrical tail. Bony spines projecting from shoulder region. Internal fertilization?	Early Paleozoic	Zero (extinct)
Cartilaginous fishes	Sharks Sting-rays	Jaws. Teeth, spines, and skin denticles the only hard parts (no internal bony skeleton). No bony scales. No swim bladder. Asymmetrical tail. Internal fertilization.	Mid-Paleozoic	0.25
Bony fishes	Ray-finned fishes Lobe-finned fishes	Bony skull, jaws, gill coverings, scales. Bony internal skeleton in this and all "higher" classes. Swim bladder or lungs. Symmetrical tail (in modern forms). External fertilization.	Mid-Paleozoic	20
Amphibians	Frogs Salamanders	Skeleton modified for support out of water. Four-legged. Gills and lungs employed in successive growth stages. Ears used for balance and hearing in this and higher classes. External fertilization in some orders; internal in others.	Mid-Paleozoic	2.1
Reptiles	Turtles Snakes Crocodilians Dinosaurs (extinct)	Lay amniote egg. Internal fertilization, as in birds and mammals. Advanced brain. Simplified set of skull bones.	Late Paleozoic	6
Birds	Ratites (example: ostrich) Passeriforms (perching birds) (And numerous other orders)	Warm-blooded, as in mammals. Wings and feathers.	Mesozoic	8.6
Mammals	Monotremes (example: platypus) Marsupials (example: kangaroo) Placentals: (most modern mammals)	Intelligent. Fur. Milk glands. Highly variable, but characteristic pattern of skeleton and teeth.	Mesozoic	3.5

* These may be subclasses or orders, or purely colloquial terms.

Not only does each successive class of vertebrates exhibit more efficient and complex organ systems, but the list almost precisely follows the order of their appearance in the fossil record (Figure 9-6). Thus an evolutionary interpretation based on comparisons of the *physical makeup* of vertebrates makes equally good evolutionary sense from the *historical* viewpoint.

If one class of vertebrates could have originated from a species within some other class, are there any fossil organisms that are intermediate between the two? These animals were once called "missing links," but nowadays many hypothetical missing links have in fact been discovered. Consider what an animal transitional between reptiles and mammals must have looked like. Did the mammalian skeleton, fur, four-chambered heart, warm-bloodedness, and milk glands appear simultaneously? That would seem highly unlikely, and in fact, the fossil record suggests that the set of characters we associate with "mammalness" was gradually acquired over millions of generations.

Not every mammal living today has a full set of mammalian characters. How should we classify the curious duckbilled platypus and spiny anteater which have milk glands, hair, and a single bone in the lower jaw after the manner of mammals, but which lay eggs like reptiles and have a reptilian type pectoral girdle?* A paleontologist does not regard the vertebrate classes as being set apart by hard boundaries. Instead, he speaks more loosely of the "reptile grade" or "mammal grade" to refer to an association of characters that are mostly, but not always totally, present. It is also apparent that *many* groups of related animals could have been evolving toward a new level of organization at the same time. The fossil record shows that mammals probably did not spring from only a single ancestral species of reptile.

Our story, then, is of how groups of vertebrates attained higher grades, both metaphorically and in the technical sense. Their evolution was marked by a number of milestones—spectacular breakthroughs that made it possible for the animals to occupy new habitats and niches. We shall see that these milestones signaled the appearance of new structures such as bone, jaws, ears, lungs, limbs, modifications of skin (hair, etc.), and the amniote egg.

ORIGIN OF VERTEBRATES

The origin of vertebrates, like the origin of the other phyla, is obscured in antiquity. Perhaps the lack of any fossil remains of the vertebrate "founding fathers" should not overly surprise us. The earliest organisms were probably soft-bodied and tiny, but even more significantly, they were few in number, hidden away in some forgotten corner of the

---

* A checklist of characters tips the balance in favor of the mammals.

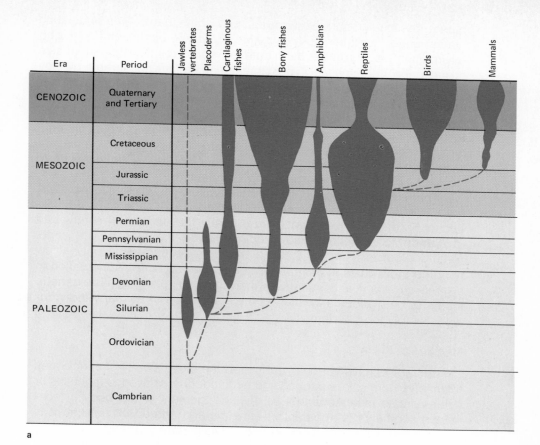

a

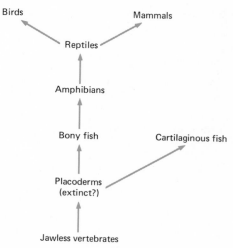

b

FIGURE 9-6 (*a*) A bubble diagram depicts how the eight vertebrate classes have fared through geologic time. The width of the shaded figure corresponds to the relative number of species within a class. The present day has been called the Age of Mammals, but according to the diagram, birds and bony fishes are thriving better, and in fact the number of mammal species has actually decreased since late Tertiary times. Probably more vertebrate species are alive today than ever before in earth history. (Nonetheless, most species of vertebrates are extinct.) (*b*) A simplified phylogeny of the vertebrates is abstracted from diagram (*a*). [(*a*) *after E. H. Colbert*, Evolution of the Vertebrates, *2d ed. Copyright © 1969. By permission of John Wiley & Sons, Inc., New York.*]

planet. The fossil record shows again and again that a *small* population placed under conditions of *environmental stress* (continually changing environment) is the proper candidate for rapid evolution. The small size of the gene pool makes it possible for mutations to spread quickly throughout a population. Genes can also be lost easily through premature death of individuals (natural selection) or by other means. One of these is a process called *genetic drift* which permits a gene eventually to be "overlooked" just in the random exchange of genetic material through normal mating encounters (Figure 9-7). Whether genes are gained or lost, rapid change over a few generations is the keystone of the evolution of small, interbreeding populations.

The rewards and perils attending a small population size are enormous. Usually the "experiment" in quickened evolution fails, and extinction is the final judgment. Occasionally the population survives to give birth to something radically new upon the face of the earth. (Anthropologists estimate that the first humans numbered only a few hundred individuals in central Africa.) In contrast, a *large* population in a *stable environment* is equated with evolutionary standstill. The genetic inertia to be overcome in altering the inheritance of so many individuals is simply too ponderous.

Affinities of Phyla

Biologists have variously proposed the close affinity of vertebrates to annelid worms, to arthropods, to echinoderms, or even to mollusks or coelenterates (see Appendix B). An arrangement of the body in connected segments is apparent in annelids and arthropods, and so it is too in the muscles of a vertebrate embryo (Figure 9-3) and in the mul-

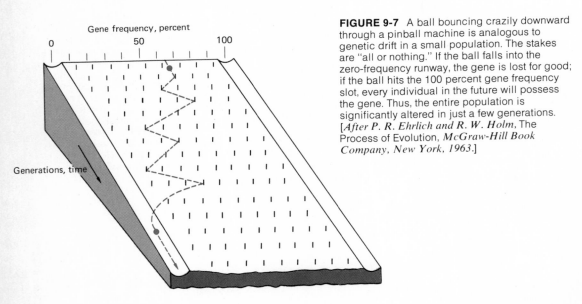

FIGURE 9-7 A ball bouncing crazily downward through a pinball machine is analogous to genetic drift in a small population. The stakes are "all or nothing." If the ball falls into the zero-frequency runway, the gene is lost for good; if the ball hits the 100 percent gene frequency slot, every individual in the future will possess the gene. Thus, the entire population is significantly altered in just a few generations. [*After P. R. Ehrlich and R. W. Holm,* The Process of Evolution, *McGraw-Hill Book Company, New York, 1963.*]

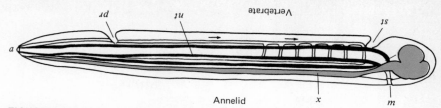

FIGURE 9-8 This diagram illustrates the supposed transformation of an annelid worm into a vertebrate. A nerve cord (*x*) runs along the underside of an annelid. The mouth of this animal (*m*) is on the bottom, and its anus (*a*) is at the end of the tail. Turn the book upside down, and now we have the vertebrate, with nerve cord and blood streams reversed. But it is necessary to build a new mouth (*st*) and anus (*pr*), and to close the old ones. The worm really had no notochord (*nt*), and the supposed change is not so simple as it seems.

[*After A. S. Romer*, The Vertebrate Body, *W. B. Saunders Company, Philadelphia, 1970.*]

tiple vertebrae of its backbone. Annelids, arthropods, and vertebrates exhibit bilateral symmetry, front and rear ends, and a central nervous system. But here the basic similarity ceases. The nerve cord is solid in annelids and hollow in vertebrates; it runs down the underside of an annelid but along the upper side of a vertebrate (Figure 9-8). There seems to be no easy way to make an annelid roll over or turn inside out to convert it into a vertebrate. Similar problems apply to obtaining the vertebrate structure from an arthropod.

Instead, modern evidence allies the vertebrates (or more precisely, the chordates) with the lowly sea urchin and his fellow echinoderms. Chordate and echinoderm larvae are astonishingly similar (Figure 9-9). The blood serum and the biochemistry of the energy cycle in muscle cells are similar in chordates and echinoderms, but different from those in the other animal phyla. This relationship does not imply the chordates *evolved* from echinoderms, but rather that these groups may lay claim to a common ancestor. The comparative studies seem to define two great mainstreams of animal phylogeny related something like a Y.

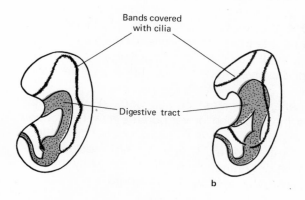

FIGURE 9-9 The plane represented by this page divides the bilateral symmetry of these larvae (greatly magnified). Larva *a* is an acorn worm, a lower chordate. Larva *b* is a starfish, an echinoderm. [*After A. S. Romer, The Vertebrate Body, W. B. Saunders Company, Philadelphia, 1970.*]

237

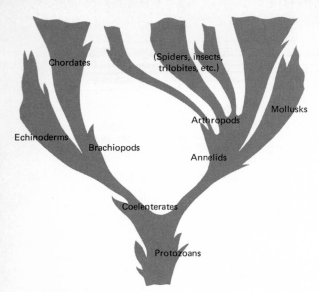

FIGURE 9-10 A simplified family tree of the animal kingdom relates the vertebrates to other phyla.

Protozoans and coelenterates are located at the juncture point of the Y. Branching out from them on the one hand are annelids–mollusks–arthropods; the other branch includes brachiopods (?)–echinoderms–chordates (Figure 9-10).

From Primitive Chordates to Vertebrates

About 98 percent of living chordates are classed as vertebrates. Still, the remaining 2 percent of the chordate phylum will help us understand vertebrate origins, for these creatures clearly hold the irreducible essentials of animal life. Whatever else an aquatic animal may possess, it certainly has to have a way to capture food and obtain oxygen from its environment. Lower chordates solve the problem in a number of ways. Some of them are sessile (attached) animals whose cilia-covered, waving arms ceaselessly wait to entrap drifting food particles (Figure 9-11a). In this respect, at least, these chordates are much like sessile crinoids (echinoderms). In other chordates, in place of tentacles we find an elaborate internal filtering system consisting of *gills* through which large volumes of water are pumped. The gills strain out food while also passing dissolved gases into and out of the body through a thin membrane.

A chordate group called tunicates (or sea squirts) shows how another important gap in the evolution of vertebrates may have been bridged. Not that adult tunicates tell us very much, for they do not look at all like a vertebrate (Figure 9-11d). They have no trace of a notochord, and they are sessile or drifting organisms. In fact, they are little more than barrels lined internally by gills and covered by a leathery

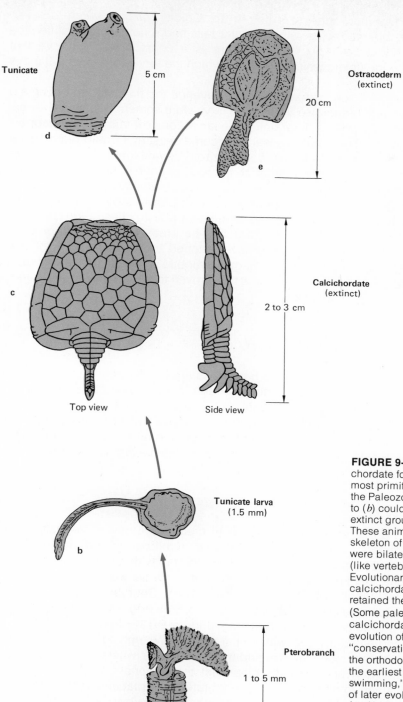

Tunicate

5 cm

Ostracoderm
(extinct)

20 cm

c

Calcichordate
(extinct)

2 to 3 cm

Top view

Side view

Tunicate larva
(1.5 mm)

b

Pterobranch

1 to 5 mm

a

FIGURE 9-11 (*a*) The branched, ciliated chordate form is considered to be the most primitive. It is suggested that early in the Paleozoic Era, a mobile larva similar to (*b*) could have given rise to a now extinct group, the calcichordates (*c*). These animals, though covered by a skeleton of echinoderm-like calcite plates, were bilaterally symmetrical as adults (like vertebrates) and not sessile. Evolutionary descendants of calcichordates lost the calcite armor but retained the notochord and muscular tail. (Some paleontologists deny that calcichordates participated in the early evolution of the chordates.) More "conservative" organisms settled down to the orthodox sessile adult life (*d*), whereas the earliest vertebrates were "off and swimming," both literally and in the sense of later evolutionary potential. Organisms (*a*), (*b*), and (*d*) are modern; they are not members of the actual vertebrate line of descent.

tunic. The innovation here lies in the tunicate *larvae*, which look like microscopic tadpoles with muscular tails and a notochord to stiffen the body (Figure 9-11*b*). Their structure is a considerable improvement over the inefficient whiplike cilia by which the larvae of Figure 9-9 move about.

Even so, how can such a larva advance the final step to become a full-fledged vertebrate? What if the larva were somehow to become sexually mature while still retaining the larval form? If this were to happen, the animal would not necessarily become sessile; it could remain ever the restless food seeker, at considerable advantage to itself. Our proposed origin of vertebrates is conjecture, of course, but it is in accord with accepted ideas about evolution. Certain salamanders, for instance, are noted occasionally to bypass an adult growth stage. Natural selection is opportunistic; it will allow the latent possibilities of *any* structure to be exploited, whether that structure be present in the adult or in an embryonic growth stage. We have reached the *first milestone* of vertebrate evolution: the development of a mobile organism from a probable sessile ancestor. In the end, this has proven to be the most consequential milestone of them all.

THE PARADE IN REVIEW

The oldest known vertebrate hard parts are locally abundant fragments of bone in sediments of Ordovician age in the United States and in Estonia. The unfolding tale of vertebrate development from that day to the present has been so skillfully reconstructed by vertebrate paleontologists, working from a profusion of fossil remains, that we can do no better than to retrace that same story with attention to some appropriate case studies.

Jawless Vertebrates

The first vertebrates were small creatures, rarely more than a few centimeters long. Essentially they were mobilized gill baskets, lacking jaws, and propelled along by a muscular tail with awkward tadpolelike motions. Unlike most modern fish, these fellows were obliged to use their gills both for feeding and breathing. Doubtless they and their contemporaries, the trilobites (Appendix B) made a living by the time-honored method of food straining (Figure 9-12). Whether the earliest fish originated in salt water or fresh is a warmly debated question answered by inconclusive evidence. No matter where they *began*, the sedimentary record makes clear that the important *later* evolution of fish that led to other groups of vertebrates took place in fresh water and eventually on land. Possibly the jawless fish were driven to inhabit

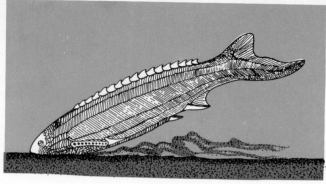

Pterolepis

Hemicyclaspis

a

b

FIGURE 9-12 (*a*) A primitive jawless fish gathers food particles by skimming through the top layer of a muddy bottom. This fish differs from modern representatives in having a bottom-heavy, asymmetrical tail and no stabilizer fins. Presumably, swimming strokes with this sort of tail kept the body pointed downward into the mud. [*After I. I. Schmalhausen,* The Origin of Terrestrial Vertebrates, *Academic Press, Inc., New York, 1968.*] (*b*) The armored body of another extinct fish looks like a miniature submarine. Its unblinking eyes stared straight upward, ever on the alert for predators. The fish even had a third light-sensitive organ in the middle of its skull. With the passage of geologic time, the vertebrate "third eye" opening in the skull tended to close up, but it was present in many fossil reptiles and amphibians, and even today in lizards. The organ is submerged in man but survives as the pineal gland. [*After J. M. Weller,* The Course of Evolution, *McGraw-Hill Book Company, New York, 1969.*]

freshwater ecological niches by competition with trilobites which remained exclusively marine. The fishes' swimming ability enabled them to cope with strong water currents.

The early jawless fish reached the *second milestone* of vertebrate evolution: the development of a bony skeleton. It was a very curious skeleton, quite unlike the sort familiar to us. In some species it consisted of a myriad of plates, large or tiny, embedded in the skin. Internal bone was absent except as a thin partition secreted around cavities and canals. Recent studies show that the skin of all vertebrates is arranged in a mosaic pattern of little polygons like the design of tile on a bathroom wall (Figure 9-13). According to one theory, the bone developed first as tiny outgrowths of the skin; with further evolution, these fused into a continuous pavement.

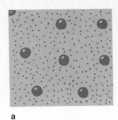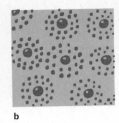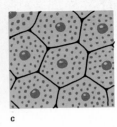

a   b   c

FIGURE 9-13 The origin of bony armor is postulated, from isolated centers of bone deposition (*a*) to a solid mosaic of plates (*c*). [*After L. B. Halstead, The Pattern of Vertebrate Evolution, 1968.*]

Given the *fact* of bone, what was its original *purpose*? Perhaps it was armor protection against the bite of the fearful eurypterids—giant predacious "scorpions" that lurked in the watery environment. Perhaps a bony cover was a means to waterproof the body. Still another possibility is that bone was a storehouse of needed calcium phosphate, a reserve against seasonal variations in the concentration of that substance in the water.

After their initial development, jawless fish went into eclipse as more advanced fish drove them out of most niches. Today, the jawless hagfishes and lampreys occupy a forlorn niche as the scavengers and parasites of the underwater world. They are not likely to become fossilized, as they embody soft cartilage but no bone.* Their existence reminds us that jawless vertebrates have probably survived continuously since the disappearance of the armored varieties from the fossil record in mid-Paleozoic time.

Placoderms

Around 50 million years after armored jawless fish had evolved, the *placoderms* began to be fossilized. Even though placoderm remains are plentiful, no one has yet been able to establish for certain where these outlandish fish came from, or what their descendants are. Evidently placoderms evolved from jawless vertebrates, but exactly which ones? Their ultimate extinction probably was not outright, but instead, it was a result of their continuing evolution into living higher grades of fish organization. (Some authorities speculate that the chimaera, or ratfish, is a relatively unmodified descendant of placoderms.) Placoderms flourished during the Devonian Period, appearing in the fossil record only slightly earlier than the bony fish and cartilaginous fishes. Placoderms were the first fishes to radiate into a host of niches that had not been available to their jawless ancestors.

---

* A well-preserved Pennsylvanian fossil is known.

The spectacular advance that made this possible brings us to the appearance of jaws, the *third milestone* of vertebrate evolution. Jaws were obviously advantageous, as they enabled the placoderms to take a satisfying bite of meaty prey, as contrasted with the mere slurping or timid nibbling habits of the jawless fish. Placoderms were undoubtedly the first vertebrate carnivores. Some of them developed the jaw theme to extravagant proportions; a man would have made a mouthful in a single crunch by one gigantic beast (Figure 9-14). In all, the placoderms are a miscellaneous group that, as the paleontologist A. S. Romer remarked, are best described as "a series of wildly impossible types which do not fit any proper pattern."

Jaws originated in a most remarkable fashion by modification of gills. The gills of a fish, arrayed along both its sides just behind the head, are reinforced by a series of V-shaped bony supports like so: (rear body)<<<<(head). Let us begin the train of events with the gills of a jawless vertebrate (Figure 9-15a). After evolution had enlarged a forward bony arch (b) and had developed in it a freely moving hinge action, behold, jaws were the result! Further evolution strengthened the jaw structure by anchoring another gill arch as a supportive prop against the jaw hinge (c).

The embryology and nerve pattern of modern fish, and findings in the fossil record, support this interpretation of the origin of jaws. We have here an example of *preadaptation*. That is, a structure already present (gills) was modified and pressed into a new form of service (jaws). However, we must not confuse the idea that gills were pre-

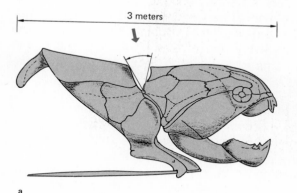

a

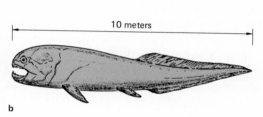

b

FIGURE 9-14 (*a*) The jaws of this placoderm, the first giant vertebrate, were armed with awesome cutting edges (not teeth). The hinged joint (arrow) permitted its skull to tilt *up* as the lower jaw tilted *down* and thrust forward. What sort of prey would have necessitated such a fantastic gape? [*After E. H. Colbert,* Evolution of the Vertebrates, *2d ed. Copyright © 1969. By permission of John Wiley & Sons, Inc., New York.*] (*b*) The entire fish grew as much as 10 meters long. [*After A. S. Romer,* The Vertebrate Body, *W. B. Saunders Company, Philadelphia, 1970.*]

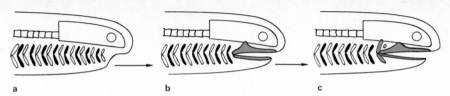

a   b   c

FIGURE 9-15 Stages in the evolutionary development of jaws are suggested here. Fossils exhibiting stages (*a*) and (*c*) are well known. Unfortunately, stage (*b*) is plausible but only conjectured, as the critical fossils are yet unknown. The placoderms deserve much continued study. [*After J. M. Weller,* The Course of Evolution, *McGraw-Hill Book Company, New York, 1969.*]

adapted to form jaws with the mistaken notion that they were pre*ordained* to do so. Gills did not originate with a mystical assurance that some day a new thing would be made from them. For that matter, fish did not develop jaws merely because they wanted to. An easier explanation is that the fish that *inherited* jaws simply ate up most of the hapless fish that did not.

Bony and Cartilaginous Fish

The *bony* fish—that is, fish that possess a complex internal skeleton of bone—appeared in the Devonian Period slightly before the *cartilaginous* fish. At last, the common sorts of fish had emerged from the shadowy underworld of grotesque placoderms to assume undisputed rulership of the waters. No sensational new milestones of evolutionary development have marked the progress of the two higher fish classes. They have capitalized, instead, on doing well what they know best, by radiating out into every conceivable underwater niche and habitat. What, then, are some of the "progressive" features that distinguish the higher fishes?

Skeletons Changes of the *external* skeleton were among the more noteworthy developments of continuing fish evolution. Both the bony fish and cartilaginous fish (sharks and the like) have shed the massive armor plate that must have been so burdensome to their placoderm ancestors. The reduced vestiges of armor appear as tiny denticles in shark skin (imparting to it a sandpaper texture), and as *scales* in the bony fish. The fossil record nicely documents a further evolution of fish scales. Bony fishes living in the Paleozoic Era were covered by row upon overlapping row of thick scales capped by shiny, enamellike outer surfaces (Figure 9-17). With the passing of the Mesozoic and Cenozoic Eras, fish scales became progressively thinner and more flexible. (Some modern fish are completely naked.) What the fish lost in armor protection they made up for by increased agility of movement.

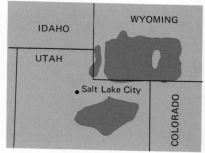

FIGURE 9-16 Because the soft cartilage of a shark or sting-ray (shown above) quickly rots, a completely preserved fossil skeleton of a cartilaginous fish is one of the rarest of finds. During the Eocene Epoch, a group of freshwater lakes in Wyoming and Utah (inset map) supported a balanced fauna and flora. Some bedding planes in the lake sediments contain numerous fish of all kinds, ages, and sizes. Evidently they were killed catastrophically. By counting fossils in the mass-death communities, paleontologists have established that herbivorous fish greatly outnumbered the carnivorous fish, just as expected. [*American Museum of Natural History photograph.*]

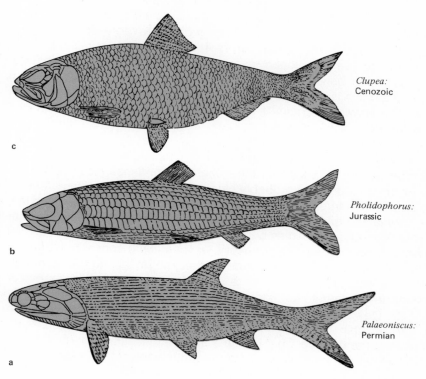

Clupea:
Cenozoic

c

Pholidophorus:
Jurassic

b

Palaeoniscus:
Permian

a

FIGURE 9-17 Three stages, or grades, of bony fish evolution show that with time the scales became thinner and more rounded, the tail became symmetrical, the jaws shortened, and the skull became extensively modified. Out of some 20,000 modern fish species, only a handful survive as living fossils representing types (*a*) and (*b*). [*After E. H. Colbert, Evolution of the Vertebrates, 2d ed. Copyright © 1969. By permission of John Wiley & Sons, Inc., New York.*]

The tail tended toward a symmetrical shape, paired fins fore and aft became the rule, and many species developed streamlined body shapes.

Teeth were present in a few placoderms, but in the higher fish they became standard equipment. Predatory fish developed the familiar rows of pointed, needlelike teeth. Other fish were endowed with pavements of numerous flattened teeth suitable for crushing mollusk shells.

The evolution of the *internal* skeleton followed an interesting divergent pathway. Certain of the jawless fish contain only cartilage: springy material of the same type that maintains the shape of a person's nose and ears. Possibly sharks and other cartilaginous fish inherited the soft internal skeleton which they have retained without much change to this day. Another possibility, in view of their later debut yet general similarity to bony fish, is that cartilaginous fish are degenerate cousins of the "ruling" fish class. The fossil record is moot on this point. The "feel" of a shark body is similar to that of an inflated rubber tire. And just like a quality tire, one of these incredibly tough citizens of the shark clan is "good for thousands of miles." Sharklike fishes evolved considerably during the late Paleozoic Era, but from the Mesozoic to the present the established body styles of various shark species have remained conservatively stable.

The other class, bony fish, ushered in at least a minor breakthrough with the appearance of an ossified (bony) internal skeleton. The braincase became a complicated assortment of bones representing ossified cartilage, united with bones appropriated from the ancestral armor plate. Vertebrae developed as simple "spools" enclosing the notochord. Other bones—hundreds of them—took shape, as anyone who likes to eat fish is aware. Bony skeletons proved to be a huge success, and the everwidening family tree of bony fish (Figure 9-6*a*) suggests that the group will continue to expand long into the distant future.

Ray-finned fish and lobe-finned fish   To continue our story of the vertebrates, we must now focus on a character of fishes that in the early Devonian Period would probably have been counted as an insignificant detail. The fin structure of bony fishes had evolved along two lines that have remained distinct right up to our own time. *Ray-finned* fishes bear a set of controlling muscles buried inside the body wall. These muscles operate a fan-shaped arrangement of spines (rays) connected together by a webbing of soft tissue (Figure 9-18*a*). The ray-finned fish, although highly successful, have proven to be but a side issue as far as evolutionary development into other animals was concerned.

The Devonian *lobe-finned* fish were especially well adapted to life in a freshwater environment. The fin rays of these fish projected, not directly out of the body, but from the tip of a muscular lobe containing stout internal bones (Figure 9-18*b*). A prominent group of early lobe-

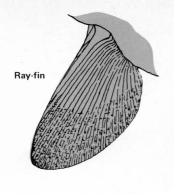

Ray-fin

a

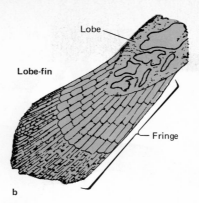

Lobe

Lobe-fin

Fringe

b

FIGURE 9-18 (*a*) Ray-fins connect directly to the body wall, whereas (*b*) lobe-fins include a fleshy appendage containing bones. [*After* The Fossil Book *by Carroll Lane Fenton and Mildred Adams Fenton. Copyright © 1958 by C. L. F. and M. A. F. Used by permission of Doubleday & Company, Inc., Garden City, N.Y.*]

finned fish were the *crossopterygians* (Figure 9-19). The sediments that contain fossil crossopterygians hint that these fish lurked in ambush for unwary prey under the cover of weeds or rocks in muddy water. The nerve pattern in the skull indicates that they had a keen sense of smell but very weak eyes. (After all, what special benefit were sharp eyes in the midst of a weed patch?) Moreover, in crossopterygians a service-able lung had evolved. Lungs were a handy item in the Devonian Period, a time of harsh climatic extremes. When a riverbed dried up, leaving only a few pools crowded with gasping, suffocating fish, the crossopterygians could interrupt their feasting by an occasional trip to the surface for a gulp of fresh oxygen. Possession of a lung was by no means unique to crossopterygians; lungfish today inhabit tropical water-courses that habitually go dry in midsummer.

Amphibians

The gradual evolution of certain of the crossopterygian fishes into true amphibians is one of the most clearly documented case histories in vertebrate paleontology. It is an example of slow attainment of a higher evolutionary grade over the span from the late Devonian to the

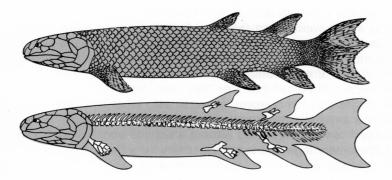

FIGURE 9-19 A late Devonian crossopterygian fish had median lobe-fins along its back and paired fins on its underside. All the fins were equally well preadapted to evolve into legs, but the median fins never did so. Again we see that a preadapted organ is not preordained to serve the new function. [*After I. I. Schmalhausen,* The Origin of Terrestrial Vertebrates, *Academic Press, Inc., New York, 1968.*]

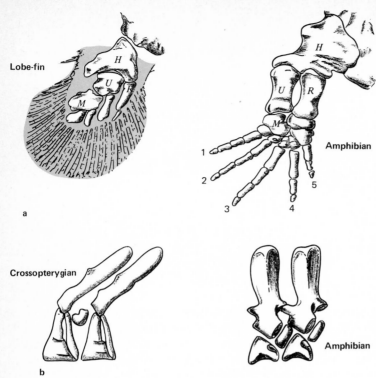

Lobe-fin

Amphibian

H

U R

M

1
2
3 4
5

Crossopterygian

a

b

Amphibian

FIGURE 9-22 (*a*) The bones of the crossopterygian lobe-fin (humerus, ulna, radius, metacarpus) were homologous with the leg and foot bones of its amphibian descendants. The five-digit scheme of most vertebrate hands and feet was established at the very outset. [*After C. L. Fenton and M. A. Fenton,* The Fossil Book, *Doubleday & Company, Inc., Garden City, N.Y., 1958.*] (*b*) Projections on the vertebrae of a crossopterygian became more robust in amphibians. Powerful back muscles, anchored to the projections, were needed to support the mass of a body no longer buoyed up in water. [*After A. S. Romer,* The Vertebrate Body, *W. B. Saunders Company, Philadelphia, 1970.*]

original form. (Why?) Williston's law applies only when *multiple* parts combine to perform the same or a similar function; it does not imply that the function itself is lost or diminished.

The evolution of the vertebrate skull furnishes a good illustration of this law. Skulls of certain of the jawless vertebrates were covered with a mosaic of innumerable tiny bones. Skulls of bony fish contain around 150 bones, and as each new class of vertebrates appeared, the number of bones was further decreased (Figure 9-23). With continuing evolution, a trend toward a dwindling number of bones persisted *within* each vertebrate class as well. Clearly, the function of a skull was not impaired as the bones were eliminated.

As amphibians invaded the land environment, their eyes and ears experienced major adjustments. The skull of a fish transmits into the region of the brain roughly 10 percent of the sound-wave energy reaching its outer surface (the remaining energy being reflected back into the water). But a skull surrounded by air might directly transmit an insignificant 0.003 percent or less. A way was needed to channel and amplify the sound before it was to be translated into nerve impulses. The amphibians witnessed the beginning stages of evolution of the middle ear, the *fifth milestone* of vertebrate evolution. The marvelous adaptation of this organ commences with the bone that props the jaw

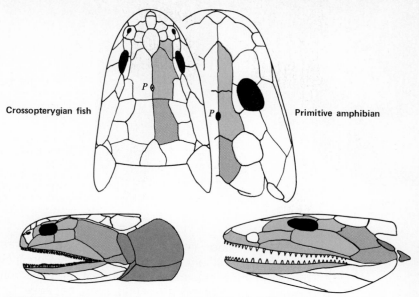

FIGURE 9-23 Shading denotes homologous bones in a crossopterygian fish and the primitive amphibian, *Ichthyostega*. The 50 bones visible in the top view of the fish skull have decreased to 35 bones in the amphibian. The side views show that homologous bones differ impressively in size and shape. The opening for the third, pineal eye (*P*) persisted in the first amphibians. [*After E. H. Colbert*, Evolution of the Vertebrates, *2d ed. Copyright © 1969. By permission of John Wiley & Sons, Inc., New York.*]

hinge of fishes against the braincase (Figure 9-15*c*). That figure reminds us that the prop had begun its illustrious career long before as one of the gill arches. Now it was to be changed once again just slightly and pressed into its *third* form of service as a conductor of sound. Preadaptation indeed! One end of the bone penetrated directly into the inner ear cavity, while its other end lightly touched an outer covering of stretched skin—a sort of primitive eardrum (Figure 9-24*b*). The bone was suspended in an open channel borrowed from what formerly was a gill chamber. Mammals in their turn made further refinements of the middle ear, as we shall see.

We know, too, that the vertebrate eye had to be altered if it were to continue in use. That is because light rays bend (refract) only slightly as they pass from water into the cornea of a fish's eye, whereas light refracts considerably when passing from air into an amphibian's eye. Moreover, amphibians needed tears to protect and lubricate the eye, and special muscles to turn it. We shall never know just how the eye structure evolved to take full advantage of the unlimited transparency of the air medium. Eyes were not preserved in the fossil record, and neither was the evidence for many another magnificent evolutionary breakthrough.

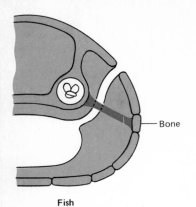

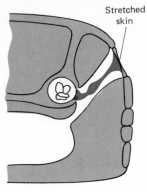

Stretched
skin

Bone

Fish

Advanced amphibian

FIGURE 9-24 Sound energy conducts into the inner ear of a fish immersed in water more easily than it enters the inner ear of an amphibian immersed in air. Animals that operate in the air medium require special sound-amplification devices in the middle ear. [*After J. M. Weller, The Course of Evolution, McGraw-Hill Book Company, New York, 1969.*]

Reptiles

The transition from amphibians to reptiles was another prolonged affair lasting millions of years in the early Pennsylvanian Period. What characteristics set apart the earliest reptiles from their amphibious predecessors? That is a hard question, because the critical distinctions between these two vertebrate classes were seldom preserved in fossil form. True, the number of bones in the reptilian skull was reduced yet further (as expected), and the skull shape trended away from the broad, flattened amphibian profile to become narrow and deep. The limb bones began to be twisted and drawn more directly underneath the body (Figure 9-25). The rowing style of limb action that amphibians inherited from the fish was thus modified by the reptiles into a more efficient, straightahead walking motion. All these skeletal changes were minor, however, and their combined effect became evident only after reptilian evolution had long been underway.

Up from the water Much had happened during the heyday of amphibians. These creatures could walk on land, and they could see, hear, and breathe in the air, but they were still chained to the watery environment. The fatal flaw of amphibians was the vulnerability of their eggs. These were shed into the water by the thousands in anticipation that a few, at least, would escape hungry predators or desiccation, an everpresent threat on the land. Reptiles bypassed the demand for water breeding when they introduced the amniote egg, the *sixth milestone* of vertebrate evolution. And what a liberating innovation it was! Now the embryo was protected until it could mature to the point of self-sufficiency, surrounded the while by a private microenvironment that supplied every necessity of life (Figure 9-26). Contained in the amniote egg is a yolk for food supply, a watery replica of the ancestral pond held within the amniotic membrane, a porous shell that "breathes"—and even a waste disposal unit!

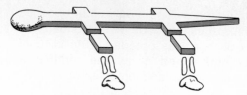

Primitive terrestrial vertebrate

Mammallike reptile

FIGURE 9-25

What selection pressure forced the reptiles to invade the unfilled dry land niches and habitats? None did, right away. The record shows that the earliest reptiles continued to slop happily about in the swamps along with their amphibious cousins for tens of millions of years. Only when the world climate began to break up into more varied zones in the Permian Period were reptiles thrust out onto dry land. Nor did every reptile forsake its ancestral habitat. The crocodile lineage (while not the most ancient of them all) at least has remained water-loving ever since it appeared. Then why did the amniote egg originate in the first place? In retrospect, we see that it appeared in the same environment that had

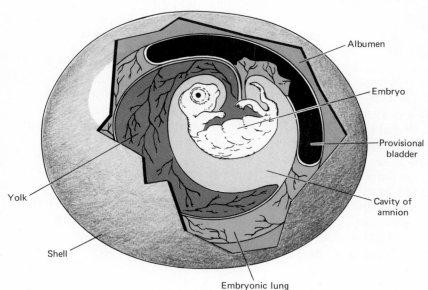

Albumen

Embryo

Provisional bladder

Cavity of amnion

Embryonic lung

Shell

Yolk

FIGURE 9-26 The amniote egg can survive a considerable drought. At the moment of hatching, the infant reptile is fully prepared for life on its own. [*After A. S. Romer,* The Procession of Life, *The World Publishing Company, Cleveland, 1968.*]

long been inhabited by amphibians, but because it had survival value, it became an established mode of reproduction. Only later did it develop that the amniote egg was preadapted to be useful in a range of drier environments undreamed of by the reptiles who "invented" it. Of course, this is not an explanation of the complex biochemical steps that led to its appearance.

The reptiles illustrate an example of a continuing debate about evolutionary pathways. Certain paleontologists believe that the fossil record shows amphibian evolution proceeding along a broadly advancing front. Some even claim that as many as four independent lines of descent led from the amphibian to the reptilian grade. If that were so, the reptiles had a *polyphyletic* (with multiple phylogenies) origin (Figure 9-27*a*). Other paleontologists, impressed by the complexity of the amniote egg, insist that this egg could have evolved only once. They affirm that reptiles radiated out from a single ancestral stock in a *monophyletic* (with one phylogeny) origin (Figure 9-27*b*). Which idea is correct? Ultimately the question will be resolved (if ever) by appeal to fossil evidence that is presently unavailable. None of a number of proposed classification schemes throws much light on the problem of reptile origins. For example, one classification, based upon the number and position of openings in the skull, is only a way to *describe* different reptiles (living and fossil) while failing to *explain* their evolutionary relationships.

The appearance of reptiles upon the dry land late in the Paleozoic Era triggered revolutionary changes in the ecological balance. Prior to that time, many vertebrates (including the "amphibious" reptiles) made a living by eating fish or invertebrates such as mollusks and insects. Predatory vertebrates, in turn, ate the invertebrate feeders, and whenever they could, the predators ate one another. From today's vantage point, this pathway through the food chain appears to be extremely primitive and unbalanced. No vertebrates were taking advantage of the abundant energy present in green vegetation.

The first strict vegetarians were land-based reptiles. Modifications required in the evolution of a successful herbivore are considerable. The animal's teeth and jaws must be outfitted to crack open tough plant materials. The gut must become greatly elongated to handle large volumes of low-calorie roughage. Protective devices must be evolved if enough individuals are to survive predation to be able to perpetuate the species. These changes were perfected over the Permian and Triassic Periods—a span of some 100 million years. Gradually the modern situation emerged in which most animals are herbivores, a few are invertebrate feeders, and a tiny minority are predators that feed on other animals but not on one another (except in extremely adverse circumstances).

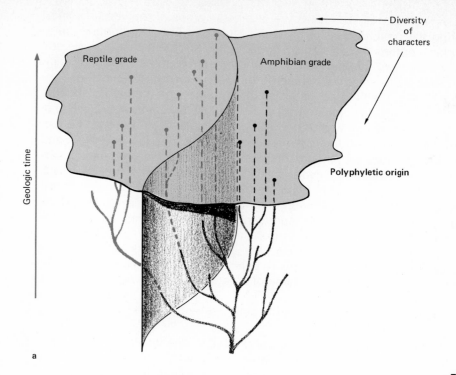

a

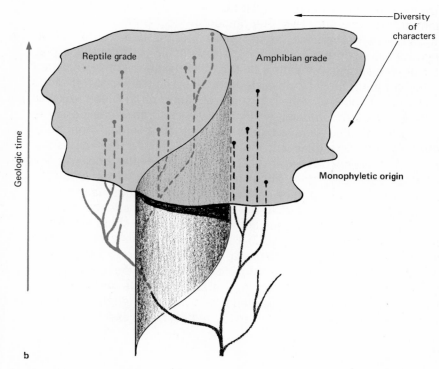

b

FIGURE 9-27 (*a*) A three-dimensional diagram similar to Figures 7-11 and 7-24 shows the passage of geologic time on the vertical axis. In any single time plane (horizontal direction), adjacent areas in a diversity map mark off the amphibian and reptile grades. Since reptiles and amphibians also existed in past geologic time, the grades actually become solid figures, containing volume. In a polyphyletic origin of reptiles, *several* groups of amphibians reach the reptile grade in different ways and at different times. (*b*) According to a monophyletic origin, *only one* amphibian species was able to breach the barrier to the reptile grade. Some paleontologists have scornfully referred to a unitary origin of higher categories of organisms as an "Adam and Eve complex," but of course verbal jousting is no way to settle a scientific dispute!

A reptilian roll call Numbers of diverse types of reptiles flourished during the peak of their reign in the Mesozoic Era. The importance of modern reptiles (snakes and lizards, turtles, crocodilians, and the shy tuatara of New Zealand) is but a pale reflection of the glory that once was theirs. We shall note some of the more interesting reptile "experiments," all of which led to extinction through one means or another.

Mammallike reptiles Among the earliest to appear was a group of reptiles well on their way to becoming mammals. In the Permian Period, the *mammallike reptiles*, while retaining the reptilian jaw and skull, had already developed a limb structure and the diversified set of teeth characteristic of mammals. (Note the latter features in that strange beast, *Dimetrodon* (Figure 9-2).) Even the bizarre sail (held rigid by vertebral spines) on the back of *Dimetrodon* may have been a step in the direction of mammal status. Measurements of various specimens indicate that as the skeleton became larger, the sail became larger even faster.

The size relationship of the sail to the main bulk of the animal poses the problem of the *square-cube* law. By analogy, consider the formulas for the area and volume of a sphere of radius r:

$$\text{Area} = 4\pi r^2 \qquad \text{Volume} = \frac{4\pi r^3}{3}$$

If, for example, the radius were doubled, the area would increase by a factor of 4 (or 2 squared), but the volume would increase eightfold (2 cubed). A large increase in volume accompanies only a modest increase in area. Now, *Dimetrodon* did not have a spherical body, but our reasoning holds nevertheless. The amount of heat produced internally in its body was proportional to its *volume*, but the heat could be dissipated only through the skin which covered its *area*. *Dimetrodon* may have been the subject of nature's first primitive experiment in regulated body temperature: the warm-blooded condition so well known in birds and mammals. The sail upheld a sheet of skin, rich in blood vessels, that helped *Dimetrodon* catch a little more warmth as it basked in the sun, or that radiated away excess heat at other times. The heat-exchange function explains why the size of the sail had to vary disproportionately to keep pace with volume changes of the body proper.

Further tantalizing data about mammallike reptiles are supplied from the *Karroo Series*, a thick sequence of Permian and Triassic river-borne sediments in South Africa. Not a single egg has ever been found in Karroo beds containing fossils of these animals! We could argue that eggs were simply not fossilized, but then how is it that other delicate structures, such as infant skeletons, plant material, and even insects, were preserved in abundance? More plausibly, the young were born alive and cared for by their parents. The same animals may have had hair. Impressed upon their cheekbones are traces of numerous nerve pathways that were probably linked to tactile hairs like the familiar whiskers of dogs and cats. In view of all this evidence, whether we should class these transitional creatures as mammals or reptiles seems to be an arbitrary decision.

The mammallike reptiles met a peculiar fate. Some species became extinct outright, but others lost identity by evolving into true mammals. For some reason, the reptiles-become-mammals were not able to gain command of the land surface. If they had, the history of vertebrate life might have been short-circuited at least 100 million years, and those magnificent beasts, the dinosaurs, would never have trod the ground. As it actually happened, the mammals went into eclipse as (from their point of view) the long, cold night of reptile domination drew on.

The terrible lizards   The Greek root of the word "dinosaur" means "terrible lizard," but in reality dinosaurs are not related to lizards except in a general way. Dinosaur lore has become so popular (thanks to elementary school teachers and the wealth of excellent dinosaur books) that we shall need to touch upon only a few ideas. Who does not indeed already know about the armor-plated *Stegosaurus*, the long-necked "brontosaurus," rhinoceroslike *Triceratops*, and the fearsome *Tyrannosaurus* (Figure 9-28)? Considering the amount of effort that has been spent studying dinosaurs, surely we could expect the major features of their life and times to be well understood. Quite the contrary, some of the most basic assumptions about dinosaur behavior are being challenged in the light of new evidence. Dinosaur research is one of the more active fields of vertebrate paleontology today (Figure 9-29).

Dinosaurs were so successful that they have been called the ruling reptiles of the Mesozoic Era. When they appeared about the middle of the Triassic Period, they already had evolved into two distinct orders that were to persist until the time of their ultimate demise. Dinosaurs

Cretaceous

Triceratops

Tyrannosaurus

Jurassic

Stegosaurus

"Brontosaurus"

FIGURE 9-28 Dinosaurs radiated into a wide variety of niches and habitats, as the dissimilarity of their profiles suggests. Some popular accounts show all the famous dinosaur types in one scene, locked in a grisly combat for survival. Nothing could be farther from the truth. The well-known dinosaurs, pictured here according to their relative sizes, lived in different places and were separated by tens of millions of years.

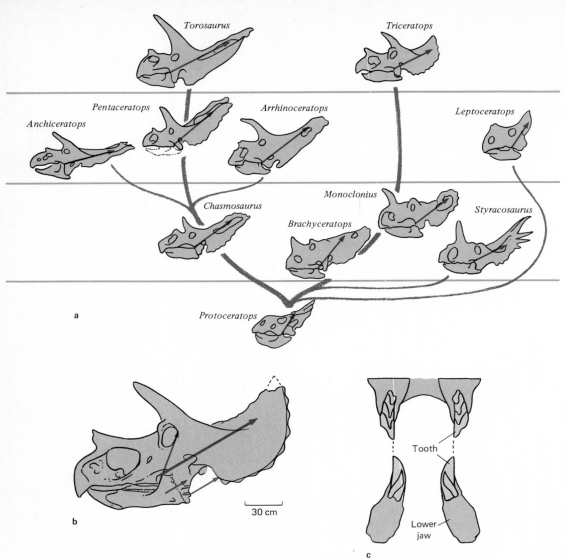

FIGURE 9-29 (*a*) An evolutionary family tree for the Late Cretaceous, ceratopsian (horned) dinosaurs contains two main branches, each with minor offshoots. Horizontal lines are approximate time markers. With time, the ceratopsian species became larger and developed enormous bony "frills" at the base of the skull. (*b*) The protective shield function of the neck frill has long been recognized, but recent studies suggest that its chief use was to anchor the powerful jaw muscles. Arrows point out the attachment and suggest the relative strength of various jaw muscles. (*c*) Behind the horny beak lay rows of imposing teeth, shown here in an end view. The fit of the bite indicates that the teeth were useful *only* for shearing motion—not for grinding. The ceratopsian juggernauts probably fed on tough, fibrous material such as cycad leaves. [*After J. H. Ostrom, "Functional Morphology and Evolution of the Ceratopsian Dinosaurs," Evolution, vol. 20, no. 3, 1966.*]

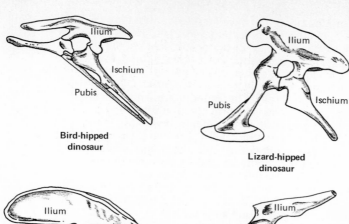

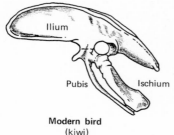

Bird-hipped dinosaur

Lizard-hipped dinosaur

Modern bird (kiwi)

Modern lizard (iguana)

FIGURE 9-30 Not only were the three pelvic bones oriented in different directions in the two dinosaur orders, but other features, such as the teeth and foot construction, were diagnostic as well. The resemblance between these bones in dinosaurs and in modern birds or modern lizards is unmistakable. [*After A. S. Romer, The Vertebrate Body, W. B. Saunders Company, Philadelphia, 1970.*]

with a lizard-type pelvis (Figure 9-30) included both herbivorous and carnivorous species. Representatives with a birdlike pelvis were exclusively herbivores. A number of species of both groups inherited a bipedal (two-legged) gait which conferred upon them a distinctive birdlike appearance. The front limbs of some bipedal dinosaurs degenerated to the point of becoming vestigial organs. *Tyrannosaurus* could not even reach his mouth with his hands (Figure 9-28).

Dimensions of adult dinosaurs ranged upward from rooster-sized to include the largest land animals ever to live. The nasty tendency for volume changes to outstrip increases in area (the square-cube law) is especially troublesome to any land-living monster. Its body weight is proportional to volume, but the sturdiness of the limbs is proportional to the area of their cross section. Giant dinosaurs sacrificed a little strength for a large saving in weight by developing bones with deeply hollowed areas.

The Great Dying The most dramatic event ever documented by fossils is the abrupt and total disappearance of a host of ruling reptiles at the close of the Mesozoic Era. What factors wiped out the dinosaurs? The vertebrate paleontologist G. L. Jepsen* summed up our ignorance of the causes of dinosaur extinction thus:

* G. L. Jepsen, "Terrible Lizards Revisited," *Princeton Alumni Weekly.* Vol. LXIV, No. 10 (Nov. 26, 1963), pp. 6–10, 17–19. © 1963. Princeton University Press.

Authors with varying competence have suggested that dinosaurs disappeared because the climate deteriorated (became suddenly or slowly too hot or cold or dry or wet), or that the diet did (with too much food or not enough of such substances as fern oil; from poisons in water or plants or ingested minerals; by bankruptcy of calcium or other necessary elements). Other writers have put the blame on disease, parasites, wars, anatomical or metabolic disorders (slipped vertebral discs, malfunction or imbalance of hormone and endocrine systems, dwindling brain and consequent stupidity, heat sterilization, effects of being warm-blooded in the Mesozoic world), racial old age, evolutionary drift into senescent overspecialization, changes in the pressure or composition of the atmosphere, poison gases, volcanic dust, excessive oxygen from plants, meteorites, comets, gene pool drainage by predators, fluctuation of gravitational constants, development of psychotic suicidal factors, entropy, cosmic radiation, shift of Earth's rotational poles, floods, continental drift, extraction of the moon from the Pacific Basin, drainage of swamp and lake environments, sunspots, God's will, mountain building, raids by little green hunters in flying saucers, lack of even standing room in Noah's Ark, and paleoweltschmerz.*

Clearly, the phenomenon must be explained by what happened not only to *dinosaurs*, but to *all the other* organisms. Why did the ammonites (marine cephalopods) go extinct at the same time? Why did the surviving reptiles and numerous mammals and birds refuse to participate in the Great Dying? We don't know! Perhaps as one author† sourly commented, "The Age of Reptiles ended because it had gone on long enough and it was all a mistake in the first place!" In a more serious vein, we could retort that the Age of Reptiles was a natural event, especially in the first place.

Invaders of sea and air   In Triassic times, the ray-finned fish, long the denizens of fresh water, began to populate marine waters too. Reptiles promptly followed the fish into the marine niches and habitats. The seagoing reptiles violently disrupted the marine food chain, forcing a reorganization just as their relatives had done on land The predators that monopolized the apex of the food pyramid were sharks and aquatic reptiles, the latter more recently replaced by whales. Their prey—cephalopods and primitive bony fish, later displaced by advanced fish—have occupied the next lower trophic level. Mesozoic and Cenozoic marine sediments record a steady adjustment in relative dominance amongst members within each of these two trophic levels. Today

* Literally, "ancient world agony."
† W. Cuppy, *How to Become Extinct*, Dover Publications, Inc., New York, 1944.

cephalopods are minor, and marine reptiles and primitive bony fish are insignificant, but in the Mesozoic, the balance of nature had strongly favored these groups.

Prominent among marine reptiles were the highly specialized, fish-shaped *ichthyosaurs* (see Figure 8-12). Their immediate ancestry cannot be traced, for when ichthyosaurs suddenly appeared in the fossil record they were already full-fledged. (The same is more or less true for the origin of seagoing reptiles and mammals of all types.) Ichthyosaurs splendidly illustrate a generalization known as Dollo's law, which asserts that *evolution is irreversible*. Everything that happens in history is a cumulative product of what happened before. Therefore, no event can repeat precisely. The shape of ichthyosaurs harked back to that of fishes in response to the requirements of an aquatic environment. But at the same time, ichthyosaurs kept the reptilian skeleton and the use of lungs, not gills. In view of these observations, we might broaden Dollo's law to state that organisms never wholly return to their ancestral condition, *nor* do they ever wholly depart from it.

In the Jurassic Period, some of the reptiles passed the *seventh milestone* of evolution by developing the power of flight (as opposed to soaring or gliding). The force of gravity has always confronted the energetic vertebrates with difficult problems. Amphibians, you recall, required special adaptations of the limbs and limb support before they were to be capable of walking on land. Conquest of the air was a more formidable accomplishment because, according to the square-cube law, the weight of the body is proportional to its volume, but the strength of wing muscles is proportional only to their cross-sectional area. To a degree, the flying reptiles offset the weight problem by possessing light, but fairly strong, hollow bones. This and other adaptations symbolized the advance of certain specialized reptiles into the grade of organization known as birds.

The wings of the carnivorous pterosaurs (flying reptiles) were membranes of skin stretched from the body across to an enormous extension of the fourth finger. The skeleton of *Rhamphorhynchus* is convincing testimony to the four-footed posture of its ancestors (Figure 9-31). Pterosaurs became extinct about the time that birds became abundant. Their extinction was outright, probably through losing too many niches to the superior birds. Pterosaurs were not the immediate ancestors of birds.

Birds

Those highly adept creatures, the birds, have been called "glorified reptiles," and so they are indeed. Bird fossils are notoriously poorly and infrequently preserved. By fortunate coincidence, a tropical sea occupied the region of Bavaria, Germany, during the Jurassic Period

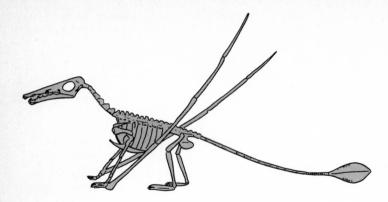

FIGURE 9-31 The claws of fish-eating *Rhamphorhynchus* were probably used as hooks for roosting in trees or on cliffs. The delicate skin membrane of the wing (whose former existence is amply proved by impressions preserved with the bones in the sediment) was extremely vulnerable to tearing. The unlucky victim, shorn of its power of flight, was probably condemned to die of starvation. [*After E. H. Colbert*, Evolution of the Vertebrates, *2d ed. Copyright © 1969. By permission of John Wiley & Sons, Inc., New York.*]

when birds were emerging from reptiles. There, birds were occasionally entombed in the carbonate mud (now a fine-grained limestone). Quarrying operations at Sölnhofen have uncovered four fossils of the earliest known bird, *Archaeopteryx* (Figure 9-32*a*). This animal is a near-perfect transitional form between reptiles and birds (Figure 9-32*b*). Without question, *Archaeopteryx* would have been identified as a relative of bipedal dinosaurs except that along with its bones are preserved the imprint of feathers! Useful in the beginning as heat insulation, feathers were preadapted to become structures necessary for sustained flight. *Archaeopteryx* could run and flap about like a chicken, but its body was probably too heavy and its breast muscles were probably too weak to enable it to fly.

Birds became completely modernized by the beginning of the Cenozoic Era. One consequence of adaptation for flight was a reduction of the original reptilian "rudder" tail to a mere nubbin in birds. Loss of a tail brought about superior maneuverability and lessening of body weight, but at the same time, birds were compelled to make unceasing corrective motions in order to keep their balance. Quick reflexes were needed and accomplished by an enlargement of the midbrain, which became highly sophisticated in modern birds. Birdbrains definitely merit our admiration!

Mammals

What characteristics distinguish a mammal? It is true that most species possess hair, sweat glands, and external ears and sex organs (in males), and that they maintain constant body temperature and give direct birth to young for which they provide loving care. However, isolated exceptions to every one of these attributes can be found among modern representatives of the group. Mammary glands (source of the term *mammal*) are not fossilized at all, nor are any of the other features mentioned above. Even though the fossil record is a complete

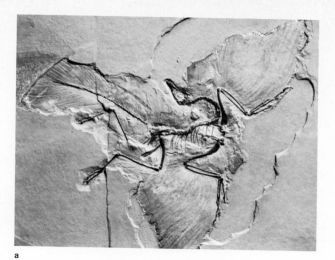

FIGURE 9-32 (a) Only two good skeletons and two incomplete skeletons of crow-size *Archaeopteryx* have been found after more than a century of quarrying at Sölnhofen, Germany. In view of their rarity, the immense amount of information that they contain, and their vital importance to the history of birds, these fossils are easily the most precious known to geology. This complete skeleton shows the bird sprawled out in its death throes. [*American Museum of Natural History photograph.*] (b) Corresponding structures in *Archaeopteryx* and a modern pigeon (not to the same scale) are emphasized in black. In every instance, the skeletal features of *Archaeopteryx* are more nearly reptilian. Many dinosaurs had hollow bones, as do all modern birds, but *Archaeopteryx* did not. [*After E. H. Colbert*, Evolution of the Vertebrates, *2d ed. Copyright © 1969. By permission of John Wiley & Sons, Inc., New York, 1969.*]

a

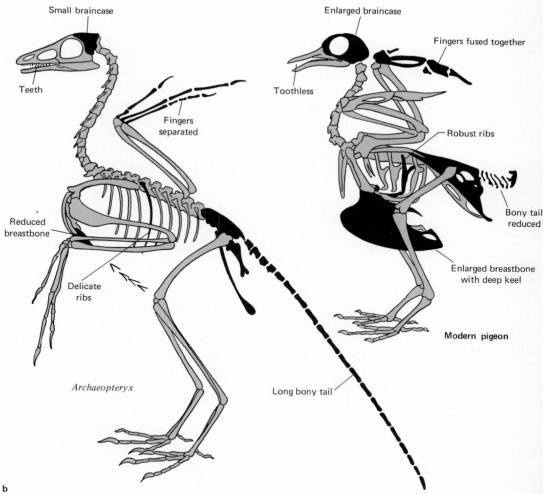

b

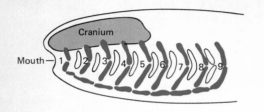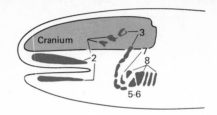

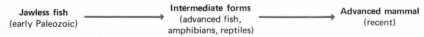

| Jawless fish | Intermediate forms | Advanced mammal |
| (early Paleozoic) | (advanced fish, amphibians, reptiles) | (recent) |

FIGURE 9-33 Nine primordial gill arches, present in ancient jawless fish, were preadapted to serve other functions in advanced mammals. In one current view, the arches became: (1) lost; (2) lower and upper jaw—the latter divided further to become incus and malleus (middle ear); (3) jaw prop—later to become stapes (middle ear); (4) lost; (5, 6) thyroid cartilage of larynx; (7) epiglottis (protecting larynx during swallowing); (8) cartilage rings supporting trachea; (9) lost. The third gill slit ultimately became the eustachian tube that connects the ears to the throat.

blank on these matters, paleontologists have had greater success reconstructing the history of mammals than of any other vertebrates.

One reason is that mammal-containing Cenozoic rocks are well preserved because of their comparatively recent deposition. Skeletal structures can richly inform us about the life style of mammals. The record shows that as they emerged from mammallike reptiles, the suspension of the *lower jaw* was modified by fine degrees into an arrangement that granted it a flexible stroking or sliding motion. This critical development enabled the infant mammals to nurse—a pleasure denied to reptiles. (Can you imagine a crocodile trying to suck?) As these changes occurred, the jaw was simplified to become a single bone. As bones were eliminated from the jaw, they migrated into the middle ear to become the delicate incus ("anvil") and malleus ("hammer") (Figure 9-33).

Mammal *teeth* were reduced to 44 or fewer (in contrast to the 2000 teeth that some vegetarian dinosaurs had). Not only did mammal teeth diversify into molars, incisors, and the like, but as species radiated into different niches, they evolved highly specialized whole sets of teeth. So well are teeth preserved, and so useful are the clues afforded by dentition, that many mammal paleontologists are primarily experts in "toothology." A hundred million years from now, a paleontologist could easily tell that shearing and stabbing teeth signified a carnivorous diet for cats, that the four giant molars of an elephant's dentition were handy for crushing twigs, and that humans were omnivores. In fact, if we understand "primitive" to mean the opposite of "specialized," then human teeth, devoted as they are to coping with a wide variety of food, are among the most primitive that are known.

However, the most telling quality of mammals, the one that sets them above all their predecessors, is their *restless, intelligent activity*. Compared to its reptilian forebears, even the most stupid mammal is an intellectual giant. Brain size is an imperfect, but nonetheless useful, index of intelligence. Paleontologists have noted the gradual evolution of the mammalian brain as it expanded relative to the size of the whole animal. In particular, the decision-making forebrain (cerebrum) of later mammals became spectacularly enlarged.

A thumbnail history The earliest descendants of mammallike reptiles known to us were tiny creatures, hardly the size of a rat. Likely they were timid, nocturnal animals that carefully avoided the terrifying ruling reptiles of that day (Figure 9-34). Dominant among Mesozoic mammals were insect eaters; thus in respect to diet as well, the mammals steered clear of direct competition with the ruling reptiles. The difficult era of reptile domination was not necessarily wasted time to these mammals, for many evolutionarily superior traits were probably perfected in them during the Mesozoic.

After giant reptiles became extinct at the close of the Cretaceous Period, mammals promptly filled the vacated niches. Cenozoic mammals developed in a succession of two evolutionary waves. The Paleocene Epoch ushered in a host of heavily built, rather clumsy, and (if brain size is indicative) dull-witted "experimental models." The next wave, which rolled in during the Eocene Epoch, brought a replacement of archaic animals by an explosive radiation of modern mammal types. The oldest known bats, whales, true primates, dogs and cats, horses, rhinos, pigs, and numerous others are represented by Eocene fossils. As time went on, these basic stocks specialized to become the highly adapted animals that we know today. The fossil record began to look more and more like the modern world with one important exception.

FIGURE 9-34 Size is not everything, of course, but it is easy to see why a typical Mesozoic mammal would have felt intimidated at the feet of a ruling reptile, *Tyrannosaurus*. [*After D. Cohen, The Age of Giant Mammals, Dodd, Mead & Company, Inc. New York, 1969.*]

Throughout the Cenozoic, world climate gradually became stratified into contrasting zones, finally culminating in the cruel advance of major ice sheets across much of the continental surface. Here was an environmental challenge to survival certain to have far-reaching consequences. The Pleistocene Epoch witnessed a muted version of the Great Dying reminiscent of the passing of the dinosaurs. This time, prominent victims of extinction were splendid large mammals, among them mammoths, saber-tooth cats, North American horses and camels, and the giant sloth. Had not Africa remained a natural wildlife refuge during this time, our acquaintance with nearly all the large mammals might have gone scarcely beyond a legacy of dry bones and ceremonial cave paintings. A question widely debated is the relative importance of human activity or climatic factors in the Pleistocene extinctions. Once of minor significance, man's influence over the delicate balance of mammal survival is expanding with alarming speed. We must look forward confidently (and with sadness) to the demise of yet more large mammals—whales, lions and tigers, elephants, and others. Is the Age of Mammals drawing to a swift close?

While the expansion and decline of mammal species were going on in the connected continents, South America and Australia were shunted aside into the isolated, half-forgotten backwaters of mammal evolution. The latter two continents became the habitat of the more primitive of the three groups of modern mammals. Most backward, and hardly known at all from fossils, are the egg-laying *monotremes*, represented today only by the platypus and spiny anteater. Somewhat more advanced are pouched mammals, the *marsupials*, typified by the kangaroo. The young of marsupials, born looking like half-matured fetuses, must experience a lengthy development in the mother's pouch, from which the infant slowly gains independence. By far the most successful mammals (95 percent of modern species) are the *placentals*. The active and rather able young of placental mammals are unmistakable miniature replicas of the parents at the time of birth.

Especially in Australia, marsupials were left unmolested to evolve along distinctive pathways. This land was isolated so perfectly that bats and rodents were the only placentals ever known there until man came, accompanied by his predacious dogs. The native Australian fauna provides the most celebrated case of convergent evolution known to paleontology. There are close marsupial equivalents to wolves, wolverines, groundhogs, moles, and others (Figure 9-35). The convergence strongly suggests that the final outcome of prolonged adaptive radiation is governed, not so much by the available type of animal (whether marsupials or placentals in this example), but rather by the *nature of the niches waiting to be filled*. Because the niches in Australia are similar to those in other continents, marsupials that adapted to them necessarily came to resemble a variety of familiar placental mammals.

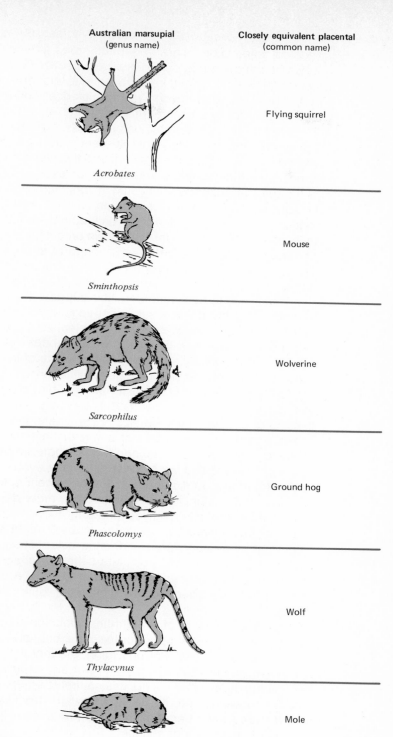

Australian marsupial
(genus name)

Closely equivalent placental
(common name)

Flying squirrel

Acrobates

Mouse

Sminthopsis

Wolverine

Sarcophilus

Ground hog

Phascolomys

Wolf

Thylacynus

Mole

Notoryctes

FIGURE 9-35

A case h(orse)tory   The story of horses is the most fully documented case history of any vertebrate group. When little "eohippus" (or dawn-horse) appeared during the great Eocene burst of mammal evolution, it bore only a faint resemblance to modern horses. Indeed, we might have failed to recognize the historical connection between eohippus and today's species if we did not also have a plentiful fossil record of transitional forms. Eohippus was a generalized, rather hunchbacked animal standing about 40 centimeters (16 inches) high at the shoulders. Its limbs and teeth indicate the niche it occupied, and also supply prophetic clues to what later horses were to become. Already, eohippus had departed to some extent from its more primitive ancestors, which possessed five digits on each foot. Its toes, reduced to four on the front feet and to three on the hind, ended in small hoofs, but most of the animal's weight bore down on a thick pad covering the sole of the foot. Its low-crowned teeth, though of an herbivorous type, were definitely not suited for chewing abrasive, silica-rich grass. Brains were not its strong point, either; eohippus would have been quite untrainable.

These and other characteristics suggest that the earliest horses were adapted to browsing succulent vegetation in a woodland environment. The protohorses were pursued objects of prey, just as zebras and wild horses are today. Eohippus was doubtless accomplished in the hazardous art of dodging, especially by virtue of those long toes which supplied traction for quick changes in direction.

The progression of the horse lineage, beginning in the Eocene, is often cited as a simple, ideal model of evolution (Figure 9-36). With the passing of geologic time the size of the animal increased. The side toe bones shrank, first to a state of being distinct but nonfunctional, and finally to mere splinters attached to the single large toe of the modern species. The feet and wrist bones gradually evolved from a flat-footed design to an elongate, spring-toed arrangement able to flex only in the fore-and-aft direction. Horse teeth became more deep-rooted, and they developed complexly wrinkled chewing surfaces capped by cement. These changes were so systematic that we might be tempted to conclude that an evolutionary trend, once established, must continue forward in a straight line (a postulate called *orthogenesis*). According to this interpretation, horses will proceed to grow ever larger until some aspect of the square-cube law limits their bulk and the animal goes extinct.

But is the concept of straight-line evolution actually justified by the data? There *is* good support for the idea, but several important qualifications are demanded. For one thing, the fact that today there is only one horse genus (*Equus*) consisting of six closely related species is likely to misguide our thinking. It is natural to emphasize only the evolutionary pathway aimed like an arrow toward *surviving* horses. The fossil record shows a number of extinct side branches of the horse phylogeny leading off in various directions.

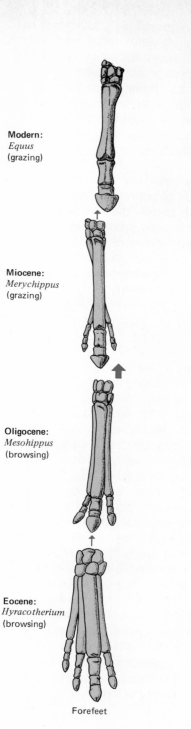

Modern:
Equus
(grazing)

Miocene:
Merychippus
(grazing)

Oligocene:
Mesohippus
(browsing)

Eocene:
Hyracotherium
(browsing)

Forefeet

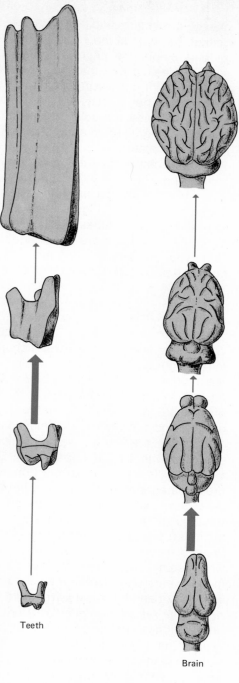

Teeth

Brain

FIGURE 9-36 Many aspects of horses evolved together, but not all at the same rate or at the same time. Thickened arrows indicate the times of most rapid change of a particular feature in the mosaic evolution (see text) of horses [*After J. M. Weller,* The Course of Evolution, *McGraw-Hill Book Company, New York, 1969.*]

In addition, horse evolution would have been far less noteworthy if changing climate had not exercised a powerful influence over its course. Throughout the Tertiary Period the woodland habitats gradually yielded before the encroachment of open grassy plains. The fossil record of teeth shows that the surviving horse lineage was the one able to switch from a browsing to a grazing diet. Very few vertebrates have successfully negotiated such a dramatic transformation in their food supply. Likewise, the development of long limbs having single spring-action toes signified an adaptation to distance running over firm ground in the open plains.

Not all aspects of the horse (or any other animal) evolved at the same pace. The most striking modification of brain size took place late in the Eocene, but for the teeth and limb bones, the crucial time was during the Miocene (Figure 9-36). Consequently, from today's vantage point the earlier species display strange mixtures of primitive characters residing in the same beast together with other characters that were already far advanced. The uneven timing of the changes (*mosaic evolution*) has been noted in many other fossil sequences.

A POSTSCRIPT

We owe the finis to these several chapters on the geologic history of life to a biologist of the previous century, T. H. Huxley. He likened the outreach of evolution to filling a barrel with apples. First, one heaps in the fruit. Still there is room among the apples for small pebbles to be placed, but the barrel is not yet full. Sand may be settled down into the depths, and finally the remaining pores may be saturated with water.

Just so, the first life was microscopic, each cell self-contained but endowed with limited capabilities. By the time the major phyla took form early in the Phanerozoic Eon, most of the marine "apples"—the broad categories of niches and habitats—had been filled in by living organisms, but the land surface and the air were yet barren. Slowly, pebbles and sand and water were added: organisms radiated until they occupied previously undreamed-of niches, and by their very existence created even more niches. Sometimes the filling process faltered, as when ichthyosaurs became extinct. Their niche lay unexploited for 20 million years until marine mammals recolonized it. But these setbacks were minor and temporary. We might say that *the universal law of evolution is the relentless tendency of life ever to expand and diversify* until (in the biblical metaphor) it is "pressed down, shaken together, running over."* That is the most sublime thought in paleontology, if not in all of science. Will the only species able to comprehend the richness of this heritage learn some day how to live in harmony with it?

* Luke 6:38.

READINGS

Burian, Zdeněk, 1972: *Life Before Man*, Thames and Hudson Ltd., London, 228 pp.

Colbert, E. H., 1951: *The Dinosaur Book*, McGraw-Hill Book Company, New York, 156 pp.

Colbert, E. H., 1969: *Evolution of the Vertebrates*, 2d ed., John Wiley & Sons, Inc., New York, 535 pp.

Jepsen, G. L., 1964: "Riddles of the Terrible Lizards," *American Scientist*, vol. 52, no. 2, pp. 227–246.

*McAlester, A. L., 1968: *The History of Life*, Prentice-Hall, Inc., Englewood Cliffs, N. J., 151 pp.

* Available in paperback.

chapter 10

Fields

Geophysical instruments (not to scale): (*a*) mass
spectrometer, (*b*) gravity meter, (*c*) seismometer,
(*d*) strainmeters, (*e*) computer. [*Judy Camps.*]

Our world includes a great variety of physical objects. Not so apparent are the several kinds of *force* that are also present in and about us. We can readily sense the force of gravity, but others (such as magnetic force) cannot be detected without the aid of instruments. The effect of these forces on the earth varies too. For example, the role of gravity in holding the earth together, in controlling sedimentary deposition, and in making rivers flow downhill, is well known. Magnetic force, on the other hand, has no sizable influence on geologic processes. Although magnetism does little to *shape* the earth, magnetic *information* has been of immense value to geologists, as we shall see.

A convenient way to describe force is through the notion of a *force field*—a region of space in which the force is present. From point to point within this volume, the intensity (or strength) of the force, and the direction in which it is exerted, may vary. In short, the field has a "shape." Earth's gravitational and magnetic fields penetrate clear to the center, and extend (at least in principle) to the infinite recesses of the universe. On the other hand, the earth's stress field (a description of forces that, among other things, cause earthquakes) is contained entirely within the solid earth. We can also broaden the idea of a field to describe the variation in space of just about any property of matter. For example, the earth's temperature field in part governs the manner in which heat can flow. Suppose that an object placed within a field is allowed to move in response to the force (for example, a compass needle within the magnetic field is caused to rotate). When the object moves, *energy* is released. We may express these relationships in algebraic form as

$$\text{Energy} = \text{force} \times \text{distance}$$

What can we learn about the earth from the interplay of mass, energy, and force fields? By considering these subjects, which make up the branch of geology called *geophysics*, we may look at the earth

273

from a very special viewpoint. In one sense, geophysics is a highly exact science based upon information gathered by sophisticated instruments. And yet, even after complex mathematical processing, the data may lead to uncertain conclusions. All we can know about the inaccessible depths of the earth comes from the force fields that penetrate it, or from internal energy that reaches the surface. Because the data are so indirect, the geophysicists (and most other geologists) must ever be designing mathematical models. Aspects of reality are always lost in a model. Worse yet, many geophysical models are ambiguous; some are "solved" by an infinite number of mathematically correct answers! Fortunately, a little geological common sense and additional independent data can often be used to narrow these possibilities down to a conveniently small number.

SHAPE OF THE EARTH

Let us begin our examination of geophysics by considering the shape of the earth. The curvature of the earth's shadow as it advances across the face of the moon during an eclipse, and the way that departing ships gradually disappear over the horizon (hull first, then deck, then mast-top), long ago convinced people that the earth is, broadly speaking, spherical. In 220 B.C., the Greek scholar Eratosthenes, Keeper of the Scrolls for Egypt, thought of an excellent way to determine the size of the earth. He assumed (correctly) that the sun is so distant that its rays approach the earth along essentially parallel lines (Figure 10-1). He also had observed that on the longest day of the year, the sun passes directly over Syene (near the present Aswan Dam in southern Egypt), so that sunlight reached the bottom of a deep well located there. He noted that on the longest day in Alexandria, a city to the north of Syene, the noon sun is not directly overhead; a sundial post casts a shadow whose angle with the post is about 7.2° (angle A). Finally, Eratosthenes arranged to measure a long trek between Syene and Alexandria to be made by a camel that had been carefully trained to plod along at a steady pace. The distance turned out to be 5000 stadia, or in modern units, about 926 kilometers.

With these data in hand, Eratosthenes was ready to solve the problem. Inspection of the figure shows that angle B equals angle A equals 7.2°. Therefore the distance between the two cities, compared with the earth's circumference, has the same ratio as angle B has when compared with a full circle. That is,

$$\frac{925 \text{ kilometers}}{\text{Earth's circumference}} = \frac{7.2°}{360°} \tag{10-1}$$

The answer he obtained missed the correct value of the polar circumference (39,940 kilometers) by only 12 percent. Considering the nu-

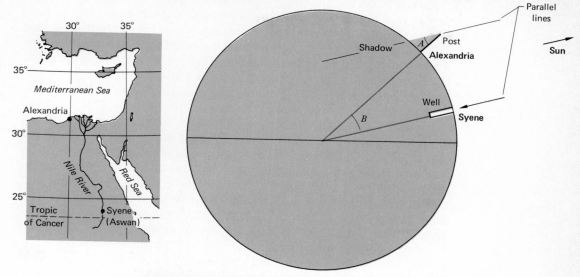

FIGURE 10-1 Eratosthenes's problem was simplified because the sun cast no shadow in Syene on the first day of summer. However, we could get an answer from measurements made at noon on any day at any two points separated by a large north-south distance. Angles and distances are exaggerated for greater clarity.

merous sources of possible error, especially in the cross-country walk, it appears that Eratosthenes must have had an astonishingly well-calibrated camel!

In modern times the shape of the earth has been measured again and again, ever more precisely. At first, the measurements were ground-based, but more recently, scientists have relied in addition upon radio tracking data from satellites that are orbiting in the earth's gravity field. Extremely accurate measurements are required because variations of the earth's shape and of its gravity field from place to place are very small. What do the findings of these years of painstaking effort inform us about the earth's interior?

More Models

Mathematical models can describe the shape of the earth to various degrees of exactness. At one extreme is a perfect sphere, a fair approximation that also corresponds to the simplest possible mathematical formula. At the opposite extreme is a formula so complicated that we can only state theoretically that it must exist. This expression would faithfully trace out a surface reproducing all the irregular ups and downs of the real earth. Which of these approximations is superior? Actually, neither of them is too useful because the object of model making is to go beyond mere description to arrive at a satisfying explanation. A successful model can *summarize* a great amount of data in compact form, and *predict* other consequences that were not obvious before the model was formulated. Such explanations are possible only if we assume that in spite of the irregularities, a few simple physical laws describe the natural phenomena that shape the earth.

Our planet would be a perfect sphere only if it were homogeneous, not rotating, and devoid of any source of internal energy. Obviously the spherical model fails to account for several factors of which we are already aware. The second model mentioned earlier fails for a different reason. It would describe the earth's surface very well indeed, but its mathematics would be so complicated that we could not understand it. We could not see the forest for the trees, so to speak. And so our first task is to "smooth" the data enough to allow the important details to show through, but not so much that the model becomes trivial. Even the most mathematical scientist must act like an artist!

Suppose that we consider the effect of the daily rotation on the shape of the earth. Both the solid earth and its gravity field become distorted. The equator is "slung" outward into a bulge, and since the volume of the earth remains constant, the two polar regions flatten as they are drawn closer to the center. A cross section through the poles, initially a circle, deforms into an ellipse in the rotating model (Figure 10-2).

The Spheroid

The shape just described is called the *spheroid*. It would result if a rotating earth, denser toward the center, were covered everywhere by

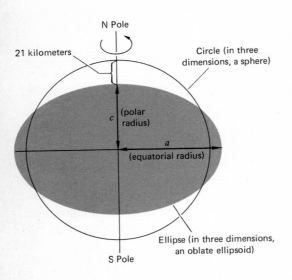

FIGURE 10-2 The eastward rotation of the earth (pictured here in polar cross section) deforms an otherwise spherical shape into an oblate ellipsoid. The polar flattening, enormously exaggerated in the drawing, is equal to $(a - c)/a = 1/298.25$. In terms of distance, the diameter is shortened about 42 kilometers. An astronaut viewing the earth from a great distance could not detect the polar flattening.

N Pole

21 kilometers

Circle (in three dimensions, a sphere)

c (polar radius)

a (equatorial radius)

Ellipse (in three dimensions, an oblate ellipsoid)

S Pole

an ocean of uniform depth. The spheroid has the great advantage of being simple mathematically, and its resemblance to the actual earth is so close that further refinements will turn out to be just tiny deviations from the spheroid. Indeed, the spheroid itself is almost a true sphere. The polar radius is a distance only 0.335 percent (or 1/298.25) shorter than the equatorial radius (Figure 10-2). Even so, this slight amount of polar flattening is highly significant. Calculations show that the observed polar flattening is a natural consequence if the earth behaves almost as a ball of rotating fluid—as though it had very little internal strength.

At first, this idea seems preposterous. Are not rocks strong enough to hold up towering vertical cliffs? They are, of course, but the question of *scale* becomes important in the behavior of solid material when it makes up a large object. Consider a small cube of rock at some depth within the earth. It will respond to *stress** applied on it by undergoing *deformation* or *strain*†: change of shape or volume or both. If stress exceeds a certain value, the strain becomes permanent as fractures develop, or as the rock deforms in a plastic (flowing) manner. Now, rocks placed under high confining pressure deep in the earth are stronger than similar rocks at the earth's surface. But at the same time, higher temperatures at depth tend to weaken the rock. As a result of all this, the rocks anywhere in the earth beneath an outer shell a few kilometers thick have no ability to hold a permanent shape. A cavity at great depth would promptly collapse through a ductile, creeping motion of its walls. Deeply buried rocks can be described as acting like putty. Left to itself, a blob of silicone putty flattens into a thin sheet as gravity pulls it closer to the center of the earth. In like manner the whole earth, "left to itself" in empty space, is drawn by gravity into a shape that occupies the smallest possible volume. Therefore the earth is essentially a sphere with a slight rotational bulge.

SEISMIC WAVES

Stress imposed upon rocks may be relieved by slow creep, but often the adjustment is sharp and sudden. To people in many areas, earthquakes are common events. Earthquakes play important and varied geological roles. Along the Pacific borderlands, and in a region extending from Java westward to Portugal, almost everyone is aware (often from firsthand experience) of the destructive power of a shud-

---

* Stress is expressed as force/area, as though the cube were under pressure exerted by the rock enclosing it on all sides.
† A person who says, "I am under a lot of strain," is incorrect. He should say, "I am under a lot of stress."

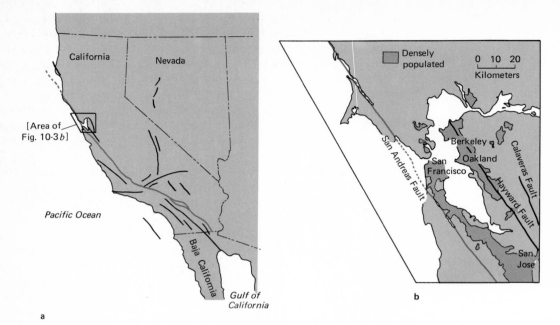

FIGURE 10-3 (*a*) The San Andreas Fault system (color line) intersects the continent at the head of the Gulf of California and continues most of the length of the state of California. It includes a number of subsidiary faults (black). (*b*) Small lakes and deeply indented bays mark the trace of the San Andreas Fault near San Francisco. Earthquake damage was greater in areas of unconsolidated fill, or "made land," which shook more violently than did tracts surfaced by hard bedrock. Since 1906 the Bay Area population has mushroomed, creating a land shortage. Developers have constructed housing subdivisions directly across the active faults.

dering earth. But earthquake waves also provide insight into processes going on at great depth, over very long time intervals. For example, they help to reveal the nature of the earth's hidden interior and the means by which mountain chains are formed.

Although earthquakes have been recorded for more than 3000 years, *seismology*, the science of earthquake phenomena, came into its own less than a century ago. An earthquake that destroyed much of San Francisco* in 1906 became the starting point for a series of careful observations that were to establish the origin of most seismic disturbances. San Francisco stands nearly astride the San Andreas Fault system, one of the largest and potentially most dangerous areas of active earth movement known (Figure 10-3). Circumstances of the 1906 disaster were particularly favorable for study, as the earthquake had just relieved the strain that had long accumulated in the rocks bor-

---

* About 80 percent of the destruction was caused by fires that could not be put out because the water mains were broken.

278 FIELDS

dering the San Andreas Fault. Triangulation networks on either side of the fault trace were surveyed and resurveyed before and after the quake, and the data were combined by the seismologist H. F. Reid into a model that he called the *elastic rebound* theory of earthquakes.

In Reid's formulation, heavy vertical lines represent the same fault as it appears in a sequence of moments *A* through *D* (Figure 10-4*a*). Let us suppose that a number of reference lines had been established across the fault—straight fence rows would serve our purpose well. Continued buildup of stress causes the rocks on either side to move slightly so that at time *B* the fences are strained into a mildly S-shaped distortion. For the time being, though, the fault remains locked together by friction or cementation. At instant *C*, stress has exceeded the restraints, and rupture begins to fling the rocks on either side violently in opposite directions. Where the reference lines are momentarily crowded together (gray areas), the earth experiences intense compression that propels the rupture still farther along the fault. A portion of the energy is released as *elastic*, or *seismic*, *waves* which travel great distances in all directions through the earth. After the quake (time *D*), the fault returns to an unstrained condition, but with the reference lines neatly offset.

A number of earthquake phenomena are explained by the elastic rebound model. Only in the San Andreas rift has the rate of strain accumulation been determined even approximately. There, a steady creep rate of around 5 meters per century (5 centimeters per year) has led to a present-day distortion pattern looking like *B* in Figure 10-4*a*. A single earthquake may create a displacement of as little as a fraction of 1 centimeter to as much as 20 meters. (In the 1906 San Francisco quake, the offsetting motion was about 5 meters in a horizontal direction. This fact, combined with the observed rate of creep, suggests that San Francisco should be "due" for another violent shock around the turn of the century. Actually, such earthquake predictions are of little value because other factors may dramatically hasten or delay the next inevitable seismic disturbance.)

The length of the break reopened by a quake can be anything from a few meters to an impressive 1000 kilometers. Once rupturing has begun (as in *C* of Figure 10-4*a*), it propagates along the break typically about 3 to 4 kilometers per second. Thus, fracturing that starts at one end of a long fault could take as much as 5 minutes to reach the other end, but an observer stationed, say, at the midway point would experience tremors lasting only a few seconds as the rupture hastened by. Clearly, a disturbance so prolonged and extensive will generate a rather complex series of seismic waves. (The simplest seismic wave signature comes from an instantaneous "point source" such as an underground explosion.) In spite of these complications, it is usually possible to locate a small volume of rock, called the earthquake *focus*,

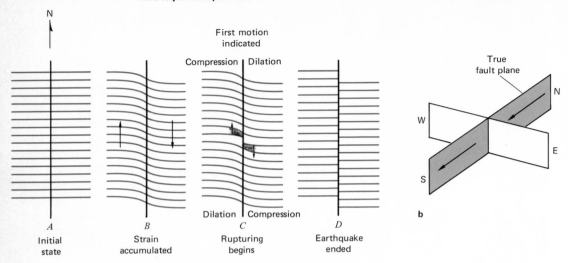

FIGURE 10-4 (*a*) For simplicity, the fault pictured in plan view is a north-trending vertical plane, but the elastic rebound theory holds equally well for slippage in any direction on a surface having any arbitrary orientation in space. Orientation of the fault surface can be inferred even at very remote distances. When rupture begins (stage *C*), the first motion of the ground is compressional, or *toward* observers located in the southeast and northwest quadrants. In the northeast and southwest quadrants the first motion is dilational, or *away* from the observer. From this information alone, we can determine that the fault is headed either north-south or east-west (but not in any other direction). (*b*) Of these two possible fault planes, one of them (not shaded) is eliminated by additional geologic data. [(*a*) *after H. Benioff, "Earthquake Source Mechanisms," Science, vol. 143, pp. 1399–1406, 1964. Copyright 1964 by the American Association for the Advancement of Science.*]

where movement began (Figure 10-5). If the focus lies at depth, we may further designate the earthquake *epicenter,* the point on the earth's surface directly above the focus.

Types of Seismic Wave

Seismologists can pinpoint the location of an earthquake focus within a few kilometers, even though the quake may have occurred beneath the ocean floor thousands of kilometers from the nearest observer. How can this remarkable feat be accomplished? To understand the procedure, we must inquire briefly into some features of seismic waves and the means of recording them.

We may begin by noting what happens when a pebble is dropped into a quiet pond. From the point of impact, ripples spread outward, wrinkling the water's surface into a series of ridges and troughs (Figure

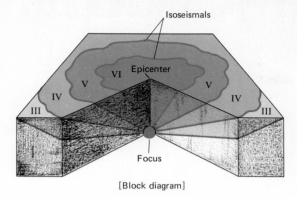

Isoseismals

VI Epicenter

V

IV   V

III   IV

III

Focus

[Block diagram]

FIGURE 10-5 Isoseismals, or lines of equal earthquake intensity, show how intensity steadily diminishes away from the epicenter. A scale (roman numerals) rates the intensity according to whether sleepers are awakened, chimneys collapse, railway lines are twisted, etc. Earthquake intensity scales, based as they are upon a blend of physical effects and the psychological reaction of people, are practical for reporting news but of limited value to a seismologist. Recently, seismologists have devised a quantitative magnitude scale based upon actual energy release. [*After Arthur Holmes*, Principles of Physical Geology, *2d ed.*, The Ronald Press Company, *New York, copyright* © *1965.*]

10-6). Water ripples, a type of surface wave, show quite literally the motion of *wavefronts*. Wavefronts connect all the parts of the disturbed medium that are momentarily in the same "phase"—for example, all the wave crests, or all the wave troughs.* Since the ripples travel at constant speed, the wavefronts are concentric, ever expanding circles. We could also draw any number of *ray paths* (in the pond example, radii of the circles) that intersect the wavefronts everywhere at right angles.

* Therefore, wavefronts are spaced one wavelength apart. See Figure 3-5 for a review of wave terminology.

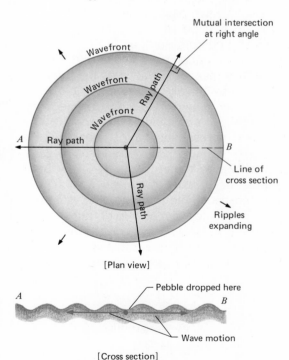

Mutual intersection at right angle

Wavefront

Wavefront

Wavefront

Ray path

A   Ray path

B

Line of cross section

Ray path

Ripples expanding

[Plan view]

A

Pebble dropped here

B

Wave motion

[Cross section]

FIGURE 10-6 Three representative ray paths are shown out of an infinite number of possibilities. A ray path is the line along which an infinitesimally small "packet" of wave energy travels away from the source of disturbance.

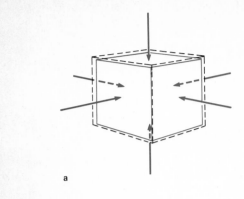

FIGURE 10-7 (*a*) If "pure compression" is acting on an object, its size is changed, but its shape is not. (*b*) As the card pack is sheared, each card (and hence, the entire pack) maintains constant volume. Shape is changed, but size is not.

Seismic waves too can be described by wavefronts and ray paths, even though the vibration is unlike the motion of water in a ripple. Just how is matter deformed as a seismic wave passes through it? First, we could think of force exerted upon a cube of matter from all sides, causing the cube to shrink (become compressed), as in Figure 10-7*a*. An original cube (broken lines) simply becomes a smaller cube (solid lines) after being compressed in this manner. Sometimes the pressure is called hydrostatic pressure because this type of compression would act upon a cube placed underwater. Some materials can be compressed more easily than others. A quantity called bulk modulus (*B*), which may be thought of as an incompressibility factor, describes this resistance.*

The other type of deformation is pure shear (Figure 10-7*b*). A pack of playing cards lying on the table nicely illustrates shear deformation. A push parallel to the top face can cause the cards to slide past one another, deforming the pack. The resistance of matter to deformation by shear stress is described by another quantity, *G*, called the modulus of rigidity. The larger the value of *G*, the more rigid a substance is.

Now, all types of seismic vibration are simply combinations of

---

* Bulk modulus *B* is given by

$$B = \frac{\text{change in stress}}{\text{change in strain}} = \frac{\text{increase in hydrostatic pressure}}{\text{reduction in volume per unit volume}}$$

The larger the value of *B*, the more incompressible the material is.

compressive and shearing motion. *Compressional* seismic waves are a series of condensations and rarefactions progressing through the earth (Figure 10-8a). These alternating zones mark the traveling wavefronts; they are analogous to the crests and troughs of water ripples. Individual particles vibrate in the direction of wave propagation* (that is, *parallel to* a ray path). Solids, liquids, and gases can transmit compressional waves.

As *shear waves* penetrate the earth, individual atoms vibrate in a plane *perpendicular to* the ray path (Figure 10-8b). The atoms interact by way of a sideways motion; they do not "bump into" one another in a straightahead direction, as in a compressional wave. We would therefore predict that shear waves are sustained only if the atoms in a material are securely coupled together by stout chemical bonds. Experiments confirm that shear waves are conducted through *solids only*.

Only compressional and shear waves can be generated in a homogeneous substance having no boundary surface. These are sometimes called *body waves* because they can penetrate into the depths of the earth. However, we noted in Chapter 2 that the earth is *not* homogeneous. Indeed, we are about to consider the seismic evidence for its distinctly layered structure. This layering permits body waves to pass through, but it also gives rise to *surface waves*, which are represented by an assortment of types having various complicated motions. As body waves expand downward and outward from the earthquake focus, their intensity decreases as the wave energy becomes distributed throughout a larger and larger volume of rock. Because surface waves are "channeled" into the crust and uppermost mantle, they retain a much greater intensity (and destructive capacity). Surface waves may circle the earth several times before friction finally damps them out.

Seismic wave *velocities* supply another vital line of evidence to our sounding of the depths of the planet. Happily, the velocity is governed by only three variables, the elastic properties (summarized by the terms B and G), and density [designated by the Greek *rho* (ρ)]. Compression is not a factor in shear waves, whose velocity (denoted as v) is simply

$$v_{\text{shear}} = \sqrt{\frac{G}{\rho}} \tag{10-2}$$

When shear waves meet a liquid or a gas, what happens to their velocity? Therefore, what is the value of the modulus of rigidity (G) for a liquid or gas?

---

* As the wave advances, individual particles vibrate briefly, then settle to rest in their original positions. That is, the *form* travels but the *substance* remains, on the average, stationary. In a constant current, say, of water, the substance flows but the form persists unchanged.

283   SEISMIC WAVES

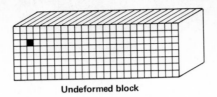

Undeformed block

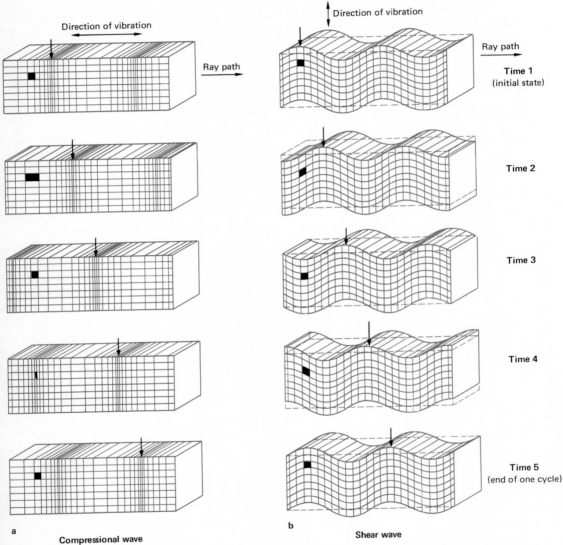

Direction of vibration

Ray path

Direction of vibration

Ray path

Time 1
(initial state)

Time 2

Time 3

Time 4

Time 5
(end of one cycle)

a

Compressional wave

b

Shear wave

FIGURE 10-8 (*a*) As a compressional wavefront (vertical arrow) moves to the right, a small reference cube (black) undergoes alternate expansions and contractions. Both its volume and shape vary as the wave form sweeps by. (*b*) The shape, but not the volume, of a small reference cube changes when it is deformed by a shear wave. [*After O. M. Phillips*, The Heart of the Earth, *Freeman, Cooper, & Co., San Francisco, 1968.*]

The velocity expression for compressional waves is only slightly more awkward:

$$v_{\text{compressional}} = \sqrt{\frac{B + \frac{4}{3}G}{\rho}} \qquad (10\text{-}3)$$

All the quantities in these formulas have algebraically positive values. Therefore, which travels faster through a given solid—shear waves or compressional waves?

Seismologists have called compressional waves *P waves* because they arrive first at a distant point and are hence "primary." Shear waves are "secondary" or *S waves* because they arrive somewhat later. Lastly, the various types of surface wave arrive in a confused jumble of overlapping signals. Table 10-1 organizes the names applied to seismic waves.

TABLE 10-1
Seismic Waves

Body waves
   P (primary, also compressional, push-pull)
   S (secondary, also shear, shake)

Surface waves
   L (for *l*ong wavelengths; includes a number of wave types)

Seismographs and Seismograms

To our knowledge, the first earthquake wave detector was invented about A.D. 132 by a Chinese man of genius, Chang Heng (Figure 10-9). Although it was a crude affair that responded only to violent motions, at least it could indicate an approximate direction toward the epicenter. Only much later did *seismographs*—devices that write a permanent, continuous record of earth motion—begin to be perfected. However complicated the modern instrument may be, its principle of operation is extremely simple. Central to every seismograph is a mass that is suspended in space but free to move relative to the ground (Figure 10-10). Its supporting framework is bolted to a heavy concrete slab or directly into bedrock. When seismic waves arrive, everything shakes except the isolated mass which hangs steady by virtue of its mechanical inertia. Thus a relative motion is set up between the steady mass and everything else. A drum wrapped in a paper sleeve can serve as a recorder of this motion. A pen scribes a visual record (*seismogram*) in the form of a helix round and round the drum. More sophisticated instruments record the data as a series of digits on magnetic tape for direct computer processing. Still another type of instrument is shown in Figure 10-11.

Since seismic waves may arrive from many angles, the details of the tremors can be worked out only by comparing the records from

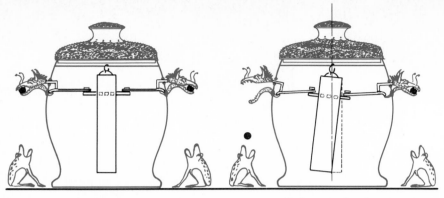

At rest   During an earthquake

FIGURE 10-9 Inside a bronze jar was suspended a pendulum attached at right angles to an eight-spoked wheel. When an earthquake shook the jar, the pendulum moved the spoke levers, causing a ball to drop out of the mouth of the dragon into the eager waiting mouth of a frog below. This device indicated the first motion of the ground when shaken by an earthquake (see Figure 10-4a). [*After J. T. Wilson, "Mao's Almanac,"* Saturday Review, *Feb. 19, 1972.*]

FIGURE 10-10 Mechanical isolation of the steady mass in a seismograph is carefully engineered. A mechanical or electronic linkage (not shown) greatly amplifies the image of the displacement of the heavy mass. A restraint is actually required to damp the pendulum, which otherwise would continue to swing almost indefinitely at its natural frequency. Specialized long- and short-term instruments respond to different seismic wave frequencies. These instruments are suited to frequencies of a few seconds to a fraction of a second. [*After A. N. Stahler,* The Earth Sciences, *2d ed., Harper & Row, Publishers, Incorporated, New York, 1971.*]

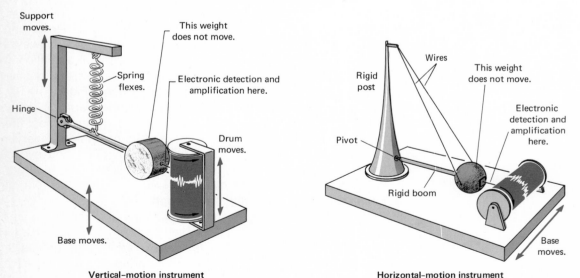

Vertical-motion instrument   Horizontal-motion instrument

FIGURE 10-11 Another instrument, the strainmeter, is best suited to measure very long period waves that require, say, 300 to 3000 seconds for a single vibration. These vibrations, known as *free oscillations*, cause the entire earth to "ring like a bell," often for days following a major earthquake. A detector records changes of as little as 1 part in 1 billion in the distance between two piers. A long quartz rod is attached to one of the piers, and the motion is measured between the other end of the rod and a reference point on the second pier. [*Teledyne Geotech photograph.*]

three detectors: one that responds to horizontal vibration in an east-west direction, another for the north-south direction, and a third that senses vertical motion.

Suppose that an earthquake happens somewhere within a distance of 1100 kilometers from a seismograph station. P, S, and L waves start out together from the focus, but the slower S waves and even slower L waves promptly begin to lag behind. Their imprint on a seismogram appears as a series of jogs.* At greater distances from the focus, the seismic signature becomes more complex. Interfaces between deep layers partially reflect P and S waves up toward the surface, and a wave reaching the surface may actually be reflected back down again. Identifying each of numerous wave arrivals becomes a difficult challenge to the seismologist. Many seismograms display arrivals whose origin has never been positively identified.

Locating Distant Earthquake Epicenters

The preceding information prepares us to understand a clever scheme for locating earthquake epicenters. Over the years, seismologists have used the occasional earthquakes whose locations and times of occurrence are well known as a means to determine how rapidly the seismic waves travel. They have constructed a *time-distance graph*, which in turn can be used to locate an earthquake whose

---

* P and L wave signals are generally quite distinct, as the seismograph is quietly resting before triggered by the P wave, and L waves are relatively intense. Picking out the arrival of the S wave from a noisy record often requires the perceptive eye of a trained seismologist.

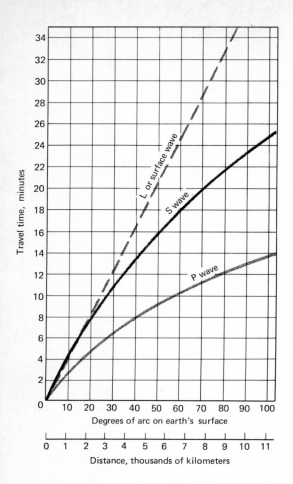

Travel time, minutes

Degrees of arc on earth's surface

L or surface wave

S wave

P wave

Distance, thousands of kilometers

FIGURE 10-12 In a time-distance graph, the origin (zero-zero, lower left-hand corner) refers to the time and location of the seismic disturbance.

epicenter and moment of occurrence are initially unknown (Figure 10-12). The curves show that as distance between the quake and a seismograph station increases, the P, S, and other waves pull farther apart, and so the time interval between arrivals also grows larger.

We begin with a seismogram, for example the one pictured in Figure 10-13. This seismogram from a long-period instrument in Alert, northern Canada, on May 18, 1962, indicates the "first arrival" of the P wave at 15:09:06 Greenwich Mean Time. (Significant wave motion continued to be recorded for more than 2 hours. Each succeeding wave shows a sharp first arrival, followed by an extended *wave train* because the rupturing took place over a prolonged time and distance, not just a brief instant at an infinitesimally small point.) The time intervals separating the arrivals of various waves are carefully measured and plotted on a movable scale. To find the station-to-quake distance, we can slide the time scale (while holding it vertical) out along the curves

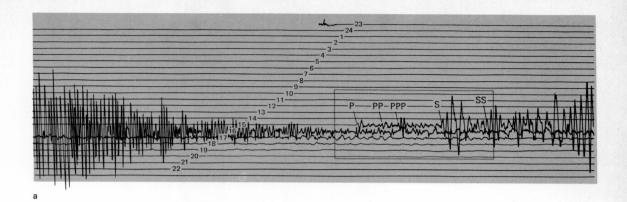

a

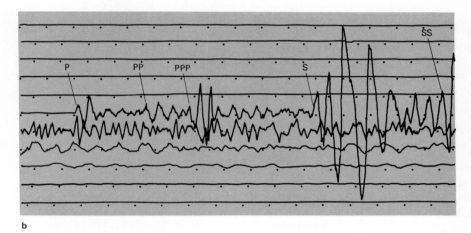

b

FIGURE 10-13 (*a*) Uncurled, the trace on the drum chart appears as a series of parallel lines, one for each hour of the day. The earthquake record is superimposed upon ever present restless background movements called *microseisms*. These result from storms at sea, tides, temperature changes, and other causes. (*b*) Wave arrivals are more distinctly seen in an enlarged part of the record (enclosed by the box above). Tick marks divide the hourly records into minutes. *PP*, *PPP*, and *SS* are waves that previously reached the earth's surface at a distant point, only to be reflected down and back up again one or two times. [*After J. H. Hodgson*, Earthquakes and Earth Structure, *Prentice-Hall, Inc., Englewood Cliffs, N.J., 1964.*]

of the time-distance graph (Figure 10-14). The plotted marks will fit the curves at only one position. It corresponds to the unique distance at which the delays of wave arrivals are the same as those seen in the seismogram (Figure 10-13). Having found this distance, how can you deduce the *time* of the earthquake?

So far, we have learned distance but not direction. Finding the geographic location of the epicenter requires data from three or more stations. At each station, a distance obtained in the manner just described

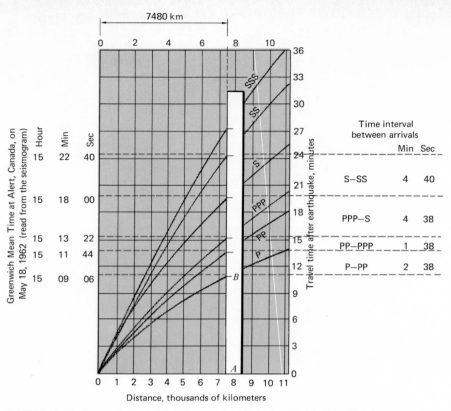

FIGURE 10-14 The seismogram of Figure 10-13 is interpreted in this slightly more elaborate time-distance graph. Intervals between wave arrivals (right-hand margin) were read from the seismogram and plotted on the vertical movable scale. Only at a distance of 7480 kilometers do these intervals match the time-distance curves. The origin of the graph is "floating" in the sense that the instant of an earthquake (time zero) may be any arbitrary time of the day. What is the significance of the time interval *A-B* (10 min, 53 sec)? At what time of day did this earthquake take place?

is scaled off as the radius of a circle drawn about the station (Figure 10-15). These circles intersect close to a single point—at least, they should if the records were interpreted correctly! Large earthquakes may be registered at 100 or more seismograph stations. The wealth of data from these big disturbances makes it possible, in effect, to locate the epicenter over and over, thereby providing a statistically accurate "fix" of its location.

Perhaps the most remarkable feature of the time-distance plot is that it is valid within narrow limits, no matter where the epicenter and seismographs may be located. We conclude that the earth is highly symmetrical in all directions about its center. Recently, large arrays of seismographs have been used to investigate small regional deviations

290 FIELDS

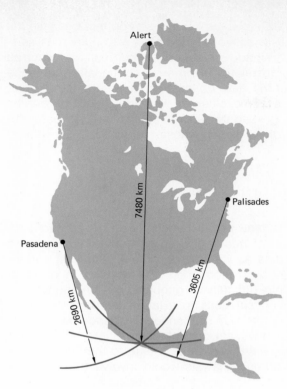

FIGURE 10-15 Stations in Alert, Canada, in Pasadena, California, and in Palisades, New York, recorded the May 18, 1962, shock. Circles, whose radii correspond to distances calculated in a similar manner for each station, intersect in west-central Mexico (a region of active faulting and volcanic activity). This unusually severe earthquake killed 3 people, injured 16 others, and caused extensive property damage. [*After J. H. Hodgson,* Earthquakes and Earth Structure, *Prentice-Hall, Inc., Englewood Cliffs, N.J., 1964.*]

from the standard curves. The largest such local array is a battery of 525 seismographs distributed over a 200-kilometer region of Montana. The worldwide network is a giant array of more than 1000 additional stations. Sophisticated observations reveal slightly different travel speeds through oceanic and continental paths, both in the crust and in the mantle beneath. Although these ultimate refinements in the art of seismology were developed as a means to pinpoint clandestine nuclear tests, our understanding of the earth's structure has also benefited from the findings.

Deep-Focus Earthquakes

In considering the spheroid, we noted that deeply buried rocks should yield gradually, as a plastic mass, rather than as a rigid object possessing strength. And since earthquakes result from *sudden* (elastic) release of accumulated strain, can we conclude that seismic disturbances are not possible at great depth? For years, geologists were convinced that they are not, but some peculiar seismic signatures required their attention. These records showed little or no surface waves, and distance calculations revealed that *no* station was very close to these distinctive earthquakes—surely an odd fact because the quakes kept taking place in active areas dotted with numerous seismographs. Damage to man-made structures, although slight, occurred over a wide area. Only by postulating *deep-focus* earthquakes could the geologists account for all the data collected. We have since learned that deep-focus quakes occur exclusively in narrow belts at depths down to, but not exceeding, roughly 700 kilometers. Deep-focus earthquakes are an important phenomenon whose significance is explored in following chapters.

PROBING THE DEEP INTERIOR

The earth's interior, so near and yet so remote, is an unapproachable frontier. In no other aspect of geology must we rely so heavily upon models fitted to indirect data as we must in probing this vast subterranean bulk. Different kinds of information are required in order to shrink the limits on that old difficulty of geophysics—ambiguous answers that are equally good solutions to a large number of models. Let us look first at scattered bits of miscellaneous evidence before concentrating on the more definitive, seismic information.

The earth's *moment of inertia* provides a clue to the internal distribution of its mass. A rigorous analysis would involve tricky mathematics which we shall sidestep by way of an umbrella analogy (Figure 10-16). Spinning the umbrella about its long axis is much easier when it is closed than when it is open, even though the mass of the umbrella stays the same. These two possibilities are described by the moment of inertia, a property that takes into account not only the mass of the umbrella but its distribution about the central axis.

And so it is with the whole earth. If it were a homogeneous sphere, its moment of inertia would be $0.4MR^2$, where M is mass and R is the radius. The moment is a sort of average result obtained by multiplying the mass of innumerable tiny "elements" of volume by the square of the distance, R, of each element from the center, and summing up. The moment of inertia of the real earth, calculated from certain characteristics of its spinning motion, is only $0.33MR^2$—a significantly smaller value. This situation can result only if the earth's mass is highly concentrated toward the center, as we might expect it to be.

Average *rigidity*, calculated from the manner in which the solid earth is deformed by tidal forces, turns out to be approximately that of steel. If the whole earth were composed of material similar to rock exposed at the surface, its average rigidity would be much greater. Therefore the large contribution by rigid outer rocks must be compensated by an interior region where rigidity is much less.

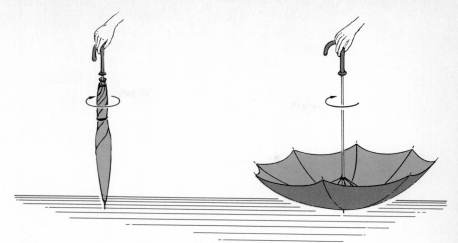

Umbrella analogy

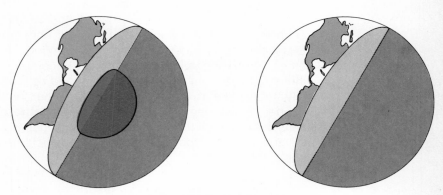

Corresponding density distribution in the earth

FIGURE 10-16 When an umbrella is opened, part of its mass is redistributed further from the axis of spin, and its moment of inertia increases. It requires a greater torque, or twist, to produce the same rate of spin in a given time. [*After O. M. Phillips*, The Heart of the Earth, *Freeman, Cooper, & Co., San Francisco, 1968.*]

Seismic Solutions

Meticulous probing of the earth's internal structure by seismic waves has proven to be a difficult technique, but rewarding in the detailed information it provides. Studies of the deep earth continue to be an important theme in modern seismology. Our best introduction to this topic is through the now familiar concept of ray paths and wavefronts (see Figure 10-6). Fronts of P and S waves, traveling into the body of the earth, become downward-expanding, three-dimensional "shells," in

293 PROBING THE DEEP INTERIOR

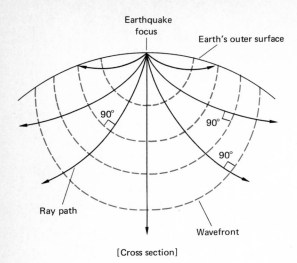

Earthquake focus

Earth's outer surface

90°

90°

90°

Ray path

Wavefront

[Cross section]

FIGURE 10-17 Wavefronts in the earth are elongated downward; ray paths are gentle curves.

contrast with the mere two-dimensionality of the pond wavefronts (Figure 10-17). As they pass downward, the seismic waves initially encounter rocks that are increasingly rigid, incompressible, and dense. Equations (10-2) and (10-3) show the competitive effect of these properties: The rising value of rigidity and incompressibility (numerator of the equations) causes the waves to speed up, and the increasing density (denominator) slows them down. The terms in the numerator happen to increase more rapidly, with the net result that the deep portions of the wavefronts travel fastest. Consequently, the fronts assume a distended egg-shape and the ray paths, which are perpendicular to the fronts, become gentle curved lines.

Cores, inner and outer In 1906, that momentous year of the San Francisco earthquake, the seismologist R. D. Oldham published the first hints that the earth has a core. However, it remained for Beno Gutenberg, another of the great men of seismology, to obtain the first accurate estimate of the core's radius. The core signifies its presence in an unmistakable manner. Body-wave arrivals are registered up to a distance of about 103° (or 11,500 kilometers along the surface) from an earthquake epicenter. But within a short distance beyond this limit, the waves fail to appear! Evidently the missing seismic waves have run into an obstruction (the core) that either blocks, or reflects, or refracts the ray paths, much as a lens alters the course of a ray of light.

What happens to these obstructed waves? Except for a tiny residue that "leaks" around the core perimeter, S waves are reflected or converted into P wave energy. Put another way, the S wave velocity falls abruptly to zero because the waves have encountered material of zero rigidity (G, Equation [10-2]). (The earth's rather low average rigidity had already hinted this possibility.) Therefore, the core must be liquid. At least it exhibits certain properties of commonplace liquids. Curiously, even under such enormous pressure,* the density† of the core is only in the range of 10 to 13 grams per cubic centimeter, which is less than the density of liquid mercury and other exotic heavy metals.

Considerable debate about the chemical composition of the core has revolved around the density question. As a first guess, the best candidate appears to be nickel-iron alloy. Meteorites comprised of this high-density substance are known to be abundant in the solar system, and presumably similar material was accreted into the earth. Suggestive data are provided by a series of shocking (!) experiments, in which an explosive charge creates million-atmosphere pressures for a thousandth of a second or so. Unfortunately, the measurements seem to disqualify the clear-cut choice of meteoritelike material. They show that an alloy consisting of iron (atomic number 26)‡ with a few percent nickel (atomic number 28) under these high pressures is more dense than the core is. Therefore, iron and nickel in the core must be alloyed with a lighter element such as silicon (atomic number 14) or even sulfur (atomic number 16). Here is more evidence for the uniqueness of the earth; its substance cannot be a simple accumulation of the common stony and nickel-iron meteorite types.

What happens to the P waves that are intercepted by the core? Compressional waves, you recall, do travel through liquids, but since the term G in Equation (10-3) is zero for liquids, the wave slows down as it enters the core. The ray path that just grazes the core is partly reflected and partly refracted sharply downward into the deep interior, finally to emerge in a region approximately opposite the epicenter (Figure 10-18a). Refracted rays, intersecting the core more nearly at right angles, follow other complicated paths that also eventually reach the surface again within a circular region between 142° and 180° from the epicenter. Between 103° and 142° is a ring-shaped *shadow zone* (Figure 10-18b) that would have received seismic energy had not the core "stolen" it. Like a lens, the core refracts the waves into a smaller area, causing them to arrive at the region opposite the epicenter with abnormally high energy.

Years after the core was discovered, improved seismographs demonstrated that seismic waves are not quite perfectly blocked from entry into the shadow zone. P waves, very weak and late in arrival, were getting through. Ray-path calculations point to the existence of an *inner core* that is solid, owing to the intense pressure in the ultimate depths of

---

* Pressure of 1.4 million atmospheres at the core boundary, increasing to 3.6 million atmospheres at the center.
† Distribution of density in the core is obtained from seismic data, including a study of free oscillations (see Figure 10-11).
‡ Atomic number is equal to the number of protons in a nucleus.

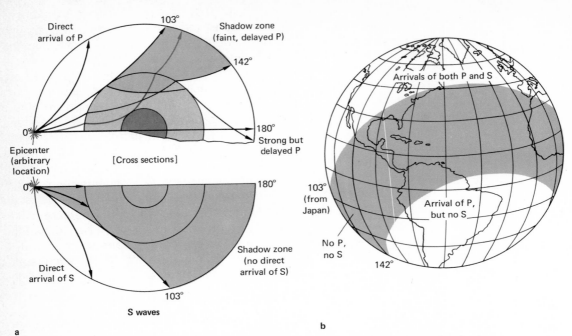

FIGURE 10-18 (*a*) Cross sections picture certain of the P and S seismic ray paths coming from a focus at the surface of the earth. (The half of the earth not shown in each cross section is understood to be symmetrical with the half that is shown.) Ray paths that intersect the core are either stopped (S waves), or diverted from the shadow zone except for very weak P waves (color) that are refracted by the inner core. (*b*) The earth's core cast this shadow zone in the path of seismic waves originating in Japan. [(*b*) *after Arthur Holmes,* Principles of Physical Geology, *2d ed., The Ronald Press Company, New York, Copyright © 1965.*]

the earth. Presumably the inner core is made of the same material as the outer core, but the outer-core liquid blanket so effectively shields the inner core that the seismic probe gives only vague information. Seismologists estimate that the volume of the inner core is only about 5 percent of the volume of the total core. The radius of the total core, in turn, is about 3475 kilometers, or 55 percent of the radius of the earth.

Mohorovičić discontinuity About the time that geophysicists first recognized the dramatic evidence for the core's existence, confirmation of another definite layer was published by the Yugoslav seismologist A. Mohorovičić (Mo-hoaro-VEE-chik). He noted that a shallow earthquake in northwestern Yugoslavia gave rise to peculiar "double arrivals" of P and S waves at some of the nearby stations. These waves, he reasoned, must have "echoed," or reflected off, an interface (internal surface)

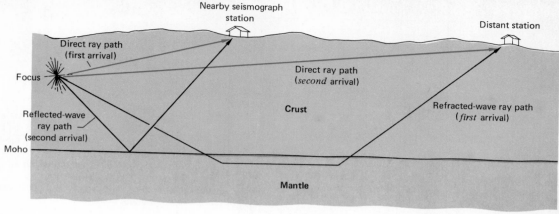

[Cross section]

FIGURE 10-19 A station near the earthquake first receives the waves that traveled by a direct path, then it picks up the reflected waves whose travel path is considerably longer. At greater distances, the seismic record is more complicated. Waves that travel through the crust, then through the upper mantle, and back up through the crust arrive before the waves whose ray path is direct through the crust.

lying at a depth of some 50 kilometers (Figure 10-19). By today, tens of thousands of seismograms, many of them energized by explosive charges, have contributed to further exploration of this feature. They show that the *Mohorovičic discontinuity* (the tongue-catching term for the interface, soon shortened into "Moho") is a warped, buried surface of nearly worldwide extent. As far as present knowledge goes, the discontinuity could be knife-sharp, but more likely it is a transitional zone perhaps as much as 0.5 kilometer thick.

By convention, the vast bulk of the earth between the Moho and the core is designated the mantle, and the outermost, relatively thin "rind" above the Moho is the crust. However, the Moho is only a seismological marker for the base of the crust.* Possibly the rock chemistry and mineralogy in lower crust and upper mantle are also different, but we have no direct information on this subject. The deepest borehole fails by a wide margin to pierce the mantle. A series of holes through the Moho† would shed light on some of the most outstanding problems in geology. In the 1960s the United States government funded a project to drill these superholes, then canceled it in an early stage. A more modest deep-ocean drilling program that fortunately took its place has given us the first detailed look at the sediments lying on the ocean floor.

---

* P-wave velocity jumps from about 7 kilometers per second in the lower crust to 8.2 kilometers per second in the uppermost mantle.
† They must be "Moholes," of course!

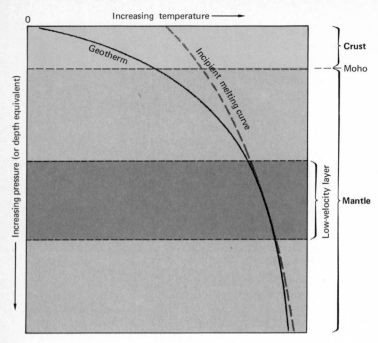

FIGURE 10-20 The geotherm is a curve that describes the increase in temperature with increasing depth in the earth. Near the surface, the temperature rises rapidly, but at greater depth the rate of increase becomes less. A broken curve shows that the beginning of melting of crust and mantle rocks occurs at higher temperature with increasing pressure. In the low-velocity zone, the two curves barely graze one another: a small fraction of the mantle actually melts. At still greater depths the pressure effect dominates, and the mantle is completely solid.

Mantle, upper and lower According to seismic evidence, the boundary between liquid outer core and solid lower mantle is clear-cut. At last, the mantle too is beginning to yield some of its internal secrets, but more grudgingly because here the contrast is only between different regions of solid material. Seismologists have concentrated on looking for abrupt changes in wave velocities and other subtle indications. They have discovered the mantle to be somewhat inhomogeneous horizontally, but even more noticeably varied with depth.

The uppermost mantle is rigid and relatively cool. Immediately below the uppermost zone, P and S wave velocities sharply decrease in a *low-velocity layer* lying at a depth between about 100 and 250 kilometers. Competing effects of temperature and pressure are believed to account for this layer (Figure 10-20). Increasing temperature, acting alone, would cause the mantle to melt, but the intensified pressure at depth encourages it to remain solid. In the low-velocity layer these factors hang in delicate balance; model calculations suggest that perhaps 1 percent of the rock may be a liquid film coating the grains. Even this trivial extent of partial melting is sufficient to reduce the amplitude of the S waves. The low-velocity zone appears to be a prominent source of basalt magma. It is also a lubricated "skid" that conceivably might permit the rigid crust and upper mantle to slide over it. And so, the low-velocity layer in some respects may be a boundary even more fundamental than the Moho.

298 FIELDS

Several minor discontinuities are found in a *transition region* between 400 and 800 kilometers deep. Most likely, they are due to high-pressure collapse of the crystal structures of silicate minerals. In ordinary silicates, you recall, four oxygen atoms tetrahedrally surround each silicon atom. Experiments suggest that under high pressure, the oxygen atoms rearrange into a sixfold coordination with silicon. A more tightly packed (more dense) silicate mineral is the result. The *lower mantle* (below 800 kilometers) probably contains these high-pressure forms of silicate and oxide minerals throughout. Its density varies smoothly in response to compaction by the weight of overlying rock.

What is the chemical composition of the mantle? Our best evidence is found in volcanic "bombs" that have been torn from the walls

FIGURE 10-21   Literally millions of mathematical models have been calculated (by computers) in an attempt to match the characteristics of the mantle and core with geophysical data. So-called Model 435002 has achieved fame as one of the most successful; it is consistent simultaneously with body-wave travel times, free oscillation and surface wave data, the earth's total mass, and moment of inertia, to within about 0.2 percent. The horizontal bar indicates the depth over which conditions in the mantle can be simulated by modern experimental equipment. [*After D. L. Anderson, C. Sammis, and T. Jordan, "Composition and Evolution of the Mantle and Core,"* Science, *vol. 171, pp. 1103–1112, 1971. Copyright 1971 by the American Association for the Advancement of Science.*]

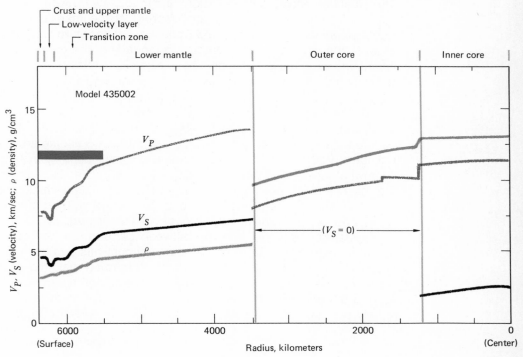

of a magma chamber and swept upward during eruption. All these volcanic inclusions consist of dark-colored, dense minerals. Olivine predominates, but pyroxene, garnet, and various oxide minerals may be present with the olivine. (Significantly, none of these minerals is hydrous.) Certain structures, called kimberlite pipes (after Kimberley, South Africa), appear to be prodigious vertical blowholes. They contain a jumble of these dark, dense rocks, plus rubble obviously torn loose from the crust, and occasionally, diamonds. High-pressure experiments prove that diamonds can form only at depths below 120 kilometers—that is, in the mantle.

Of course, some questions remain. Which part of the mantle do these wayward fragments represent? How were they changed while en route to the surface? Which of them are mantle-derived, and which came from the equally unknown lower crust? Already we know from experiments that the volcanic or kimberlitic inclusions do *not* become a basalt magma when melted. Hence a more primitive rock type, never yet seen, must lie at depth. It is this hypothetical material, when partially melted, that supplies the basalt magma so abundantly brought to the surface.

Crust, oceanic and continental From the geophysical viewpoint, the crust is hardly more than a thin, slaglike veneer overlying the mantle. The crust has accumulated throughout the ages by upward migration of mantle substances; it is truly an offspring of the mantle. What have geophysical techniques revealed about crustal structure? Superficially, at least, the most apparent contrast on the face of the earth is between *continents* and *oceans*.* Not only have seismic studies confirmed these differences; they have enhanced them by showing that, internally, the continents and oceans are highly distinct also.

Beneath the varied topography of the ocean floor lies a crust that is partitioned rather consistently into three layers (Figure 10-22). The composition of the deepest, or "oceanic," layer (about 5 kilometers thick) is unknown. Possibly it is mantle material that has been converted into somewhat less dense minerals by reacting with water. Next above lies the basement layer, uniformly about 1.5 kilometers thick. Dredge samples show that the basement is a glaze of basalt, erupted underwater and present universally throughout the oceans. A layer of unconsolidated sediment, typically between 0.2 and 2 kilometers thick, caps the sequence. Far from land, the sediment fills and overflows the irregular top surface of the basement, spreading out into vast, flat *abyssal plains*. (Nearer the continents, the sediment is much thicker. In rugged undersea areas distant from land, the sediment is absent or merely ponded in local depressions.) Finally, an average thickness of

* We commonly understand "ocean" to mean seawater. A geologist includes both the water and the oceanic crust in the term.

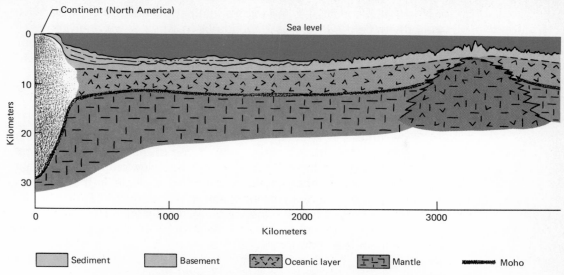

FIGURE 10-22 The western North Atlantic basin, seen in this cross section, typifies "normal" oceanic crust everywhere. Chapter 11 comments in more detail on the topographic features of the ocean floor. Vertical exaggeration: about 40 times. [*After J. Ewing, "Seismic Model of the Atlantic Ocean,"* The Earth's Crust and Upper Mantle, *Geophysical Monograph 13, p. 221, 1969. Copyright by American Geophysical Union, Washington, D.C.*]

about 5 kilometers of seawater fills the ocean basins. Thus the Moho lies around 12 kilometers below sea level, but the solid crust is only about 7 kilometers thick.

Continental crust stands in sharp contrast to the oceanic type in every important respect. Take thickness, for example. Depth to the Moho beneath the continents is about 35 kilometers on the average, though wide deviations from this figure are noted in many places. Except where sedimentary strata are at hand, seismic interfaces are elusive affairs, seemingly of local importance only. It is difficult to estimate a mean composition of the incredible diversity of continental rocks. In spite of our ignorance of many details, the result of such assays seems clear. Average continental rock, if melted, would congeal into something resembling granodiorite—an igneous rock less dense than, and chemically quite unlike, the basalt of the ocean basins (see Chapter 4).

MORE ABOUT GRAVITY

Having learned some basic ideas about the crust and the mantle, we may now return to the role of gravity. The force of gravity is a type of probe that is sensitive to the earth's internal distribution of mass. What the gravity field can tell us depends upon whether we analyze its effects on a global scale, or in a more restricted domain measuring tens to hundreds of kilometers across. Let us begin with a "broad brush" treatment: the whole earth.

Force of Gravity

How does the force of gravity vary from place to place on the earth? Part of the answer makes use of one of the simplest, yet most elegant, laws known to science. In 1686, Isaac Newton announced that "Every particle of matter attracts every other particle in the universe with a force that is directly proportional to the product of their masses, and inversely proportional to the square of the distance between them." In mathematical terms, if the masses are m_1 and m_2, the distance between them is r,* and force is F, then

$$F \text{ is proportional to } \frac{m_1 m_2}{r^2} \qquad (10\text{-}4)$$

Mass and weight should not be confused. Mass is a fundamental property of a given piece of matter regardless of where it is. Astronauts on the moon *weigh* less than they do on earth, but their *mass* is the same in both places. (Their "heft" is the same.) In ordinary earthbound situations, we think of weight as being the attractive force exerted by the earth on a mass located on its surface. Thus weight is denoted in the equation by F and mass is designated by m.

Our question becomes, "Does an object weigh more at the pole or at the equator?" Newton's formula predicts that since the polar radius is shorter than the equatorial radius, the force of gravity should be greater at the pole. In addition, the earth's rotation affects the weight of an object at the equator but not at the poles. Do the radius factor and the rotational factor work in the same direction, or do they tend to cancel one another? The sum of these effects is quite small by human standards; a man who weighs 100 kilograms would feel only 0.5 kilogram heavier as he traveled from the equator to the pole. He probably would not notice the difference, but a sensitive gravity meter operated in high latitudes would read so far off scale in the tropics as to be useless without a major adjustment (Figure 10-23).

---

* We may consider the gravitational force between one sphere (the earth) and another sphere to equal the force that would exist if the entire mass of each sphere were concentrated into a point at its center. The centers of the two spheres are thus equivalent to "particles" separated by a distance r. To a very close approximation, similar reasoning holds for a spherical earth and a small *non*spherical object located on its surface.

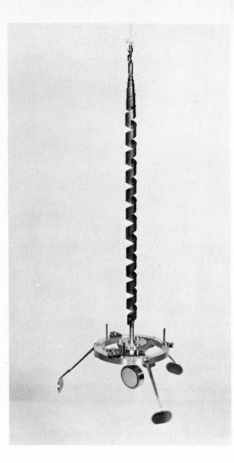

FIGURE 10-23 In this version of a gravity meter as in other geophysical instruments, springs play a major role because the distance they stretch is directly proportional to the force being measured. Here, the spring (shown isolated from the remainder of the apparatus) is a metal ribbon twisted into a helix. Variations in the force of gravity cause the spring to rotate as it stretches or contracts. An image of the rotation is greatly magnified by a light beam reflected from the mirror attached to the base of the spring. This instrument can measure differences in gravity of as little as 1 part in 10 million. [*Gulf Research & Development Company photograph.*]

The Geoid

Our approximations of earth shape have progressed as far as the spheroid. This figure corresponds to the position of sea level when polar flattening and the equatorial bulge are taken into account, but it disregards lateral variations such as continental and oceanic crust. An improved model, the *geoid*, brings us one step nearer reality. If the spheroid is an average, or *idealized*, sea level, the geoid describes the *actual* sea level. That is, it almost does. Out to sea, the geoid coincides with the water surface itself, but in the continents, the geoid at any spot is the elevation that water would seek in an imaginary slot canal connected to the sea. The spheroid is a regular figure with a simple mathematical formula. The geoid corresponds to an extremely complex formula; it rises above the spheroid in one place, and sinks beneath it in

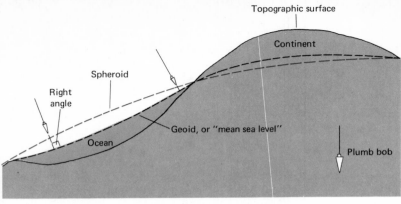

[Cross section]

FIGURE 10-24 The geoid responds to local variations in the gravity field. The spheroid is but an average surface about which the geoid deviates above and beneath.

another. And yet, the wrinkled geoid defines a horizontal surface because a plumb bob (a lead weight on a string) hangs perpendicular to the geoid at every point.

Logically, we would expect the geoid and spheroid to be related as in Figure 10-24. Continental rocks rising above sea level should exert gravitational attraction on the water in our hypothetical canal, drawing the geoid upward slightly. Conversely, the geoid should sag a bit in the ocean basins.

Artificial satellites have proven to be the ideal "gravity machine" for determining the shape of the geoid. We may think of a satellite as falling endlessly; it does not crash only because the curvature of the earth is downward, away from the path of its unrestricted fall. After many orbits, a well-placed satellite will have sampled most of the earth's gravity field. Another advantage is that a satellite orbit, being rather high in the sky, responds only to broad variations—it automatically smooths the gravity data.

Careful measurements of satellite orbits have been reduced (after *very* elaborate calculations) to a map of the geoid (Figure 10-25). As expected, the geoid turned out to be an undulatory surface, but the shape of that surface came as a considerable surprise. Its contours cut serenely across oceans and continents as though the boundaries were not there! What is the origin of the gigantic sag off the tip of India, or the large hump in the southwest Pacific? These and most of the other ups and downs correspond to no topographic feature on land or under the sea that we know of. Instead, the geoid must have a deeper source related to large-scale density variations in the mantle. Perhaps some day the geoidal shape will be explained, but for the present it remains a mystery.

TABLE 10-2
Models* of the Shape of the Earth

Model		Characteristics
Sphere		Homogeneous Not rotating
Spheroid		Internal layers becoming more dense at greater depth Rotating
Geoid		Internal layers becoming more dense at greater depth Layers inhomogeneous laterally Rotating

Increasing degree of simplification and "smoothing" of gravity data variations

Increasing complexity of mathematical formulas describing the shape of the earth

* All these models are abstractions; they assume an earth covered everywhere by sea-water (unaffected by waves or tides). Departures from a sphere shown in the bottom two figures are extremely exaggerated.

FIGURE 10-25 Contour lines (10-meter interval) depict a geoid fitted to a spheroid in which polar flattening is 1/298.3. Shaded areas are depressed below the reference spheroid; clear areas lie above it. Undulations in the geoid are exceedingly minute; they differ from the spheroid by only 1 part in 2 million. That is why some 50,000 observations of satellite orbits were required in order to calculate the geoid to within ±3 meters. Area in black represents the Himalaya Mountains and Tibetan Plateau. [*After E. M. Gaposchkin and K. Lambeck, "Earth's Gravity Field to the Sixteenth Degree and Station Coordinates from Satellite and Terrestrial Data," Journal of Geophysical Research, vol. 76, p. 4869, 1971. Copyright by American Geophysical Union, Washington, D.C.*]

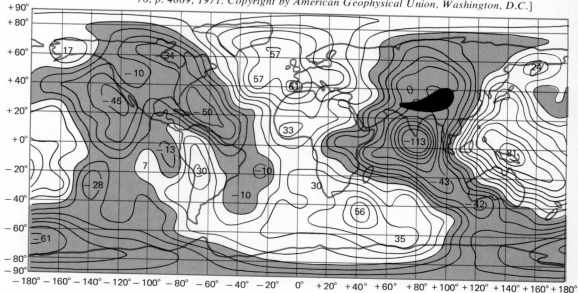

Isostasy

Still the nagging question persists: Why *doesn't* the geoid consistently rise up under continental areas, as predicted by Figure 10-24? It was a question (though posed in other terms) that first aroused notice during the Great Trigonometrical Survey of India, in the mid-1800s. British surveyors had used two methods to measure the distance between Kaliana (a village at the foot of the Himalaya Mountains) and Kalianpur, 600 kilometers to the south. A standard technique, by which a network of triangles is carefully laid out across the countryside, gave the distance correct within 0.2 meter—of that they were certain. Another procedure involved sightings in the two villages of the angle between a star and a vertical line. From the astronomical measurements, the arc of the earth's curvature (equivalent to difference in latitude between the villages) could be figured. The distance obtained astronomically was way off the mark by about 150 meters. When the Reverend J. H. Pratt, Archdeacon of Calcutta, heard of the discrepancy, he immediately spotted the difficulty: The plumb bob, used to establish a vertical line, was deflected ever so slightly northward toward the mass of the Himalayas. The effect should be strongest in Kaliana, near the mountains. [Since the plumb bob hangs perpendicular to the geoid, we could say that the geoid slopes upward under northern India (Figure 10-26)].

In a long paper filled with tedious calculations,* Pratt estimated the northward deflection according to a model that assumes the Himalayas to be perched as an "excess mass" atop the spheroid (Figure 10-26, dashed line). Much to his dismay, the computations showed that the

* Pratt made numerous references to his "computer." In his day, a computer was a man who figured with pencil and paper!

FIGURE 10-26 Diagonal ruling identifies the excess mass assumed in Pratt's first calculations. Note that the plumb bob hangs perpendicular to the geoid, not to the spheroid. Relationships of the spheroid, geoid, and surface of the earth are greatly exaggerated.

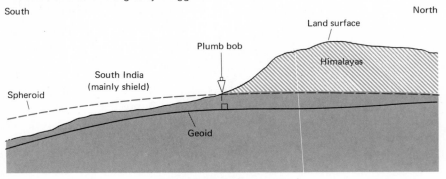

[Very large vertical exaggeration]

deflection "ought" to be 3 times greater than it actually is. Plainly, there had to be a *deficiency* of mass in the Himalayas compared to the mass he had assumed.

Pratt's report attracted the attention of the British astronomer G. B. Airy, who promptly offered an explanation. Suppose that the crust is severed by faults into large blocks that may adjust freely upward or downward. (Or, if the crust is not actually broken, suppose that it can flex like a rubber sheet.) Let us postulate, furthermore, that the crust is uniform throughout, but less dense than the substratum on which it rests. By this conception, high mountains or plateaus are supported from below by thick crust. Where there are lowlands or ocean basins, the crust is thinner (Figure 10-27*a*). In Airy's words:

It appears to me that the state of the earth's crust lying upon the lava* may be compared with perfect correctness to the state of a raft of timber floating upon the water; in which, if we remark one log whose upper surface floats much higher than the upper surfaces of the others we are certain that its lower surface lies deeper in the water than the lower surface of the others.

A thickened Himalayan "root" of low-density crust protruding down into the mantle accounts for the mass deficiency demanded by Pratt's calculation.

Pratt objected to Airy's solution, and in later papers he submitted an alternative theory. Some of the crustal blocks, said Pratt, stand high as mountains because they have "puffed up" like fermenting bread dough. Lowlands and ocean basins are underlain instead by compacted, high-density crust (Figure 10-27*b*). According to Pratt:

If these changes took place chiefly in a vertical direction, then at any epoch a vertical line drawn down to a sufficient depth from any place in the surface will pass through a mass of matter which has remained the same in amount all through the changes. By the process of expansion the mountains have been forced up. . . .

Which is correct: Pratt's variable-density crust or Airy's crust of uniform density but variable thickness? These gentlemen could not resolve the controversy, for the gravity data are equally well explained by both of them. On the broadest scale—continents versus ocean basins—we now realize that they *both* were right. High-standing continental crust is thicker than the depressed oceanic crust (Airy's hypothesis), but it is also true that granodioritic average continental crust is less dense than oceanic basalt (Pratt's theory). On a more restricted scale, as in the contrast between a shield and a mountain

* In today's terminology, the mantle.

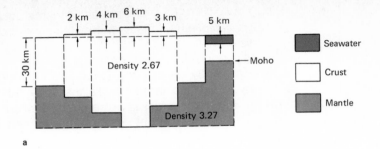

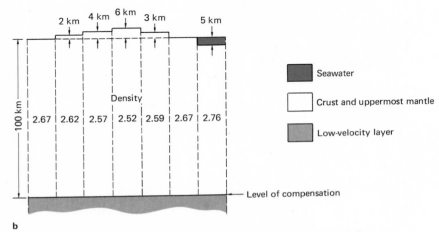

FIGURE 10-27 (*a*) According to Airy's model, shown here as an idealized cross section, crustal density is uniform but its thickness is variable. (Densities are in grams per cubic centimeter.) The mass of each crustal block varies in proportion to its thickness. (*b*) In Pratt's hypothesis, a plate of initially uniform density and thickness is permitted to expand upward to various degrees, becoming less dense in the process. The lowest-density block rises the highest, but each block retains the same mass. It is tempting to equate the rather deep "level of compensation" with the top of the mantle's low-velocity layer. [*After W. A. Heiskanen and F. A. Vening Meinesz*, The Earth and Its Gravity Field, *McGraw-Hill Book Company, New York, 1958.*]

range, the test of later findings definitely has favored Airy's model. We learned in Chapter 6 that most of the shield rock was once deeply buried within mountain systems. There is 'no reason why these rocks should have become *more* dense as erosion reduced the landscape to a low plain. If anything, the removal of overburden should have had the opposite effect.

According to Airy's scheme, the upper surface of the crust is a subdued mirror-reflection of the profile of the Moho (Figure 10-27*a*). One of the triumphs of modern seismology is the vindication of Airy's prediction. Beneath the high country of Chile, below the Himalaya Mountains and the Tibetan Plateau, and in the Pamir Range (U.S.S.R.), the depth to the Moho approaches double the normal continental value. But neither Pratt nor Airy had an inkling of why this should be so. Pratt's fermenting bread-dough analogy implies that mountain uplift is *dominantly vertical*. What could be more plausible? The true situation is now known to be far more incredible: Many of the great continental mountain chains are a contorted pile-up of rocks *shoved dominantly in a horizontal direction*. The crumpled and broken mass simply had nowhere to go except to thicken upward, and to become even thicker downward, as in Airy's formulation. (However, some mountains — the Rockies, for example — have no root zone. Density variations in the mantle help to compensate the "flotation" here. On a world average, Airy gets credit for about 65 percent, and Pratt for about 35 percent, of the total adjustment.)

Later, the term *isostasy** (eye-SAH-stasy) was coined to describe the condition of flotational balance established among large segments of the earth's crust. Isostatic approach to equilibrium is remarkable: about 91 percent complete in the continents, and 99 percent in the oceans. In hindsight, we may wonder if the train of events that led Pratt and Airy to propose the idea was more than lucky coincidence. If the Himalayas are poised in isostatic equilibrium, the plumb bob should not have been deflected at all (a fact clearly recognized by Airy). The deflection that led Pratt to an isostatic interpretation has nothing to do with isostasy! What he actually noted was the steep slope of the great Indian depression in the geoid, whose ups and downs are not of isostatic origin (see Figure 10-25).

So far, we have considered only a static situation. What happens when erosion begins to remove a mountain chain (Figure 10-28)? (Assume that the crust and mantle can adjust to the crustal "unloading" by slow creep, or flowage.) Where will the top and base of the crust be situated when the cycle of erosion and isostatic adjustment nears completion? The effect of these gradual adjustments is to prolong the lifetime of a mountain range by a factor of about 3 times.

EARTH MAGNETISM

"The earth speaks of its internal movements through the silent voice of the magnetic needle." These words, penned in 1918 by the mathematician C. Hansteen, sound strangely prophetic today. Not only does the magnetic field bear witness to present-day happenings in the earth's interior, but it is a geophysical recorder of events in earlier times. The

* From the Greek meaning "equal standing."

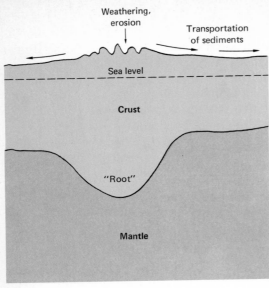

Weathering, erosion

Transportation of sediments

Sea level

Crust

"Root"

Mantle

[Cross section]

FIGURE 10-28 Airy's model is assumed in this diagram. As the land elevation decreases, so do the rate of erosion and the isostatic response to the local unroofing of the crust. The slackening of these processes can be described as a sort of "erosional half-life," analogous to the half-life of radioactive decay (see Chapter 6). Typically, about 40 million years are required to reduce 8,000-meter mountains to half that elevation, but an equal length of time is accordingly needed to reduce a shield elevation from 80 to 40 meters. Unless forces other than gravity intervene, a shield will last forever.

magnetic imprint contained in the rocks tells us a tale so astonishing that most geologists were long reluctant to accept its consequences. But that is another chapter.

Origin and Behavior of the Magnetic Field

For centuries men had pondered the mysterious force that unfailingly draws a suspended magnetic needle, regardless of how it is disturbed, back into a northerly bearing. Many thought that the North Star, "which does not move around the axis of the heavens as do the other stars," had somehow imparted "virtue" to the needle.* It remained for the British physician William Gilbert to knit together many strands of thought concerning magnetism. In his *De Magnete*, published in 1600, he asserted that the entire earth behaves as one great magnet. Since lodestone (a magnetized piece of magnetite) was the only known source of the magnetic force, Gilbert naturally assumed that permanent rock magnetism creates the force field of the earth. This seemed rea-

* Hear the song of the thirteenth century Italian poet Guido Guinicelli:
 In what strange regions 'neath the polar star
 May the great hills of massy lodestone rise,
 Virtue imparting to the ambient air
 To draw the stubborn iron; while afar
 From that same stone, the hidden virtue flies
 To turn the quivering needle to the Bear
 In splendor blazing in the northern skies.

sonable because he could demonstrate that the earth's external field resembles the field surrounding a uniformly magnetized sphere.

Gilbert's ideas soon got into trouble, however. Over the years, observations in London showed the compass direction to drift slowly but unmistakably westward. That would hardly do for a field originating in terra firma! Once the mathematics of potential theory had been invented, an analysis proved that more than 98 percent of the magnetic field issues from within the earth, especially from the region of the boundary between core and mantle. Yet the liquid core cannot be a permanent magnet despite its composition of iron. Permanent magnetism is a product of the ordered arrangement of electron "spin"—a possibility in solids only.

Indeed, permanent magnetization cannot even be achieved throughout most of the *solid* earth. Take the situation for the naturally occurring magnetic iron oxides, hematite (Fe_2O_3) and magnetite (Fe_3O_4). As temperature is raised, thermal agitation increasingly disrupts the systematic alignment of the spinning electrons. Finally, above the *Curie temperature*, permanent magnetism disappears altogether. The Curie points for magnetite (580°C) and for hematite (680°C) are far below the melting or decomposition temperatures of these minerals. (These two minerals account for most of rock magnetism, but a similar limitation applies to all other magnetic minerals.) Now consider a temperature rise of 25°C for every kilometer depth (Figure 10-20). At that rate of increase, the Curie temperature is exceeded in the vicinity of the base of continental crust. Nearly all of the mantle is too hot to retain permanent magnetism. Studies of the intensity of rock magnetism only reinforce our conclusion. Rocks lying above the "Curie depth" (that is, lying above the general region of the Moho) are far too weakly magnetized, by a factor of about 100, to account for the earth's total field strength.

Some complications How, then, does the geomagnetic field originate in the core? In order to understand the field behavior, we need first to devise a map based upon a convenient representation, the *magnetic lines of force*. These lines form a pattern that describes the *directionality* and the *intensity*, or strength, of the field. They can penetrate both the interior of the earth and the empty space surrounding it (Figure 10-29). A freely pivoting magnetic needle aligns itself parallel to the direction of the lines of force locally surrounding it. The density, or degree of crowding of the lines, symbolizes the strength of the field. On the surface, in the Northern and the Southern Hemispheres, are two magnetic *poles*. Here the magnetic lines of force tend to converge and they are directed straight downward into the earth. In contrast to gravity, which is everywhere an attractive force, the magnetic force attracts one pole, but it repels the other pole, of a magnetized object. Since magnetic poles are always paired, the overall shape of the earth's *dipolar field** is the simplest that is possible.

A closer look, though, uncovers some complications. One is the slow drift of the compass-heading already mentioned. Furthermore, the variation seen in London does not correspond exactly to the changes observed at some of the other stations. It appears that local, continually shifting magnetic fields are superimposed upon the earth's much larger

* *Di* means "two."

FIGURE 10-29   Imaginary magnetic needles trace out the contour of the lines of force. Crowding of the lines indicates that magnetic field strength at the pole is roughly twice the strength at the geomagnetic equator. In the earth's interior, the shape of the lines of force is probably exceedingly complex. According to modern theory, lines of force embedded in the core are dragged and stretched into nearly circular patterns by the liquid metal whose inner regions rotate more rapidly than the outer regions.

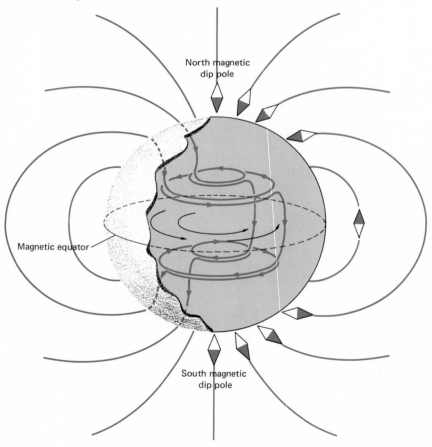

(and stronger) primary field. Suppose we subtract the symmetrical part from the total field to obtain a map of the residual component. Continent-sized *magnetic anomalies,** produced by the irregular, or nondipole, field, appear on this map. (They look similar to the drifting swirls on a weather map.) The strength of the nondipole field is significant—averaging about 20 percent of the total. And in keeping with the weather analogy, the intensities of both the dipole and nondipole fields continually change. The whole affair carries a distinct air of impermanence.

These irregularities make it necessary to use precise language in defining the magnetic pole. The location where a "dipping needle" points straight down (that is, where the lines of force are vertical) is the *dip pole*. Northern and southern dip poles are only 160° apart on the earth's surface, and *neither* of them coincides with the rotational pole, where the earth's axis of spin intersects the surface. (For example, the northern dip pole† is 17° (1900 kilometers) from the north rotational pole.) We may define another pole that best fits the symmetry of the lines of force on a worldwide average. It so happens that the "best-fit" poles lie near the dip poles, but the two kinds of magnetic pole do not exactly coincide. Like the inconstant anomaly pattern, the magnetic pole itself continues to wander about slowly. We would like to know whether the magnetic pole always stays close to the rotational pole. Their near proximity today suggests that the magnetic field may be aligned in large part by the earth's rotation.

The central dynamo Devising a theory to account for the complexities of the magnetic field has turned out to be impossibly difficult. Out of 10 or so hypotheses (some very ingenious) that have been advanced, the idea receiving most attention is based upon the *self-exciting dynamo*. Consider a metal disk fastened to a central axis and connected to a lower point on the same axle via a circular metal ring (Figure 10-30). Suppose, too, that a few stray magnetic lines of force penetrate the disk. As the disk rotates, its substance "cuts" the lines of force, causing an electric current to flow through the devious path from the disk back to the axle. Current flowing through the ring induces a magnetic field, which adds to the initial field, causing a stronger electric current to flow, which intensifies the magnetic field yet more. The device looks suspiciously like a perpetual motion machine. Actually it is not, because energy from an outside source is required to spin the disk.

So far, the calculations have proven only that a self-exciting dynamo action in the liquid core is possible, but its operation is but dimly understood. The earth dynamo must be immensely more compli-

---

* *Anomaly:* something that deviates from the common rule.
† On Prince of Wales Island, Canadian Archipelago.

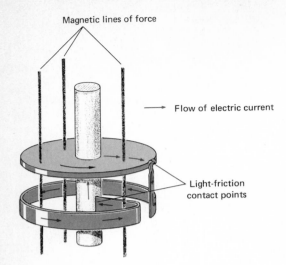

Magnetic lines of force

Flow of electric current

Light-friction
contact points

FIGURE 10-30 A laboratory self-exciting dynamo, rather like the one pictured here, was successfully operated. Because of the small size of the model, it had to be spun 2 *billion* times faster than the convective turnover time of fluid in the core. [*After T. Rikitake,* Electromagnetism and the Earth's Interior, *American Elsevier Publishing Company, New York, 1966.*]

cated than our laboratory model, for the core has no disk, and no axle or coil. It is merely a homogeneous electrical conductor. According to theory, an overturning (or *convective*) motion of the liquid core material as slow as 0.04 centimeter per second can self-perpetuate a magnetic field.

Perhaps heat from radioactive decay powers the convective overturn. Magnetic lines of force, transported along by the conducting fluid, become stretched and twisted as though they were elastic threads (see Figure 10-29). The dipole is believed to result from a large-scale motion symmetrical about the rotational axis. The nondipole field is created by minor turbulent eddies. These eddies are like whirlpools that form in a river, lazily drift downstream, and then disappear. Inside the earth, the magnetic field is probably very intense, especially within the core. Only a small fraction of it, perhaps 1 percent, "leaks" upward through the crust. Total field energy, some 10^{28} to 10^{29} ergs,* is approximately the same amount as the solar energy received by the earth in one day. We note, however, that the energy of the magnetic field is mostly *potential*. It could be transformed into other kinds of energy only if the field were to disappear. Only about 0.01 percent per year of that energy is actually dissipated.

Reading the Magnetic Record

Studies of geomagnetic field directions have been carried out for nearly 500 years, but accurate measurements of intensity began only around 1830. Human observations cover but a trifling part of the full his-

---

* A dead fly falling off the table will have acquired an energy of motion of several ergs by the time it hits the floor.

tory of the field. Fortunately, the *remanent magnetism* present in many rocks enables us to probe into past conditions of the geomagnetic field.

Remanent magnetism may be acquired in several different ways. One means is illustrated by a basalt flow. Liquid at the time of eruption, the basalt becomes a solid mass of interlocking grains (a few of them magnetite) by the time it has cooled to 900°C. As cooling continues, the temperature eventually drops below the Curie point for magnetite (580°C). Though the magnetite crystals are not free to rotate as a compass would, they do assume a magnetic directionality parallel to the local lines of force. Magnetism in this case is related to the *thermal* history of the rock (cooling below the Curie temperature).

Sediments also can be magnetized. Could sand grains perhaps rotate quite literally into alignment with the magnetic field as they settle to rest in quiet water? In most situations, this expectation is not fulfilled because a water current forceful enough to move sand is also very turbulent. Besides, each grain is subdivided into numerous tiny "domains" whose random magnetic directions tend to cancel one another. What could be used instead are ultramicroscopic grains, each barely large enough to contain one magnetic domain. In our sedimentary example, a *chemical* remanent magnetism can be set up when an iron oxide cement is introduced by percolating water. Individual grains of the cement may be as small as 0.01 microns (10^{-8} meter) across.

The strength of remanent magnetism in different rocks varies by a factor of at least 1 million. Some basalt specimens are able to deflect the needle of a hand compass held nearby, but in general, remanent magnetism can be detected only by sensitive instruments in the secluded environment of the laboratory. As a first step, the geophysicist places a fine scratch mark on the specimen while it is still attached to the outcrop. After measuring the orientation of the scratch, he breaks the piece loose. Then he determines the remanent field direction relative to the scratch, which in turn enables him to calculate that direction relative to the surface of the earth. Next, he deliberately attempts to destroy the magnetism by heating the specimen, or by "scrubbing" it electronically in a rapidly alternating field. Many rocks contain a weak magnetic component whose direction is easily changed. Only specimens with a stable, or "hard," component of magnetism are reliable indicators of the ancient field direction.

According to Figure 10-29, the magnetic lines of force may be parallel to the earth's surface, or they may meet it at any angle up to the vertical. In like fashion, magnetic directions imprinted in rocks will assume various angles depending upon where the specimen was collected. These angles are used to calculate where the magnetic pole must have been located in order to be consistent with the magnetic direction in the rock. Take a simple case: If the remanent direction is vertical, then the specimen was located at the position of the pole at the time it became magnetized.

315   EARTH MAGNETISM

Testing the dipolar model Remanent magnetism enables us to sample the ancient geomagnetic field, and thereby to test whether the great internal dynamo was subject to fluctuations over a time scale too long for direct observation. Suppose we plot the pole positions calculated for rocks that took form during the past few million years (Figure 10-31). The distribution of this large population "averages out" the fluctuations that occur over a short time scale of only thousands of years. Significantly, the swarm of data points forms a cluster around the rotational pole, *not* the present-day magnetic pole. This pattern supports the dynamo theory which predicts that over an interval of, say, 25,000 years, the magnetic and rotational poles should occupy the same *average* position. We may picture the magnetic pole as wandering aimlessly about the rotational pole, much like a nighttime insect flittering about a candle. Shifting patterns in the nondipole field account for the erratic motion of the pole; each point in Figure 10-31 represents only an instantaneous "snapshot" of the pole position. Having established the close relationship of the two kinds of pole, we may confidently use magnetic data to infer the position of the rotational pole in the much more distant past. (There is no guarantee that long ago the pole was in the same place it is today.) Remanent magnetism provides a powerful method to study ancient climates and other aspects of earth history.

Magnetic Reversals

In the early days of magnetic investigations, an astonishing discovery was made by the physicist B. Brunhes. He accidentally discovered (in central France) a lava flow that is commonplace in every respect ex-

[Polar projection]

FIGURE 10-31 Roughly two-thirds of the data points that correspond to recently formed rocks are located within 12° (1350 kilometers) of the rotational pole. Some of the scatter is due to experimental error, but most of it unquestionably is real. A colored dot signifies the modern position of the dip pole. The tight grouping indicates that the geomagnetic field had just two poles—not four or eight, which in principle might be possible. [*After R. Doell, "History of the Geomagnetic Field,"* Journal of Applied Physics, *vol. 40, 1969.*]

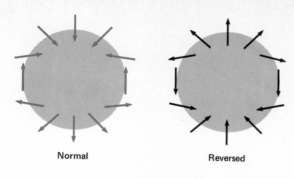

Normal   Reversed

FIGURE 10-32 Normal and reversed magnetic fields are identical in all respects except for direction.

cept that it is magnetized exactly in reverse (Figure 10-32). At first it was thought that somehow a mineral could assume a magnetic direction opposite to that of the encompassing field. Such "self-reversal" is now known to be a possibility, but evidence shows that an overwhelming number of reversed rocks actually formed in a reversed geomagnetic field. On a worldwide basis, reversely magnetized rocks are equally as common as the "normally"* magnetized sort.

The fact of magnetic reversals only places an additional burden of explanation on an already overtaxed dynamo theory. Aside from the mysteries of what might be going on in the core, what happens to the external field during a reversal? Fluctuating field *intensity* seems to

* Today's situation is normal because *we* are normal, of course! It is the other fellow (the other epoch of history, the other culture, etc.) that is "reversed."

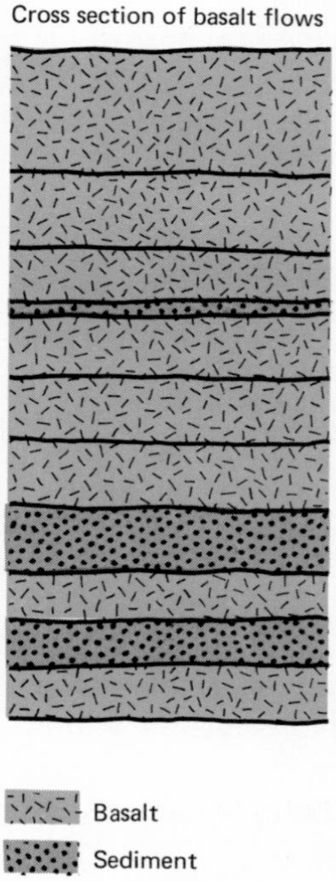

Cross section of basalt flows

N: normal R: reversed	K-Ar age, millions of years
R	0.74 ± 0.04
N	
N	0.89 ± 0.05
R	1.03 ± 0.06
R	1.03 ± 0.06
R	1.05 ± 0.07
N	1.64 ± 0.03
R	1.76 ± 0.07

FIGURE 10-33 Ages of a sequence of normally and reversely magnetized basalt flows from Madeira (a Portuguese island near the west coast of Africa) are determined by the K-Ar method. Once sufficient data are pooled, the times of magnetic reversal can be pinpointed quite accurately. Basalt is so strongly magnetized, and the remanent directions are so distinct, that simple portable apparatus is adequate to identify the polarity. [*After N. D. Watkins and A. Abdel-Monem, "Detection of the Gilsa Geomagnetic Polarity Event on the Island of Madeira,"* Geological Society of America Bulletin, *vol. 82, 1971.*]

Basalt

Sediment

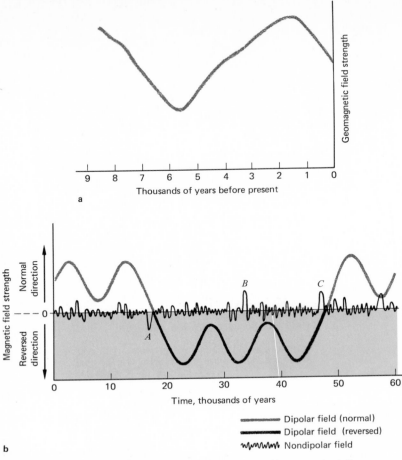

FIGURE 10-34 Measurements of remanent field intensities seem to define a smooth cycle of about 10,000 years' duration. If the present trend of decreasing intensity is not soon reversed, the geomagnetic field will go to zero in about 2000 years. (*b*) Regular 10,000-year cycles of the dipole field are superimposed upon ceaseless random noise of the nondipole field. At certain times, the nondipole field is in fact the stronger of the two. According to the model, fluctuations of the nondipole field at times *A* and *C* were able to swing the main field into an opposite direction. A large flicker at time *B*, when the dipole field was strong, was ineffectual. [*After A. Cox, "Geomagnetic Reversals,"* Science, *vol. 163, pp. 237–245, 1969. Copyright 1969 by the American Association for the Advancement of Science.*]

govern the reversal events (Figure 10-34*a*). Since 1835 the strength of the geomagnetic field has decreased by some 7 percent. Measurements of remanent magnetism in baked bricks and recent volcanic rocks of known age show that the earth's field intensity has varied smoothly down, then up, and down again during the past 9000 years. Several localities where sedimentary rocks were continuously depos-

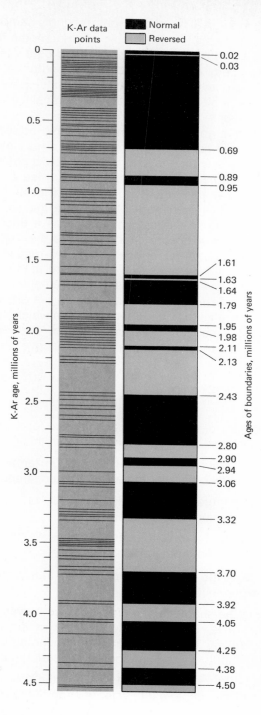

FIGURE 10-35 Approximately 150 potassium-argon ages were used to construct this time scale of geomagnetic polarity reversals. Times of magnetic reversal seem to be completely random. [*After A. Cox, "Geomagnetic Reversals," Science, vol. 163, pp. 237–245, 1969. Copyright 1969 by the American Association for the Advancement of Science.*]

ited during a reversal have been painstakingly sampled. The remanent magnetism suggests that the dipole and nondipole fields act independently. According to one model, when the dipole intensity occasionally falls close to zero, a minor jiggle in the nondipole "noise" can nudge the dipole field over into the opposite state. Reversals are sudden events on the geologic scale of time; the transition is completed in as little as 5000 years.

Potassium-argon isotopic analyses have added the crucial *time* dimension to the series of reversals. K-Ar ages determined from normally and reversely magnetized volcanic rocks around the world have established that field reversals took place at least 25 times within the past 4.5 million years alone (Figure 10-35). (Resolving the fine detail of earlier reversals is more difficult because the experimental error is greater.) We can use magnetic reversals as a tracer of earth history. You recall that evolutionary changes made the fossil record a useful index of time, but on a rather coarse scale. Organisms do not evolve or become extinct (or even inhabit) everywhere at the same time. In contrast, the field reversals fix a pattern of remanent magnetism whose boundaries are *instantaneous* and of *worldwide extent*. While rocks are forming, they are natural "tape recorders" in which the erratic temperament of the geomagnetic field is inscribed. That record has become a keynote in the most exciting geologic discoveries of modern times.

FURTHER QUESTIONS

1   Is there any other direct visual evidence for the spherical shape of the earth, besides the evidence cited in this chapter?

2   Solve Equation (10-1). What answer (in terms of kilometers) did Eratosthenes obtain for the earth's circumference?

3   Refer to Figure 10-2. What is the shape of a cross section of the ellipsoid through the equator?

4   Why does the earth's rotation affect the weight of an object at the equator, but not at the poles?

5   What is a type of compressional wave that is familiar to our everyday experience?

6   Refer to the shapes of the time-distance curves for P and S waves, Figure 10-12. Do these waves speed up or slow down as they travel farther? The straight line corresponding to surface waves indicates that they travel at constant speed. What is it?

7   Refer to Figure 10-18*a*. If the radius of the core is 55 percent of the radius of the whole earth, what fraction of the earth's *volume* does the core occupy?

8   Refer to Figure 10-18*b*. Where are L waves recorded?

9 P waves that travel through the core arrive at the surface on the far side of the earth "behind schedule" (see Figure 10-18a). What are two reasons for the delay of these waves?

10 Refer to Figure 10-19. Explain the anomalous behavior of the travel times of seismic waves refracted through the mantle.

11 At a certain place, the temperature at the surface of the earth is 5°C, and the rate of temperature increase is 25°C per kilometer depth. At what depth would the Curie temperature for magnetite be exceeded? On the continents, would this depth likely be found in continental crust or in mantle beneath? Would this depth below the ocean surface be found in oceanic crust or in mantle beneath?

12 The earth's magnetic field is a shield against high-energy particles (cosmic rays) that continually bombard the atmosphere from all directions. What might be a consequence to living organisms during a geomagnetic field polarity reversal?

13 Refer to Figure 10-36, showing a large rock lying on the bottom of a shallow bay. Its mass causes the shape of the geoid to be slightly warped directly overhead. Is the geoid warped upward (curve *A*) or downward (curve *B*)? (*Hint:* The geoid is perpendicular to the line defined by a suspended plumb bob. Would a plumb bob be deflected toward or away from the rock?)

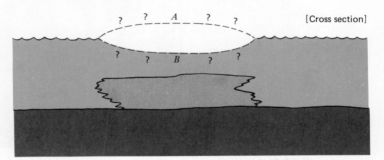

FIGURE 10-36

READINGS

* Clark, S. P., Jr., 1971: *Structure of the Earth*, Prentice-Hall, Inc., Englewood Cliffs, N.J., 131 pp.
* Hodgson, J. H., 1964: *Earthquakes and Earth Structure*, Prentice-Hall, Inc., Englewood Cliffs, N.J., 166 pp.
  Phillips, O. M., 1968: *The Heart of the Earth*, Freeman, Cooper, & Co., San Francisco, 236 pp.
  Takeuchi, H., S. Uyeda, and H. Kanamori, 1970: *Debate About the Earth*, Freeman, Cooper & Co., San Francisco, 281 pp.

* Available in paperback.

chapter 11
The Grand Synthesis: Introduction

A scene in the Alps. The Italy-Switzerland border lies on the skyline to the north. [*Courtesy Earl Verbeek.*]

A glance through old geology textbooks is a revealing experience. Facts by the thousands abound in their pages. Not surprisingly, the descriptions are mostly about the continents. In those days, little was known about the ocean basins, and even less about the earth's interior. But to anyone who was curious about the *reasons behind* the flood of details, an acquaintance with geology as it was then known was certain to be frustrating. Today, things are different — not that we have all the answers, of course. It seems that deeper explanations always lie just beneath the surface of explanations that are already understood. And yet, new understanding of the earth has led to a unifying hypothesis of such sweeping grandeur that it is able to incorporate, in one way or other, nearly every topic mentioned so far in this book.

We must acknowledge that no all-embracing theory springs up overnight. There were anticipations, early attempts, false starts, while all those necessary millions of facts continued to be gathered. So incredible did the scheme appear, when it finally unfolded, that at first there were many disbelievers. A few honest dissenters remain today. In much of future research, though, the aim will be to interpret and *appreciate* the power of the new hypothesis to explain the rightful place of each particular geologic fact in a global system.

THE WORLD OCEAN

What key discoveries brought about this dramatic new Grand Synthesis? Nearly two centuries of intensive scrutiny of the continents had failed to produce them. Instead, the decisive evidence has come from the 71 percent of the earth's surface that, until the 1950s, had remained an uncharted realm beneath the waves. Let us see what clues awaited discovery in this submarine world.

323

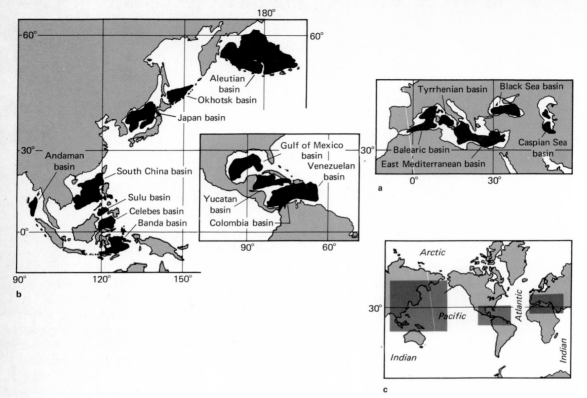

FIGURE 11-1 (*a*) In the Mediterranean region, small ocean basins (black) are virtually landlocked. (*b*) Along the western Atlantic and western Pacific they are confined between a continent and offshore island arcs. The exposed part of an island arc may be no more than a chain of volcanoes (such as the Aleutians), or it may consist of more complex, massive islands dominated by volcanic rocks (like Japan, Philippines). (*c*) Major ocean basins are the Pacific, Atlantic, Indian, and Arctic, in order of decreasing size. Shaded areas are shown enlarged in (*a*) and (*b*). [*After H. W. Menard, "Transitional Types of Crust under Small Ocean Basins," Journal of Geophysical Research, vol. 72, no. 12, p. 3062, 1967. Copyright by American Geophysical Union.*]

Some Generalizations

In reality, there is only one ocean,* whose parts connect freely. It is divided into four major bodies: the Atlantic, which is oriented dominantly north-south; the Indian, a predominantly east-west ocean; the Arctic, which caps the north polar region; and the Pacific, an immense, roughly circular ocean that alone occupies more than one-third of the earth's surface. In addition there are 15 or so *small ocean basins*, of which the Gulf of Mexico, the Aleutian basin, and the Sea of Japan are examples (Figure 11-1). These steep-sided, rather isolated depressions are nearly as deep as the "big four" basins, but they make up only

* As mentioned in the preceding chapter, an ocean must be floored by oceanic crust.

about one-fiftieth of the area of the total ocean. Because of their nearness to land, however, these small ocean basins receive one-sixth of all oceanic sediment. Also, the crust beneath them appears to be transitional between the continental and oceanic types. Are continental and oceanic crusts permanent, or is there a way to convert one into the other? This is as yet an unanswered question.

Distribution of land and sea forms a striking pattern. The Northern Hemisphere contains more than twice as much land as the Southern. However, by appropriate shifting of the poles and equator, we may define an even more distinct "water hemisphere" and "land hemisphere" (Figure 11-2). Fully 81 percent of the total land is concentrated into the land hemisphere; only 5 percent of the land is on the opposite side of the globe from other land (rather than water).* This peculiar huddling together of the continents makes us wonder if the distribution resulted purely by chance, or if something systematic is at work. Far from solving some difficulties as promised, we seem only to be collecting more unanswered questions!

Ocean-Floor Topography

In fact, a description of ocean-floor topography, to which we now turn, will for the time being only add to the confusion. The contour of the

* Contrary to popular belief, the United States is opposite the Indian Ocean, not China.

FIGURE 11-2 The pole of the land hemisphere is centered in western France; the opposite pole lies near New Zealand. Note that North America, South America, Africa, and the Indian subcontinent are roughly triangular areas pointing south like so: ▽. Is this seeming regularity a quirk, or is there a fundamental explanation? [*After L. D. Leet and S. Judson*, Physical Geology, *4th ed., Prentice-Hall, Inc., Englewood Cliffs, N.J., 1971*.]

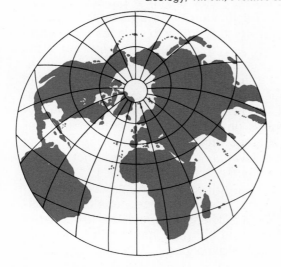

Land hemisphere

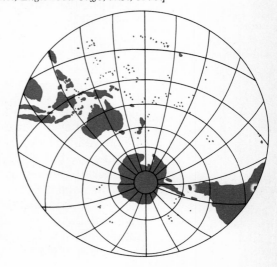

Water hemisphere

bottom is a faithful record of the processes that formed it. In many ways the underwater seascape is easier to interpret than a continental landscape whose original features may have been completely destroyed by erosion. True, a gentle rain of fine sediment may have obscured the ocean's basaltic basement in places, but even this circumstance is no obstacle to modern geophysical exploration. Analyzing the reflections of sound waves sent into the ocean sediments can solve the problem nicely (Figure 11-3a). A ship equipped with a powerful acoustic pinger (an explosive source of sound) can also measure the time required for the sound to travel downward to buried layers and return back to the surface. The deeper a layer is, the longer the delay before the sound echo is received. As the ship moves along, a profile is assembled showing the form both of the buried basement and the overlying strata (Figure 11-3b).

A certain aspect of this sedimentary carpet brings up yet another perplexing riddle. Since the ocean is the dumping ground of material shed off the continents, geologists had long hoped that one day they would find beneath the ocean a continuous record of earth history stretching back to its very beginning. Suppose we assume a sediment thickness of 900 meters, typical of the ocean floor. If sediment had been laid down during the previous 4.5 billion years, the average rate of deposition would have been only 20 centimeters per million years! Can deposition really have gone on at such an incredibly slow pace?

Methods based upon radioactivity have been developed to date these oceanic sediments (Figure 11-4). Uranium 238 and uranium 235 decay through a series of radioactive daughters on the way to becoming stable isotopes of lead (see Table 1-3). One of the daughters (in the ^{238}U decay chain) is thorium 230, which has a half-life of 75,200 years. Another daughter (in the ^{235}U decay chain) is protactinium 231, whose half-life is 32,480 years. Uranium remains dissolved in the ocean water, but thorium and protactinium stick tightly to surfaces of clay particles. These two daughter elements are constantly "milked" out of the water, to be deposited with the sediment on the bottom. The daughter isotopes decay away according to their characteristic half-lives in the sediment, but they are no longer being replenished by the decay of uranium. The older the sediment, the less radioactivity remains in it. Analyses have shown that a thickness on the order of 1 to 10 meters per million years is deposited in areas remote from land. Deposition is indeed slow, but at most only a few hundred million years, not billions of years, are needed to account for the observed thickness.

More recently, holes have been drilled through this sediment into the oceanic basement. According to fossil evidence, none of the sediment was deposited earlier than the Jurassic Period. The few reliable isotopic ages so far obtained from the basaltic crust itself are likewise in accord with the relatively young sediment ages. Of course, we could

[Cross section, not to scale]

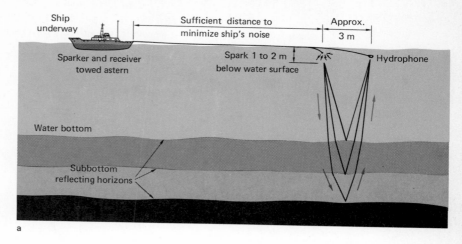

a

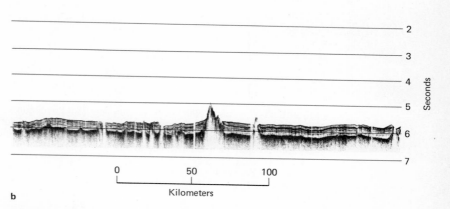

b

FIGURE 11-3 (*a*) A potent electric spark shot into the water every few seconds creates a train of steam bubbles. Sound waves from the miniature explosion are reflected off the bottom and off interfaces located as much as 2 kilometers deeper. On its return to the surface, the reflected sound is picked up by a hydrophone trailed behind the ship. (*b*) A continuous sparker profile is built up of a myriad of thin vertical lines, somewhat as a television picture is assembled by many sweeps of an electron beam. The vertical scale (highly exaggerated) is based on seconds of reflection delay time. This record from the equatorial Pacific shows a rough basement masked by strata that lap against the buried hills. Here and there, naked volcanic peaks rise above the general level of the sediments. [(*b*) *after J. I. Ewing, "History of the Ocean Basins as Recorded in the Sediments,"* Governor's Conference on Oceanography, *Lamont-Doherty Geological Observatory, Palisades, N.Y., 1967.*]

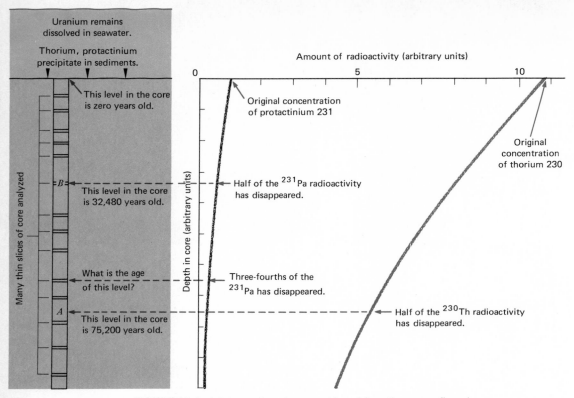

FIGURE 11-4 A long sediment core, retrieved from the ocean floor, is sliced into many small pieces that are analyzed individually. Thorium 230 and protactinium 231 are present in greatest abundance at the top of the core. Deeper sediment is older, containing less radioactive daughters than freshly deposited sediment at the top. The simple relationships pictured here may be confused somewhat by undersea erosion, changes in deposition rate, and disturbance by burrowing organisms.

insist that there was no such thing as an ocean before the Jurassic, or perhaps that the ocean refused to accept sediment for the first 96 percent of earth history. Both ideas are too absurd to dwell upon further. Nonetheless, the difficulty remains. How is it that rocks of the continental crust may be any age from recent to as old as 3.75 billion years, but that oceanic crust is youthful everywhere? Somehow the ocean floor must be destroyed and restored. How can this be accomplished?

Continental shelf, slope, and rise We would find the transition from continental to oceanic crust to be interesting and informative, if only we could see it. Unfortunately, it lies at the edge of the *continental shelf*, where it is everywhere hidden beneath sediments, and additionally by water nearly everywhere. The shelf borders the world's land masses, from which it gradually descends to an average water depth of some

130 meters (but with large deviations from that value in different places). Shelf width averages about 80 kilometers, but it may vary locally from zero to as much as 1500 kilometers (off the northern coast of Asia). In all, the shelf occupies an area about equal to that of Europe plus South America. Less than 10 percent of this extensive region can be termed well known.

Sparker profiles and samples recovered from numerous oil wells have established that the crust underlying the shelf is truly continental—internally, it is a continuation of the structure and composition of rocks found nearby on land. We could say that the ocean, being a bit too full, slops up over the edges of the continents a little. This raises the intriguing question of whether the ocean "ought" always to be so full, or whether today's situation is abnormal. The origin of the continental shelf is linked to the interplay between rising and falling sea level and the supply of detritus released from the continents: a topic for later chapters.

Seaward of the shelf edge, the imposing *continental slope* drops off into the abyss of the deep ocean. Contours of the ocean bottom in this little-explored region also vary widely. Typically, the downward angle of the continental slope is about 5 degrees, but where rivers have poured forth a great mass of sediment, as in the Bay of Bengal (between India and Burma), the shelf and the slope merge together as one enormous, gently inclined plane. An opposite extreme of topography is found off the west coast of Florida, where a rocky escarpment drops 3 kilometers in a horizontal distance of less than 5 kilometers—quite a contrast to the Floridian flatness! In many areas the slope is cut by submarine canyons, some of which outrival the largest canyons on dry land.

Most of the sediment transported across the shelf reaches a final resting place at the base of the continental slope. Canyons that cut across the shelf nearly to shoreline (as along the Pacific Coast of the United States) efficiently channel the flow of sediment into the depths. Sediment fans accumulate at the foot of the canyons, from which strong bottom currents may drift the material laterally. The result is a thick sediment wedge, the *continental rise*, banked up against the slope on one side and tapered gradually toward the thin veneer that blankets "normal" ocean floor on the other side (Figure 11-5).

Physiography of the deep ocean Within the past two to three decades our knowledge of the deep ocean, one of the last frontiers of earthly exploration, has grown with explosive speed. Best known is the Atlantic basin, but the once trackless Pacific, too, is rapidly being laced by the traverses of oceanographic vessels. Reconnaissance of the Indian and Arctic basins, though lagging far behind, will be a major objective in the future as we seek to confirm (or perhaps modify) our concept of the Grand Synthesis as it applies to these areas.

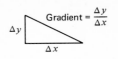

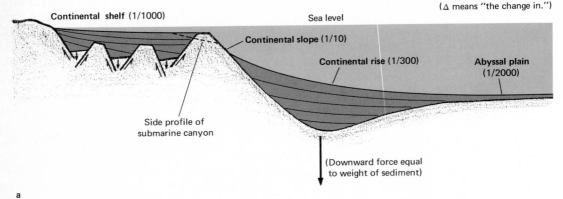

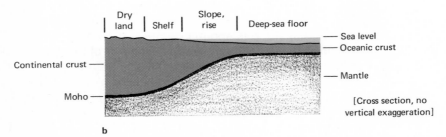

FIGURE 11-5 (*a*) A cross section (with vertical exaggeration) depicts the relationship of shelf, slope, and rise. Continental basement beneath the shelf may be a complexly faulted "borderland" (as off the coast of southern California), in part buried beneath sediment. The ocean floor beneath the rise develops an isostatic sag under an accumulation whose thickness along the U.S. Atlantic Coast and elsewhere may exceed 12 kilometers. Gradients of the submerged surfaces (in parentheses) are typical values. (*b*) Transition between continent and ocean is accomplished beneath the continental shelf and slope. The crust thins oceanward by the descent of the slope, but more especially by the rise of the Moho. (See Chapter 10 for a review of continental and oceanic crust, and the Moho.) [*After H. W. Menard*, Marine Geology of the Pacific, *McGraw-Hill Book Company, New York, 1964.*]

The ocean ridge system If continents and oceans are the largest features of the earth's crust, what would be the second largest feature? It is a system of mountainous ridges and broad, somewhat smoother "rises" whose existence long went unrecognized because only a few high points project above sea level. An imaginary journey along the 60,000-kilometer length of the *ocean ridge system* will enable us to focus on different characteristics in the areas where they are best understood.

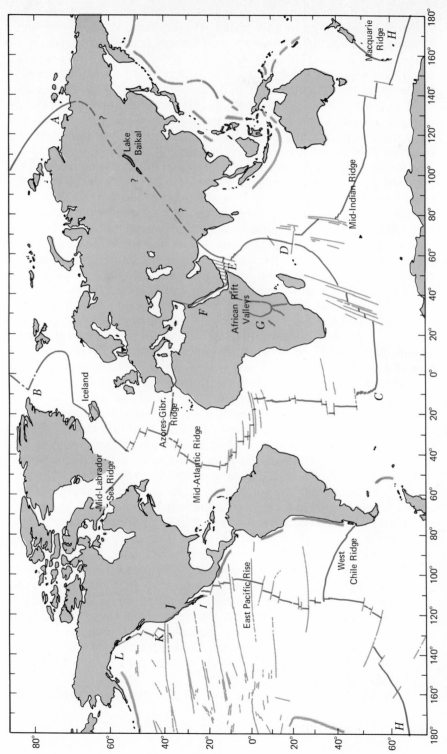

FIGURE 11-6 The world-girdling ocean ridge system is broken and laterally offset by fracture zones. However, in the Pacific are numerous long fracture zones not closely associated with the present locus of the ridge system. Ocean trenches rim the Pacific Ocean. [*After L. R. Sykes, "Seismicity of the Mid-Ocean Ridge System," Geophysical Union Monograph 13, p. 149, 1969. Copyright by American Geophysical Union.*]

We begin where the ridge intersects the wide continental shelf north of Siberia (Figure 11-6, point *A*). The ridge crosses the Arctic Ocean above the top of the map, and swings southward to extend the length of the Atlantic (*B* to *C*). Along that entire distance, the ridge is a symmetrical chain of rugged highlands occupying the central third of the width of the Atlantic basin, and rising to some 2 kilometers above the floor on either side (Figure 11-7*a*). Incised into the very crest of the ridge is a continuous deep cleft—the *median rift* (Figure 11-7*b*). Iceland is the largest island situated atop the ridge; farther south, the only exposed part is a sprinkling of minor volcanic islands.

Probably the most remarkable quality of the Mid-Atlantic Ridge is the midway course that it steers between the Old and New Worlds (Figure 11-6). Near the equator, where the trend of the Atlantic is sharply kinked, the ridge crest does not accommodate the bend by curving smoothly. Instead, it is severed into linear segments that are stepped off laterally by a series of *fracture zones* (Figure 11-8*a*) Topographically, the fracture zones are imposing undersea escarpments; samples dredged from the precipice walls are the customary oceanic rock types, but intensely sheared (by faulting) and in some places, metamorphosed. The central Atlantic fractures swing through gentle arcs that are portions of circles within circles. The common center of these nested circles is a distant point just off the tip of Greenland (Figure 11-8*b*). This concentric arrangement implies that somehow the curvature of the earth's surface has influenced the style of the faulting.

Precise seismic data from the ridge, interpreted at face value, seem both to clarify and to deepen the mystery of its origin. Earthquakes are frequent and shallow—always within the crust or uppermost mantle. The direction of movement, obtained from first motion studies (see Figure 10-4), comes as a complete surprise. Topography of the ocean floor looks as though the ridge were once a continuous strand that broke into small blocks that were driven apart. And yet, the seismic movements imply that the sundered blocks are approaching one another (Figure 11-9). How can the conflicting topographic and seismic evidence be reconciled? We shall see that this paradox is resolved (in the following chapter) only if new crust is created in the ridge—a process of crucial importance in the Grand Synthesis.

Similar offsetting fault scarps are distributed in large numbers along the ridge system throughout the world ocean. One fracture zone in the Pacific can be traced around a quarter of the earth's circumference! What awesome dislocation of the earth could have created such a feature? If crustal blocks move sideways, as the patterns suggest, what force drives them? What happens where the front of the block encounters something else, and what manner of "hole," if any, is opened up behind the trailing edge?

Passing below South Africa, the ridge enters the Indian Ocean

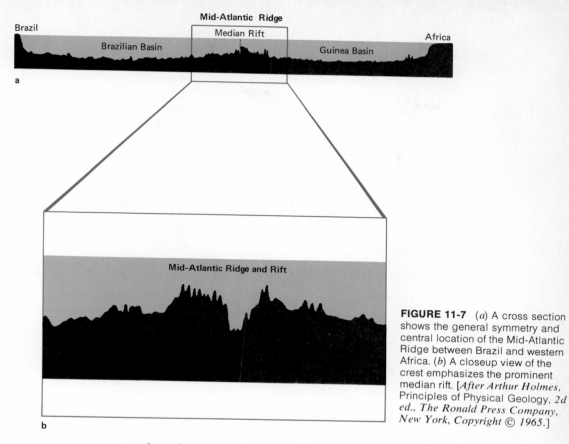

Brazil

Mid-Atlantic Ridge

Median Rift

Brazilian Basin

Africa

Guinea Basin

a

Mid-Atlantic Ridge and Rift

b

FIGURE 11-7 (*a*) A cross section shows the general symmetry and central location of the Mid-Atlantic Ridge between Brazil and western Africa. (*b*) A closeup view of the crest emphasizes the prominent median rift. [*After Arthur Holmes*, Principles of Physical Geology, *2d ed.*, *The Ronald Press Company*, *New York, Copyright © 1965.*]

where it splits into two branches (Figure 11-6, point *D*). One offshoot enters the Gulf of Aden (*E*) where the axis forks yet again. The median cleft is prominent in a northwestern continuation of the ridge along the troughlike Red Sea (*F*). To the south, another section of the normally oceanic ridge system actually runs onto the African continent where it appears as a series of *rift valleys* (*G*). Nestled in these valleys are lakes, some of them with bottoms deeper than sea level. In the same region are numerous volcanoes that pour forth unusual low-silica lavas believed to have originated deep in the mantle. A faint suggestion of yet a third branch strikes out northeast across Asia to rejoin our starting point off Siberia (*A*).

A relatively unexplored Mid-Indian Ridge commences at *D*; it continues eastward midway between Australia and Antarctica, then it enters the South Pacific. Its topographic form in the Pacific is quite unlike that in the other three ocean basins. The gently sinuous path of the East Pacific Rise occupies no midway position where it wanders northeast toward the coast of Mexico (*H* to *I*). No median rift appears, and

333 THE WORLD OCEAN

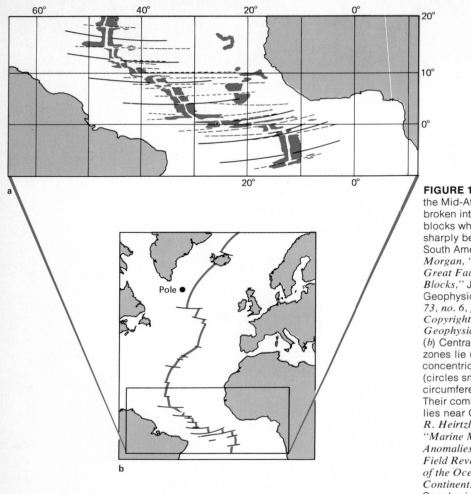

FIGURE 11-8 The crest of the Mid-Atlantic Ridge is broken into many small blocks where the ridge bends sharply between Africa and South America. [*After W. J. Morgan, "Rises, Trenches, Great Faults, and Crustal Blocks,"* Journal of Geophysical Research, *vol. 73, no. 6, p. 1964, 1968. Copyright by American Geophysical Union.*]

(*b*) Central Atlantic fracture zones lie upon portions of concentric "small" circles (circles smaller than a circumference of the earth). Their common pole, or center, lies near Greenland. [*After J. R. Heirtzler and others, "Marine Magnetic Anomalies, Geomagnetic Field Reversals, and Motions of the Ocean Floor and Continents,"* Journal of Geophysical Research, *vol. 73, no. 6, p. 2130, 1968. Copyright by American Geophysical Union.*]

[Schematic block diagram]

FIGURE 11-9 A commonsense interpretation is that offset segments of the ridge system have moved according to the black arrows. Seismic first-motion studies indicate exactly the opposite sense of movement (colored arrows).

no rugged topography embellishes the broad flanks of the rise. (Exceptions are found in isolated volcanic seamounts and the ever present fracture zones.) The rise is more a vast low bulge of the Pacific floor about as extensive as North plus South America. At point *I*, the crest of the ridge system heads up the long slot of the Gulf of California. As in Africa, the ridge "runs aground" in California (*J*). Here, the active San Andreas Fault system (see Figure 10-3) is identified, not as a continental expression of the crest, but as one of the offsetting fracture zones. The crest itself resumes offshore in the Pacific (*K*), from whence it advances (through several more jogs) to a disappearance at the Alaskan coast (*L*).

Ocean trenches   Complementary to the ridge system—a positive topographic feature—is a system of the most spectacular depressions on earth, *ocean trenches*. These negative topographic features also play an important role in our Grand Synthesis. In the deepest trenches, the ocean floor lies farther *below* sea level than Mount Everest towers *above* it. Trenches are restricted mostly to the western, northern, and eastern Pacific basin which they encircle in nearly unbroken succession (Figure 11-6). Even the Java Trench in the far northeast corner of the Indian basin, and the two Atlantic trenches "almost" lie within the Pacific. Some of these troughlike depressions are straight, or even sharply hooked on one end, but the majority bend along gentle open curves. For this reason, the chains of volcanic islands commonly situated between a trench and the continent were named *island arcs*.

Not all trenches are bordered by island arcs; a slope ascends from the depths of the Peru-Chile Trench onto the South American continent and up to the summit of the Andes almost without interruption—through a vertical range of 14 kilometers! This trench should be subject to filling by sediment washed off the continent. Local climate in Chile varies from the world's driest in the northern desert (less than 0.01 meter of precipitation per year) to one of the wettest in the southern Andes (more than *6 meters* annually). Offshore from the great Atacama Desert, the trench is 8 kilometers deep and nearly devoid of sediment, suggesting that the climate has remained arid for millions of years. Farther south, the sedimentary infilling steadily increases to the point of obliterating the trench topography altogether.

However, no other trenches lying at the foot of a continent are filled brim full with sediment. This unexpected circumstance presents another puzzle, for it would not seem likely that all these trenches had by chance formed too recently for thick sediments to have accumulated in them. Possibly there is some way to rid the deposit from the trench floor, but where can the sediment go when it is *already* resting at the lowest point on the earth's solid exterior?

Gravity measurements suggest that the missing sediments have somehow been carried downward into the earth. We learned in Chapter

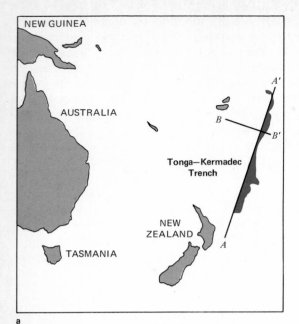

FIGURE 11-10 (*a*) An index map locates the Tonga-Kermadec Trench and the lines along which two cross sections are drawn. Section *AA'* lies along the length of the trench, and section *BB'* cuts across it. *Opposite page:* (*b*) Earthquake foci extend the entire length of the trench axis (section *AA'*), and to a depth of 700 kilometers. About half of all deep-focus earthquakes take place in the Tonga-Kermadec Trench alone. (*c*) An island arc is but weakly developed along the Tonga-Kermadec Trench. As seen in the transverse cross section (along *BB'*), the foci outline a plane descending steeply from the trench floor toward Australia. Where the north end of the trench hooks sharply westward (see map), the dipping sheet of foci likewise curls under it conformably. Do these two features share a common origin? [*After L. R. Sykes, "The Seismicity and Deep Structure of Island Arcs," Journal of Geophysical Research, vol. 71, no. 12, pp. 2982, 2995, 2994, 1966. Copyright by American Geophysical Union.*]

10 that ocean basins are about 99 percent in isostatic balance. Most of the 1 percent of *im*balance is found in the trench system. Along a narrow strip just shoreward of the axis of a trench is a large, negative-gravity anomaly (an abnormally weak gravity field) that could mean that a mass of low-density crust has displaced dense upper-mantle material.

Equally striking is the pattern of seismic data from these active regions. Earthquakes occur at the floor of the trench. Under the adjacent island arc (if present), the foci are deeper, and still farther away from the axis of the trench, deep-focus earthquakes take place at a maximum depth of about 700 kilometers (Figure 11-10). Thus the earthquake foci seem to be located along a plane that descends into the earth.

Smaller features Even the minor topographic features of the ocean floor are significant to the Grand Synthesis. The ocean contains an estimated 20,000 submerged volcanoes or *seamounts* whose relief above the ocean floor exceeds 1 kilometer. About half of these are scattered about in patches in the Pacific. In addition, the world ocean (for example, some 80 to 85 percent of the Pacific area) is studded with innumerable *abyssal hills*—elongated, irregular mounds with only a few hundred meters' relief. Abyssal hills are easily the most widespread single topographic feature on earth. Because of their relatively small size, they are also extraordinarily difficult to survey. Only recently have close-in studies been carried out by deep-submersible instrument packages ("geophysical fish").

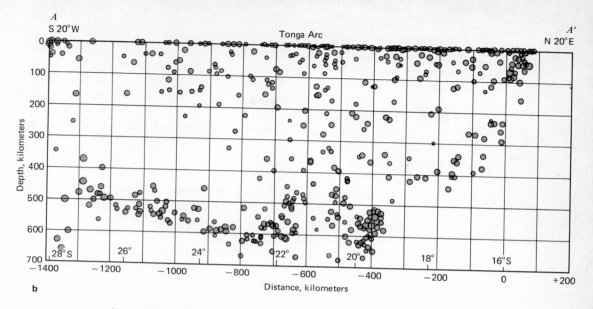

A
S 20°W

Tonga Arc

A'
N 20°E

Depth, kilometers

Distance, kilometers

b

Near-surface cross section

[Vertical exaggeration: 10 times]

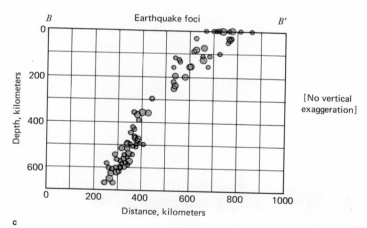

B

Earthquake foci

B'

Depth, kilometers

[No vertical exaggeration]

Distance, kilometers

c

Of more obscure origin are the strange, coral-encrusted *atolls* of the southwest Pacific. Rising steeply from the ocean floor, an atoll culminates at sea level in a broad, shallow lagoon hemmed in by a reefy barrier. Deep drilling on Eniwetok and Bikini atolls penetrated a kilometer or two of coralline deposits of shallow-water origin before entering a volcanic basement. These findings vindicate the idea of one of the first

337 THE WORLD OCEAN

men to investigate atolls, Charles Darwin. He proposed that atolls were once volcanic islands that began to subside. At first, he said, the reefs merely fringed the rocky shores of the island, but with gradual sinking the carbonate-secreting organisms (chiefly coral) were able to keep pace by building upward on the platform of their departed ancestors. Eventually, even the volcanic summit became submerged beneath a thick limestone cap (Figure 11-11).

If profound vertical movements of the ocean floor have played a role in the development of atolls, the origin of *guyots* (GHEE-oes) becomes less a mystery. Guyots, discovered during the first extensive surveys of the Pacific during World War II, are seamounts whose curious flat tops lie as much as 2 kilometers below sea level (Figure 11-11). Dredge hauls from guyots include fossils of extinct Cretaceous or Cenozoic reef organisms adapted to life in shallow water. Rounded basalt pebbles, buffeted and polished to a smooth finish, have also been recovered. Apparently the guyots too were subsiding volcanic islands on which coral did not continue to grow upward. Wave attack, which is effective down to a water depth of 100 meters or more,

FIGURE 11-11 Profiles, all to the same scale and without vertical exaggeration, reveal how puny even the greatest land volcanoes are, compared with many undersea volcanoes (seamounts and their modified kin, atolls and guyots). [*After H. W. Menard,* Marine Geology of the Pacific, *McGraw-Hill Book Company, New York, 1964.*]

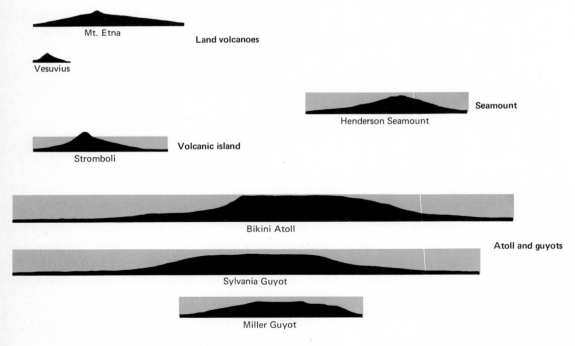

vigorously planed off the crests of the sinking islands. Guyots are clustered in the south and central-west Pacific, and in the Gulf of Alaska (Figure 11-6, near *L*).

MOUNTAIN BELTS

A successful global hypothesis must account for the major features not only of the ocean, but of the continents as well. A frequency diagram concisely summarizes the distribution of the earth's surface according to its elevation above or depth below sea level (Figure 11-12). Nearly 75 percent of the land surface lies below 1000 meters; the uplands—mountains and plateaus—are but a minor, though spectacular, part of the whole. Geologists are keenly interested in mountain chains, not just to appreciate their loftiness or beauty (as any other person would), but also to understand their complex internal structures. Thanks to the great vertical relief,* geologists can study these structures with an exactness that is not possible anywhere else. But at the same time, the erosional downcutting that etched out the relief has also

* Though mountains are generally considered to be rugged terrain, a geologist may casually refer to "mountains" in an area that has been eroded to a level plain. For example, in Late Paleozoic times, the Hercynian Mountains were a bold chain across central Europe, but today only the intensely distorted rocks of the mountain "roots" remain.

FIGURE 11-12 A frequency diagram shows how much of the earth's surface area lies at a particular height or depth with respect to sea level. By and large, the elevations of oceanic and continental crust are sharply distinguished. The two maximum points of the curve signify the most frequent (*not* the average) elevations. Ocean floor is typically about 5 kilometers below sea level and, surprisingly, more of the continents is practically at sea level than at any other elevation—testimony to the degrading power of erosion. [*After A. Wegener*, The Origin of Continents and Oceans, *Dover Publications, Inc., New York, 1961.*]

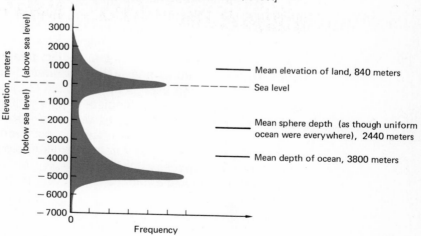

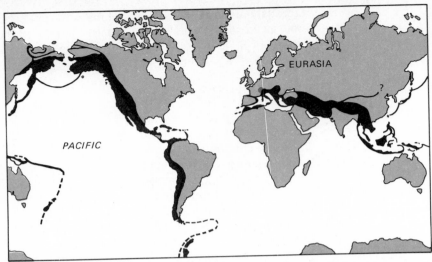

FIGURE 11-13 Mountain chains are linear both on a local and on a global scale. The two major systems trend approximately at mutual right angles. A small circle encloses Switzerland. [*After Arthur Holmes*, Principles of Physical Geology, *2d ed., The Ronald Press Company, New York, Copyright © 1965.*]

destroyed much of the original record of the rocks. Mountain chains are perpetually crumbling into ruins even while they are being uplifted.

Moreover, because of the colossal size of the subject, the organization of the rock structures may become apparent only when the results of months or even years of careful field observations are plotted. Although mapping these structures in one difficult, seemingly postage stamp-sized area after another may appear to be painfully slow progress, the effort is rewarding. A structural geologist must feel great satisfaction as he recognizes a fundamental order in what was once a bewildering confusion of peaks.

Modern high-standing mountain ranges have been uplifted within the past few million years, and in many of them, uplift is still going on. Two systems—one surrounding the Pacific and the other extending across southern Europe and Asia—dominate the mountain network (Figure 11-13). Among the "great" mountain ranges, only the Alps and the American Rockies are geologically quite well known. The structure of the Himalayas is but moderately well understood. Immense stretches of the Andes and of numerous chains in south-central and southeast Asia have been examined only from satellite photographs* supplemented by a few spot checks on the ground.

What is a mountain? In common language, a mountain refers

---

* Photographs taken from an altitude of 150 kilometers provide an excellent starting point since they record the giant structures while leaving the minor faults and folds conveniently hidden below the resolving power of the film.

merely to a prominence rising to an impressive height above the local terrain. Mountains have diverse origins—they may be volcanic (such as Fuji, near Tokyo, or Mount Shasta, California), or they may consist of crustal blocks displaced by faults (such as the Grand Tetons, Wyoming). They may be no more than isolated remnants of a high plateau already mostly removed by erosion. In each instance, whether the high elevation was achieved by volcanic eruption, by local faulting, or by a broad regional uplift, the predominant sense of motion was *vertical.* However, a simple notion of vertical uplift was soon found not to apply universally. One of the most astonishing results of structural mapping early in this century was the discovery that in many mountainous regions, *horizontal* transport of the rocks has far exceeded the vertical displacement. Could it be that wholesale lateral shifting of the ocean floor (implied by the pattern of fracture zones) finds a counterpart on the continents? Let us see how the structures in *fold mountains* are illustrated by the most intensively studied mountains of them all, the western Alps.

Some Alpine Attributes

If anyone ever were to study a mountain chain in depth, the Swiss geologists were the people to do it. In their homeland are magnificent cliffs exposing an accumulated total of 3 kilometers of vertical relief. The Swiss were well suited by temperament and national tradition for the painstaking observations that would be necessary. Switzerland is not a very large place, either—when drawn to scale, the country fits easily inside the tiny circle in Figure 11-13. Thus these skilled geologists, it seemed, could hardly fail to make short work of the endeavor. Such was not to be the case. For example, the shape of one enormous mass of deformed rocks in the southern Swiss Alps was quite unanticipated from studies of the surface. Not until the Simplon Tunnel was punched through the heart of the structure was its form correctly interpreted. Parts of the Alps remain geologically obscure today, after a century and a half of scrutiny. We shall touch upon some conclusions that seem typical of the Alps, but more important, of fold mountains in general. But we must take care not to press the analogy too far. Having formed under unique circumstances, every great mountain chain is endowed with a distinct personality.

Setting the stage Long before the Alps were uplifted, other mountains once stood in western Europe. Their life cycle had ended as erosion reduced the previously rugged landscape to a low plain. Today, small patches of these more ancient rocks are exposed within the modern Alps and farther north in France and Germany (Figure 11-14). Uranium-lead ages from zircon are in the range of 250 to 550

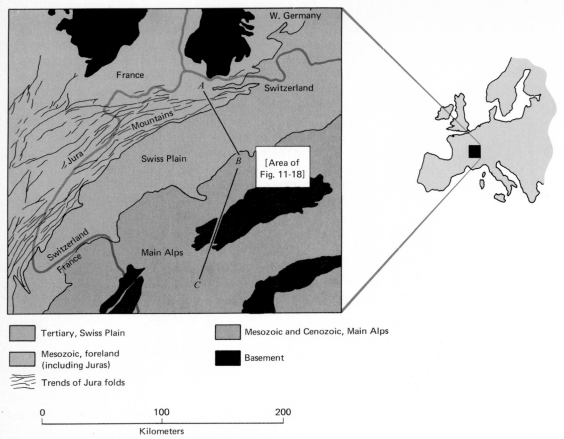

FIGURE 11-14 A greatly simplified geologic map includes the western Alps and the Jura Mountains and terrain farther to the northeast and northwest. Black areas are deeply eroded remnants of the Late Paleozoic Hercynian Mountains. In France and Germany this rock has not since been severely deformed, but similar areas within the Alps were again caught up in mountain-making movements. Sandstones and conglomerates of Oligocene and Miocene age comprise the Swiss Plain. Alps southeast of the Swiss Plain include a gigantic pile of folds draped over folds.

million years, signifying Paleozoic metamorphisms. Thus these rocks had participated in earlier orogeny, only to be involved anew in orogeny. Other fold mountains contain similar remnants of ancient basement, reminding us that in the earth's "mobile belts," the pulse of orogeny may be expressed again and again.*

By the late part of the Paleozoic Era, mountain building on the site of modern Germany and France had ceased. The area had become a

* Regional isotopic age patterns suggest that the Berkshire Highlands of Connecticut (a region of older basement in the Appalachian Mountains) has experienced at least four orogenies centered at 1150, 450, 360, and 260 million years ago.

region that was sometimes barely dry land, at other times barely awash beneath a shallow sea. To the south lay the Tethys Sea, of which the Mediterranean Sea is a shrunken vestige, according to some authorities. Sediments brought from the continent were dumped into this nearshore environment for millions of years. When the Alpine orogeny began early in the Jurassic, it raised a series of submarine ridges and island chains separated by deep troughs. (A similar situation exists today along the coasts of southeast Asia.) Occasionally the sediments would slump back into an adjacent trough. This "cannibalizing" of the sedimentary record—underwater uplift, erosion, and deposition progressing together—created an extremely complex stratigraphy.

Finally the stage was set for the main phase of orogeny. At this point, too, the Alps were similar to other fold mountains. As the world map of mountain chains implies, the edge of a continent is the most likely place for orogenic disturbance (Figure 11-13). It is this same marginal environment that is likely to contain thick deposits of sediment. So many fold mountains contain enormous sediment thicknesses that geologists had long believed that prolonged deposition was a necessary forerunner of orogeny. In the light of modern knowledge, it seems better to separate the two phenomena. For example, the Himalayas do not contain excessive quantities of sedimentary rocks. Along the United States Atlantic Coast and in the Gulf of Mexico, sediment thicknesses are equal to those found in the grandest mountain range, but there is no hint of impending orogeny. Instead, the factor that unifies the two processes of orogeny and sedimentary deposition is their mutual location at the continental border. It is an odd situation that the elevated regions of the world were substantial depressions at the outset!

The Alps take form   The orogenic force that crumpled the Alps was chiefly a lateral (horizontal) compression directed north from the Tethys Sea toward the ancestral European continent. Mechanically, this appears to be an absurd way to raise up mountains; far less energy would be expended by a simple vertical uplift. High elevations in fold mountains appear almost to be an incidental effect. Of course, this is nature's way of signifying that these deformed belts are only a surface expression of some deeper, more fundamental process. To gain an insight into what happened, let us take an imaginary journey across the Alps approximately along the line *ABC* in Figure 11-14.

Trending across the border region of Switzerland and France is a bundle of parallel valleys and ridges: the Jura Mountains (Figure 11-15a). Here, folded sedimentary strata of Jurassic* and Cretaceous age lie above Triassic beds containing the soft evaporite mineral, anhydrite ($CaSO_4$). Oddly enough, the metamorphic basement rocks un-

---

* The Jurassic Period was named after these mountains.

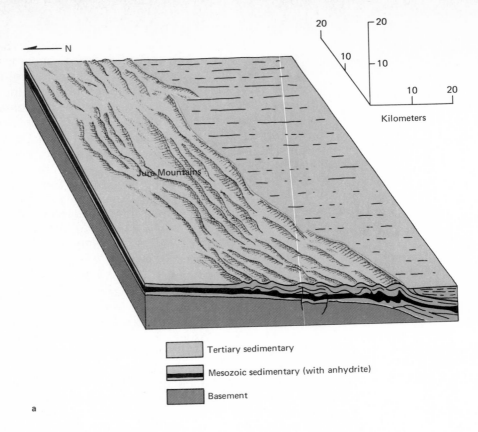

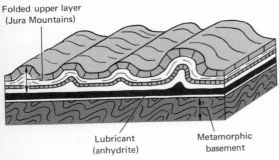

Folded upper layer
(Jura Mountains)

Lubricant
(anhydrite)

Metamorphic
basement

b

FIGURE 11-15 (*a*) A block diagram of northern Switzerland shows the Swiss Plain and the wrinkles of the Jura Mountains. (*b*) In the Jura Mountains the superficial beds became detached from the basement, then slid over it. A similar relationship between basement and covering strata is found in the folded Appalachians—the long ridges and valleys so prominent in West Virginia and Pennsylvania. [*After J. H. F. Umbgrove,* Symphony of the Earth, *Martinus Nijhoff, The Hague, 1950.*]

derneath the Triassic are nowhere to be seen, even where erosion has bitten deeply into the ridges. A wealth of data from boreholes and tunnels shows why this is so. Sedimentary strata came "unglued" and slid over the weak, slippery anhydrite substratum (Figure 11-15*b*). Basement rocks did not participate in the folding. The final result is like a rug that wrinkled as it skidded across a polished floor—but on a giant scale!

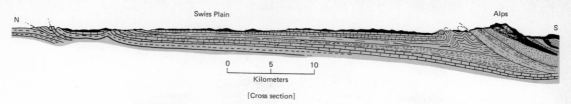

Alps

N

S

0   5   10
Kilometers

[Cross section]

FIGURE 11-16 As the giant sheets comprising the main Alps advanced northward, they crumpled and overrode their own erosional debris. [*After J. H. F. Umbgrove,* Symphony of the Earth, *Martinus Nijhoff, The Hague, 1950.*]

Adjacent to the south of the Jura Mountains lies the low Swiss Plain,* a belt about 35 kilometers wide that contains the major cities. Structurally, the Swiss Plain is a huge trough that has been filled with sandstones and conglomerates of mid-Tertiary age (Figure 11-16). These sediments were stripped off and deposited at the foot of the main Alps during an episode of intense compression and uplift.

Still farther to the south, the high wall of the main Alps juts abruptly upward. In all, the Alpine orogeny signaled a tremendous shortening and thickening of the crust. One estimate is that a region 450 kilometers across has been compressed into a belt only one-fourth as wide! A clue to the structure in this region is afforded by that famous pyramid summit, the Matterhorn (Figure 11-14, location *C*). Near its base, somewhat metamorphosed Triassic and Jurassic strata lie in an almost horizontal position. Correlation with equivalent rocks elsewhere indicates that an original accumulation of thousands of meters has been thinned and stretched to 20 to 30 meters at the Matterhorn—to only 1 percent of the original thickness. What is more, Jurassic rocks are succeeded upward by Triassic, and above the Triassic at the apex of the mountain rests a massive block of pre-Triassic basement. The entire sequence is upside down!

From this observation and hundreds more like it throughout the Alps, a unified picture began to emerge. The earth's crust, squeezed laterally as though clamped between the jaws of a vise, was buckled upward into folds tens of kilometers across. Because rocks are not especially strong on such a large scale of size, the folds tipped over and continued to creep forward as horizontal sheets, or *nappes*† (Figure 11-17). Likely the nappes flowed down a gentle slope, so that gravity aided their forward movement. Fold after fold arose until a stack several deep had formed, each later nappe extending farther to the north as an immense blanket over its crumpled predecessors. Erosion was active all the while; sedimentary debris shed off in front of the mountains was overridden by the northward advance of the piled-up sheets.

* Not all of Switzerland is mountainous.
† "Nappe" is a contraction of the French *nappe du charriage,* meaning "transported sheet."

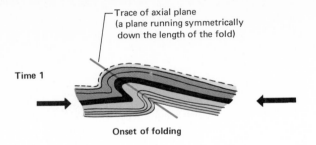

Time 1

Trace of axial plane
(a plane running symmetrically
down the length of the fold)

Onset of folding

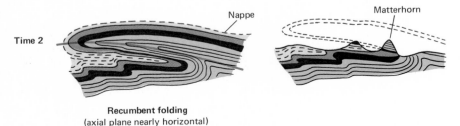

Time 2

Nappe

Matterhorn

Recumbent folding
(axial plane nearly horizontal)

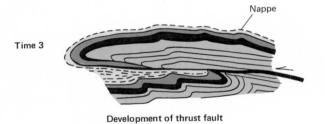

Time 3

Nappe

Development of thrust fault

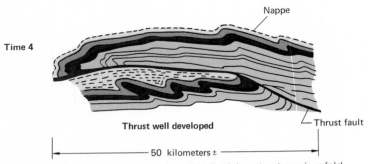

Time 4

Nappe

Thrust fault

Thrust well developed

|← 50 kilometers ± →|

FIGURE 11-17 Cross sections show that flat-lying structures in a fold mountain belt are driven by forces acting in a horizontal direction. As deformation continues, the nappe becomes smeared out and enormously elongated. In some nappes a nearly horizontal fault plane breaks through the rocks, permitting the upper slab to continue sliding over the lower. Other nappes are merely stretched thin as they are driven forward horizontally for 50 kilometers or more. In the schematic drawing (right), a nappe that once looked like stage 2 (left) has been eroded. Only the *inverted* strata are preserved in the mountain peaks (as in the Matterhorn). In such structurally disturbed regions, local stratigraphic sequences are about as likely to be found upside down as right side up.

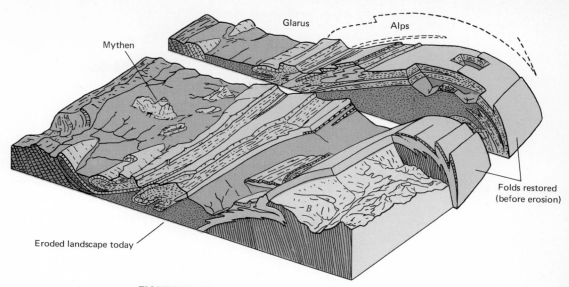

FIGURE 11-18 Mesozoic and Cenozoic sediments in the Alps were squeezed as horizontal compressive forces relentlessly crumpled the crust. This block diagram shows a region near Lucerne, Switzerland (see Figure 11-14). *B* is ancient basement rock, draped over by folds of deformed sedimentary rocks. The Mythen are gigantic blocks of limestone that were carried along within an advancing nappe. Now that the enclosing rocks have been stripped away, we find the anomalous situation of Cretaceous rocks (the Mythen) resting "rootless" upon Eocene strata. (In an undisturbed sequence, Cretaceous rocks lie *beneath* Eocene rocks.) Offset block in the distance is the region of the Glarus Alps (see Figure 11-19). Uplift of the Alps is still going on today at a rate of several millimeters per year. [*After J. H. F. Umbgrove*, Symphony of the Earth, *Martinus Nijhoff, The Hague, 1950.*]

Far to the south, on the opposite side of the heap, is a zone where once horizontal, layered rocks now stand vertically. They are the intensely deformed and metamorphosed "roots" of the Alpine nappes, now disunited from the horizontal portions of the nappes by erosion. Mangled rocks of basaltic affinity, interpreted as fragments of the ocean floor, have been interleaved among other rocks in the southern region and among some of the nappes.

In no one spot are all the Alpine nappes to be seen. Along the long axis of the Alps, the nappes and underlying basement rise and fall in broad swells (Figure 11-18). Erosion has locally opened up various structural levels to view, thereby enabling geologists in their reconstruction to project the position of rocks that are now missing (Figure 11-19). In three dimensions, the churned Alpine rocks look like infoldings of batter in a marble cake. Surely these fantastic happenings merit a prominent status in our grand hypothesis.

347 MOUNTAIN BELTS

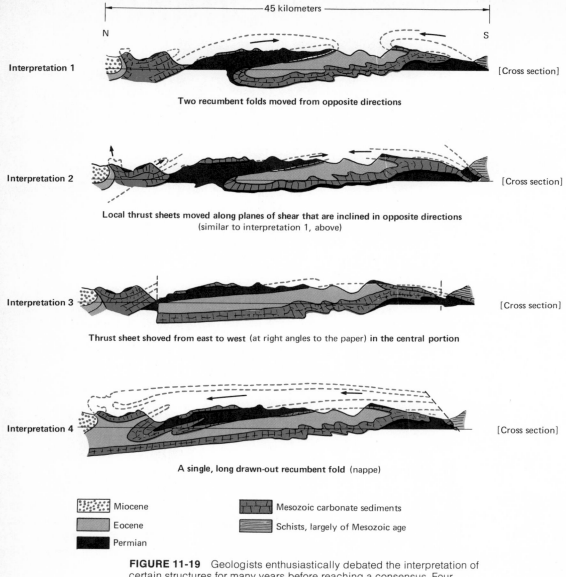

Interpretation 1   [Cross section]

Two recumbent folds moved from opposite directions

Interpretation 2   [Cross section]

Local thrust sheets moved along planes of shear that are inclined in opposite directions
(similar to interpretation 1, above)

Interpretation 3   [Cross section]

Thrust sheet shoved from east to west (at right angles to the paper) **in the central portion**

Interpretation 4   [Cross section]

A single, long drawn-out recumbent fold (nappe)

Miocene   Mesozoic carbonate sediments

Eocene   Schists, largely of Mesozoic age

Permian

FIGURE 11-19 Geologists enthusiastically debated the interpretation of certain structures for many years before reaching a consensus. Four divergent opinions, in a spectacular mountain setting in the canton of Glarus, Switzerland, were advanced. Note that rocks found at the surface are identical in all four interpretations. The disagreements were about the unseen part: structures hidden at depth, or already eroded away. Ironically, interpretation 4, which seems the least plausible because it demands that material be transported the greatest distance, turned out to be the correct one. [*After W. H. Bucher,* The Deformation of the Earth's Crust, *Princeton University Press, Princeton, N.J., 1933. Copyright 1933 by Princeton University Press.*]

Two Personalities of Solids

In view of the way that Alpine nappes are draped over one another like so many folds of cloth, it appears that the rocks flowed like putty or toothpaste (Figure 11-20). Could it be that the sediments were a weak, soft mass, not yet lithified when the folding took place? If that were so, we probably would find sedimentary objects that, *already* solid at the time of deposition, had been rafted along with the enclosing matrix but not deformed themselves. Instead, the Alpine rocks contain strongly distorted, though still recognizable, fossil shells. Quartzite cobbles in conglomerate, initially spherical or egg-shaped, have been rolled out into long rods resembling walking sticks. If this can happen to quartzite, mechanically the strongest and chemically the most inert of all

FIGURE 11-20 Intensely contorted sedimentary rocks in this cliff near Interlaken, Switzerland, make typical Alpine scenery. Size of the folds can be judged by comparison with the fruit trees in the foreground. In many settings, the rocks have experienced "penetrative deformation"; distortion is apparent on *all* scales of size. Microscopic kinks cut across individual crystals; these dislocations are superimposed upon larger corrugations visible in a hand-sized specimen. Centimeter-sized crinkles are part of folds like those pictured below, which in turn are perched on a recumbent fold several tens of kilometers across.

common rocks, small wonder that some of the more submissive beds have been flattened out to 1 percent of their original thickness! No—the deformed pebbles and fossils, and many other lines of evidence, establish that the rock was fully a solid even as it yielded to stress by flowing.

A ductile behavior of solids is foreign to our everyday experience. After all, rocks shatter when hit sufficiently hard with a hammer. When a blow is inflicted, a *brittle* or *elastic* response is manifested—the rock promptly returns to its original size and shape when the stress is removed, or if the stress exceeds a certain value, the rock simply breaks. Mountain ranges carry abundant testimony to the brittleness of rock as recorded by numerous faults and joints.*

These very same rocks may bear the imprint of a *plastic flow*, in which deformation is distributed throughout the material. In this situation, the strain becomes permanent, even after the stress has been released. Some rocks are free of strain internally, though broken through cleanly by joint fractures (brittle rupture). Others are penetrated by numberless local planes of flowage. In the ultimate stage, every layer of atoms in every crystal has glided a short distance past its neighboring layers of atoms (plastic flow).

Ductile behavior becomes more pronounced as temperature, confining pressure, and amount of water in pore spaces increase. The most crucial factor, though, is the duration of the applied stress. Ductile flow of the Alpine nappes was brought about by stress sustained over millions of years. All the while, the Alps must have been active earthquake country, in which rocks ruptured in a brittle manner.

Some Generalities

Our Grand Synthesis must take account of several more characteristics shared not only by the Alps, but by ancient and modern fold mountains generally. As already noted, great thicknesses of sediments are involved in the folding. (Volcanic rocks were also deposited, only a little in the Alps and Himalayas, but a very large quantity in the Andes.) Where fold mountains have been deeply eroded, as in the Appalachians and especially in shield areas around the world, metamorphic terrains liberally invaded by granodiorite and other intrusive igneous rocks are brought to light.

Another striking regularity of fold mountains is their organization in long linear belts. "Linear" does not mean "straight," for mountain ranges may bend through graceful arcs or even through sharp hairpin loops (Figure 11-21). The structures have a pronounced asym-

* Fractures along which no significant later movement has occurred.

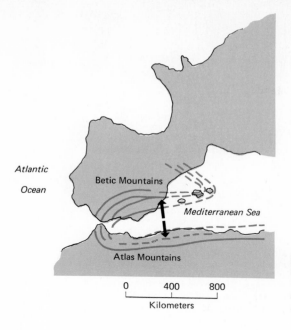

Atlantic

Ocean

Betic Mountains

Mediterranean Sea

Atlas Mountains

0   400   800

Kilometers

FIGURE 11-21   During an orogeny, rocks were shoved in the directions of the arrows. [*After J. Gilluly, "Tectonics Involved in the Evolution of Mountain Ranges,"* The Nature of the Solid Earth, *McGraw-Hill Book Company, New York, 1972.*]

metry—folds have been overturned and shoved in the direction away from the ocean, toward the continent. For instance, in the Betic Mountains of southern Spain, transport was toward the north, away from the Mediterranean (Figure 11-21). Opposite Spain, in the Atlas Mountains of Morocco, great slices of rock were thrust southward, likewise away from the Mediterranean. The reasons behind this relationship between deep ocean basin and mountain structures on land are not known.

Isotopic ages and fossil and stratigraphic evidence furnish yet another useful set of data. Tens of millions of years may pass by while a mountain chain is being uplifted, but in the larger perspective of time, an orogeny is rather sharply episodic. An amazing lot can happen in a brief period, geologically speaking; the Himalayas and the Andes have soared upward to their present grandeur mainly within the past 2 to 3 million years! Far longer intervals are required to erode mountains away, especially in view of their progressive renewal by isostatic uplift (Chapter 10).

Finally, Paleozoic fold mountains on opposite sides of the North Atlantic are an excellent illustration of the most astonishing relationship of them all. At the northern end of the Appalachians, the structural trends (folds and faults) head straight out to sea where last seen on the Newfoundland coast (Figure 11-22). Likewise, the Caledonian Mountains appear to be abruptly snubbed off where they meet the western

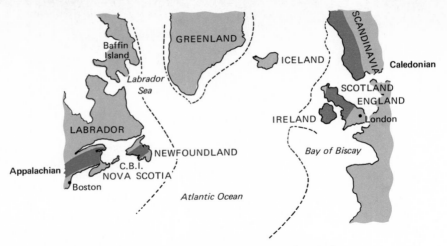

FIGURE 11-22 Early- to Mid-Paleozoic Appalachian and Caledonian trends appear to be mutual continuations of one another across the Atlantic. Similar observations pertain to other deeply eroded mountain systems (some older and some younger than those shown here) which intersect both the Atlantic Coast and the Appalachian-Caledonian belt. Dashed lines denote the edges of continental shelves. [*After M. Kay, "Continental Drift in North Atlantic Ocean,"* American Association of Petroleum Geologists Memoir 12, *1969.*]

shores of Ireland and Scotland. Three subbelts run parallel along the length of the major chains. Within each zone, the rock types, fossils, times of sedimentary deposition, trends and styles of deformation and of igneous intrusion—all are broadly similar to those in an opposite belt across the Atlantic. The correspondence is not exact, nor would we expect it to be in view of the fact that these same structures extend across the continental shelves an additional 400 kilometers on either side. Thus, about 800 kilometers of the mountains remain uncharted, concealed beneath a veneer of shelf sediments.

The Appalachian-Caledonian match is not unique; the same Caledonian trend jumps from Norway to Spitzbergen (a group of islands in the Arctic Ocean). Eroded vestiges of a mid-Paleozoic mountain range barely graze the tip of South Africa, and the range seems to resume its course in Argentina, 7000 kilometers to the west. More ancient structural trends can be matched between Australia and Antarctica, and between West Africa and Brazil.

But how can fold mountains of one continent harmonize so perfectly with mountains of some remote land mass? Years ago, geologists proposed that once connecting the continents were ancient "land bridges" that have since foundered into the ocean depths, but recent geophysical exploration fails to support this suggestion. Not a trace of continental crust is to be found in the ocean basin between the severed ends of the mountain chain. The idea that the continents were once joined together, later to split and drift apart, is even more absurd—or is it? The Grand Synthesis, as we shall see, can put together not only facts and observations, but in a more literal sense it will show how the great land areas of the world were once put together.

chapter 12

The Grand Synthesis: Conclusion

In southern California, the spectacular San Andreas Fault marks the boundary between two moving tectonic plates. [*U.S. Geological Survey photograph by Robert Wallace.*]

The preceding chapter has raised a number of profound but unanswered questions about the earth. Among these are the puzzle of the land and water hemispheres and the mysteries of the absence of thick sediment in the mid-ocean and the origins of the ocean ridge system, the trench system, and the fracture zones. What causes earthquakes and volcanism? What is the origin of the peculiar structures present in the Alps and other fold mountains?

Often in the progress of science, the safe, conservative interpretation must finally give way to evidence favoring a bizarre explanation. One of these unlikely possibilities is the notion that the continents in fact are not terra firma, forever fixed in their positions. But then, what is the evidence for such an incredible assertion, and what is its place in the Grand Synthesis?

WANDERING CONTINENTS

Sir Francis Bacon launched a historic controversy when in 1620 he pointed out the conformity in outline between Africa and South America: a natural consequence of rifting apart a former supercontinent. Later authors speculated that Noah's flood, or some other mighty water current, must have effected the breakup. (In those days, the doctrines of catastrophism and uniformitarianism were still being argued.) As the uniformitarian point of view slowly won adherents, the idea of continental drift fell into disrepute along with its catastrophic overtones.

Early in this century, the question was reopened in a uniformitarian context by two enthusiastic advocates of continental drift, the German meteorologist-explorer Alfred Wegener, and the South African geologist A. L. du Toit. Wegener was intrigued by a result of two surveys showing that a certain landmark in Greenland had moved westward 1190 meters

355

in a 137-year period.* He and du Toit called attention to the close similarities in fossil assemblages and in stratigraphic successions, and to evidence for Paleozoic glaciations spread across continental edges that supposedly were once joined together. At that time, geologists could not identify the forces responsible for continental drift. (These forces are still largely a mystery to us.) Wegener believed that the Americas had "sailed" actively over the mantle, which he offered as explanation for why mountains are wrinkled up along the western (forward) edge. His critics pointed out that a continental crust thousands of kilometers wide is too weak to act as a rigid plate. Wrinkles should have been heaved up all throughout, not just in the western area. Other proposals were tested and found wanting. And so once again the hypothesis lost favor, except among geologists in the Southern Hemisphere where the evidence is more convincing than it is in the northern continents. Since the underlying causes of continental drift could not be identified, for most people it was safest to conclude that drift had not happened.

Wegener and du Toit had only fragmentary information, suggestive but nonconclusive in itself. Then, beginning in 1956, a third revival of interest came in with a flood of new data representing 30 or more different and independent lines of evidence. Here we shall document how some of the more impressive data have converged upon the issue.

Support of the Hypothesis

Improved mapping of the ocean-bottom contours has strengthened the argument first raised long ago by Bacon. Surprising, perhaps, is the excellent fit not only between Africa and South America, but also between Europe and North America (Figure 12-1). Note that this amazing harmony is achieved only by matching the true continental boundaries: that is, along the shelf edge, or better yet, along a depth contour located about halfway down the continental slope. (Present-day shorelines on opposite sides of the Atlantic fit together only rather poorly.) The parallelism of continental margins is our most substantial evidence for rifting of the great land masses.†

* More refined measurements invalidated this conclusion. Wegener's idea of continental drift, even if correct, is not supported by some of his supposed "evidence."
† India, Madagascar, Antarctica, and Australia can be joined to Africa and South America to compose a supercontinent. Arrangement of these southern puzzle pieces is somewhat ambiguous because the land masses do not have such distinctive shapes, and the geology (especially of Antarctica) for the time being is poorly known.

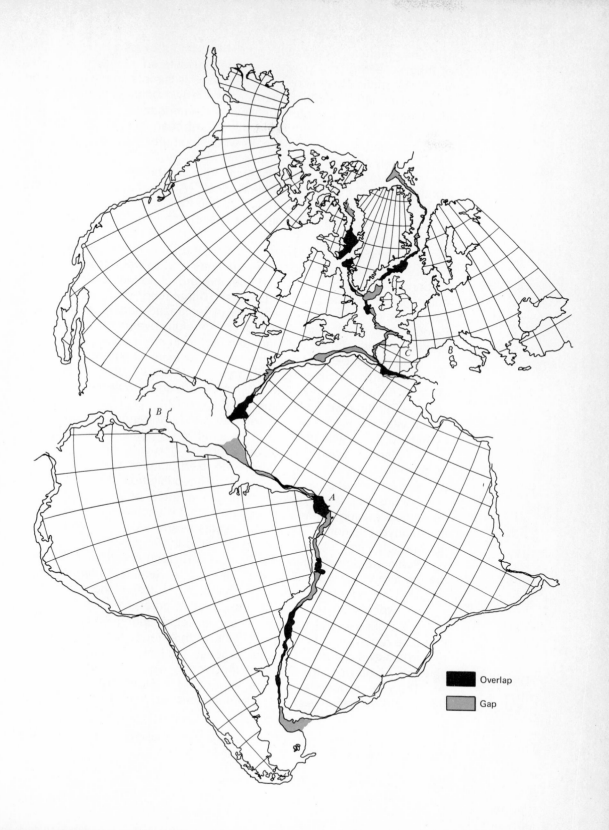

Overlap

Gap

Some difficulties are still apparent from Figure 12-1, however. One of them is a region (*A*) where Africa (in restored position) overlaps South America. This modest discrepancy is simply the delta of the Niger River, which built out long after the two continents had pulled apart. Region *B*, the junction between the Americas, is left blank on the map. Mexico, Central America, and the Caribbean islands seem to be a jumbled collection of minor blocks whose complex pattern of motion is only vaguely understood. Still another peculiarity of the restoration is the rotation of about 30° necessary to make the Iberian Peninsula (*C*) fit tightly against the Newfoundland shelf. Although this part of the scheme may appear somewhat forced, the indications of earth magnetism also suggest that Iberia actually has rotated in an arc, as the hands of a clock would do.

That evidence, an interpretation of the earth's ancient magnetic field (*paleomagnetism*), applies equally as well to whole continents as to Iberia. It becomes our second line of approach to the continental drift question. You will recall from Chapter 10 that magnetic pole positions can be calculated from an analysis of remanent magnetism in rocks. Paleomagnetic determinations must be carefully evaluated, as later disturbances have altered the magnetic record in many specimens.

Data from rocks that pass the laboratory tests for magnetic stability plot as a fairly consistent pattern on a global map. From the viewpoint of North America, around 2 to 3 billion years ago a pole lay in what is now the Pacific near South America (Figure 12-2). During the following billion years or so, the pole seemed to wander through three large serpentine loops, eventually to lie in the southwest Pacific at the beginning of the Paleozoic Era. Which actually moved—the pole, or the continent, or both? We cannot tell, for there is no fixed point of reference. What if we were to look at polar positions simultaneously from the viewpoint of *two* continents? If they were drifting, the trace of polar wandering throughout geologic time would be different for each continent because the relative position of all three subjects keeps changing. Whether or not the pole itself moves is not important to the argument.

Polar wandering curves for Europe and North America coincide at the modern North Pole (Figure 12-3). That is, remanent magnetism in recently formed rocks from both continents indicates the same pole, just as expected. The locus of more ancient pole positions traces out divergent curves. The part of the curves representing Paleozoic time could be reconciled if we were to shift North America (and its magnetic record) eastward about 30°. In the process, the North Atlantic, that great yawning rift between the Old and New Worlds, would be nearly closed.*

The *rate* of polar wandering, suggested in Figure 12-2, is instructive. Rapid migration (say, 3 to 6 centimeters per year) along a straight-

---

* However, the paleomagnetic data can be equally well explained by movements other than a westward drift of North America relative to Europe.

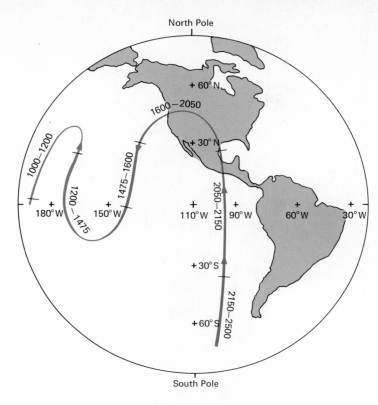

North Pole

+60°N

1600—2050

1000—1200

1200—1475

1475—1600

+30°N

2050—2150

+180°W   +150°W   +110°W   +90°W   +60°W   +30°W

+30°S

2150—2500

+60°S

South Pole

FIGURE 12-2 Rocks from North America contain a record of the changing position of the magnetic pole throughout the immense duration of Cryptozoic time. Ages (in millions of years) indicate approximately when the pole lay in a given region. [*After H. Spall, "Precambrian Apparent Polar Wandering: Evidence from North America,"* Earth and Planetary Science Letters, *vol. 10, North-Holland Publishing Company, Amsterdam, 1971.*]

away was followed by a dramatic slowdown for 200 million years or so, then a resumption of rapid motion in some different direction. Do these fits and starts actually indicate episodes of drift of North America? The periods of near stagnation at the bends of the hairpin turns were also times of intense orogeny. Does this signify that North America had collided with some other land mass?

Isotopic age patterns furnish a third line of evidence for continental drift. If continents are like pieces of a jigsaw puzzle, the age patterns are the picture printed on its surface. Both the pieces and the picture should reassemble into a continuous whole. The most decisive results are obtained where a large area of basement rock in a shield, characterized by a uniform isotopic age, is located next to another large region of the shield in which the age is sharply different. An excellent example of this is found in the Brazilian and West African shields. Here, the boundaries between the age provinces are perpendicular to the coast, and the continental shelves (an inaccessible part of the record) are narrow. In both regions, a belt that was metamorphosed around 600 million years ago cuts across a larger tract in which 2-billion-year ages are preserved. Boundaries between the age provinces cross smoothly from Africa into South America when the continents are placed together, as in Figure 12-4.

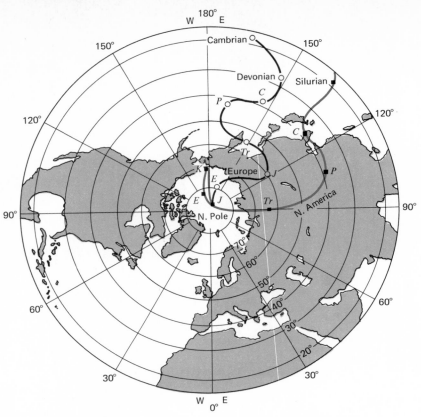

FIGURE 12-3 From the Cambrian Period to the present day, the magnetic pole has migrated generally northward toward its modern position. Apparent paths of polar wandering from the viewpoints of Europe and North America are not the same. A scatter of results has been averaged to a single point for each geologic period: C = Carboniferous, P = Permian, Tr = Triassic, J = Jurassic, K = Cretaceous, and E = Eocene. [*After Martin H. P. Bott, The Interior of the Earth, St. Martin's Press, Inc., Edward Arnold & Co., New York, 1970.*]

A fourth argument for the reality of continental drift is implicit in the distribution of fossils. They played an important part in the earlier debate led by Wegener and du Toit, but today we have far more convincing data. The argument here is based upon the occurrence of fossils of the same species of organism on two continents now separated by thousands of kilometers of ocean. (To be useful as evidence, the organism must not have been able to fly, swim, or drift across the ocean.) Fauna and flora living today are profoundly influenced by geography. For example, species of vertebrates are quite similar across the northern land masses where interchange is rather free. In South America (barely connected via the Isthmus of Panama), in Africa south of the Sahara desert, and in the Orient south of the great Himalayan barrier, we find other unique assemblages. Most distinctive of all is the

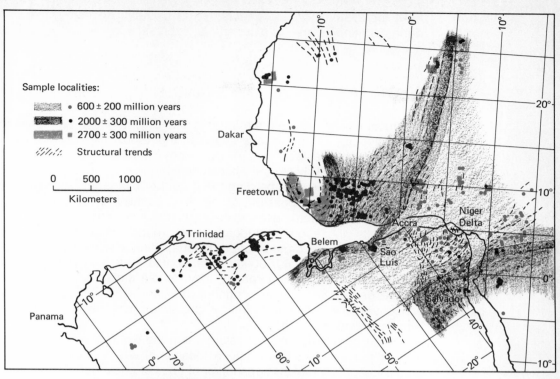

FIGURE 12-4 Isotopic ages in the Brazilian and West African shields signify times of metamorphism and orogeny. A once continuous mountain belt, traced by the distribution of 600-million-year isotopic ages, was broken apart by continental drift. In this regard, these ancient mountains are like the Appalachian-Caledonian chain mentioned in Figure 11-22. [*After P. M. Hurley and others, "Test of Continental Drift by Comparison of Radiometric Ages,"* Science, *vol. 157, no. 3788, pp. 495–500, 1967. Copyright 1967 by the American Association for the Advancement of Science.*]

totally isolated Australia with its odd assortment of marsupials.* (No one knows for sure how the marsupials got there.)

Abundant fossil remains of a grotesque mammallike reptile, *Lystrosaurus murrayi*, have been known for years from the Karoo beds of South Africa (Figure 12-5). And then, in 1969, the identical species was discovered in Antarctica! Both localities contain entire faunal assemblages (dominated by *Lystrosaurus*) that are nearly identical. Therefore the suture between the two continents must have been firmly knit over a great distance; it was not a tenuous connection like the Isthmus of Panama. The latter is more a "filter bridge" that permits some animals to cross readily while effectively excluding others.

When these and other approaches to the question are combined, the argument favoring continental drift becomes impressive indeed. Still,

---

* See Chapter 9 for a review of marsupials.

what force drives the continents apart? Could it be that entire wandering continents are but a superficial expression of some deeper phenomenon?

PLATE TECTONICS: THE GRAND HYPOTHESIS

The revolutionary concept that promises to draw together so many facets of geology began to take shape in the 1960s. Its essential detail now seems well established, even if many shortcomings of the theory call for further study. A key observation is the striking distribution of seismic activity around the globe (Figure 12-6). Modern seismological techniques make it possible to pinpoint earthquake epicenters with great precision. Nearly everywhere, epicenters are sharply confined to linear belts whose form and location are already familiar to us. Shallow earthquakes (in the crust or upper mantle) thinly trace the ocean ridge system throughout its 60,000-kilometer course. Dense clusters of points boldly outline the circum-Pacific; they denote the seismic zones that plunge deeply toward the continent from an offshore trench. A third concentration of epicenters follows the swirling trends of the Alpine-Himalayan mountain chain.

For a beginning, let us postulate that the weblike seismic pattern divides the earth's surface into a mosaic of seven immense plates, reminiscent of a cracked eggshell or of floes of pack ice* (Figure 12-6). Largest of these is the Pacific plate (*P*) bounded by the East Pacific Rise on the south, and by trenches around most of the remaining distance. This plate is almost entirely oceanic. In contrast, the African

* In addition, 20 or so minor plates are known.

FIGURE 12-5 Positioning of the external nostrils high on the skull suggests that *Lystrosaurus* (seen here in restoration) was a water-loving beast, perhaps like the modern hippo. Nonetheless, his aquatic habits scarcely would have equipped him to swim the thousands of kilometers of stormy waters currently separating South Africa from Antarctica! [*After* The Age of Reptiles *by Edwin H. Colbert. By permission of W. W. Norton & Company, Inc. Copyright © 1965 by Edwin H. Colbert.*]

plate (*A*) embraces parts of the Atlantic and Indian Oceans, and all of Africa. (Therefore the plates must be considerably thicker than oceanic or continental crust. Perhaps the base of a plate lies at a depth of 100 to 300 kilometers, in the low-velocity zone of the mantle.) Plate boundaries may even cut through a continent; the subcontinent of India is not part of Asia proper, but rather is attached to a plate (*I*) that also includes Australia and part of the Indian Ocean. Another example is western California (including Los Angeles), which is a thin sliver of continent attached to the Pacific plate, not the North American plate.

Clearly, the earthquakes are the result of recurrent motion—they show us that the "action" is confined mostly to the margins of a plate whose interior is relatively inert. We may further postulate that the motion of the plates is a lateral creep, as attested by the passive drift of continents embedded in them. Apparently the drift is taking place in random directions. Deformation and volcanism occur on a grand scale where plates are in contact. Geologists have applied the term *plate tectonics* to these interactions.

If these plates are in motion, what happens where they are in con-

FIGURE 12-6   New, refined data (a compilation of some 29,000 earthquakes reported between 1961 and 1967) emphasize how conspicuously narrow the seismic belts actually are. Volcanoes densely populate the seismic zones (see Figure 4-4). Except in Africa, shields are nearly devoid of both active volcanoes and seismic activity. Less than 6 percent of the earthquake energy is released in the ridge system. More than 90 percent is dissipated in the trench system. [*From M. Barazangi and J. Dorman, "World Seismicity Maps Compiled from ESSA, Coast and Geodetic Survey, Epicenter Data, 1961–1967,"* Bulletin of the Seismological Society of America, *vol. 59, no. 1, 1969.*]

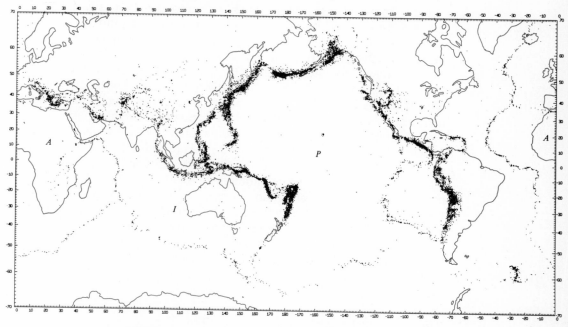

tact? How are they made, and destroyed? We could imagine three possible types of plate interaction. One sort is in progress where plate edges are pulling apart causing zones of tension. Zones of compression are located where plates are approaching one another. A third possibility is a zone in which two plates are slipping past one another by way of lateral, or shear, displacement.

Oceanic Magnetic Anomalies

Earth magnetism provides an important clue to the origin and ultimate destruction of these plates. When ships and planes equipped with magnetometers began to track the oceans, a peculiar magnetic pattern soon became evident. Plots of the earth's magnetic field strength along the surface prove to be smooth curves upon which numerous minor wiggles are superimposed, suggesting that two sources of magnetism are contributing to the field. In oceanic regions, about 97 percent of the total field strength (the smoothly varying part) is generated in the earth's core. Magnetic intensities of the residual 3 percent of the field vary irregularly from place to place. These small deviations, or magnetic anomalies, owe their origin locally, to permanently magnetized minerals in the basalt of the ocean floor.

An intensively studied segment of the Mid-Atlantic Ridge southwest of Iceland makes a good case study of the magnetic anomalies (Figure 12-7a). There, the anomaly pattern is organized into an astonishing regularity. Perpendicular to the crest of the ridge, the anomaly curves form mirror images on either side. Moreover, the highs and lows of each data-taking traverse across the ridge can be easily correlated with similar ups and downs in neighboring curves (Figure 12-7b). Later surveys established the symmetry of magnetic anomalies about the ridge throughout much of the world ocean. What could all this mean?

The answer to that question is probably the most shrewd interpretation in the Grand Synthesis, for it opened the door of understanding to all that was to follow. Ocean-floor basalt is a jumble of underwater flows laced by dikes and sills. In spite of the local complexity, broad strips of the ocean floor parallel to the ridge crest show a predominant direction of remanent magnetism (Figure 12-8). Basalt that congealed and acquired its magnetism during earlier episodes of earth history when the magnetic field was normally directed (like today's field) forms positive magnetic anomalies. Permanent magnetism of these rocks is added to (acts in the same direction as) the stronger magnetic field originating in the core. Located between these strips of basalt are other strips that froze during times when the magnetic field was reversed.* Remanent magnetism locked into these rocks is acting in a direction opposite to

---

* See Chapter 10 for a review of magnetic reversals.

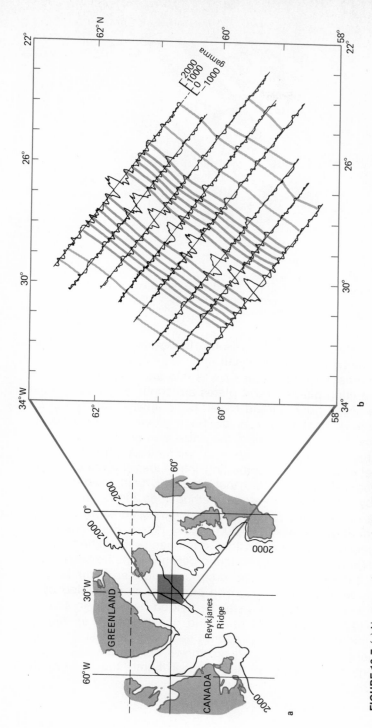

FIGURE 12-7 (a) Numerous traverses back and forth across the Mid-Atlantic Ridge (locally named the Reykjanes Ridge) were made in order to measure magnetic field intensities in the area of the square. Contours outline a water depth of 2000 meters.[*After F. J. Vine, "Spreading of the Ocean Floor: New Evidence," Science, vol. 154, no. 3755, pp. 1405–1415, 1966. Copyright 1966 by the American Association for the Advancement of Science.*] (b) Correlated profiles south of Iceland indicate that whatever is responsible for the magnetic anomalies is arranged in strips parallel to the ridge crest. A gamma is a unit of magnetic field strength. [*After M. Talwani and others, "East Pacific Rise: The Magnetic Pattern and the Fracture Zones," Science, vol. 150, no. 3700, pp. 1109–1115, 1965. Copyright 1965 by the American Association for the Advancement of Science.*]

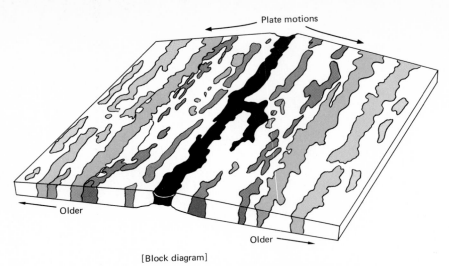

Plate motions

Older

Older

[Block diagram]

FIGURE 12-8 Normally and reversely magnetized strips of basaltic ocean floor run parallel to the ridge axis. In all, nearly 200 magnetic anomalies have been identified in the floor of the world ocean. Colored areas are normally magnetized basalt; clear areas are reversely magnetized.

the general field direction. The net result is to weaken the field strength above the reversely magnetized basalt.

We may further assemble these ideas into an interpretation of the motion of the drifting plates. These plates are continually pulling apart at the midline of the ridge. Liquid basalt wells upward into the crack, freezing there and "healing" the break. Each renewed opening of the rift tends to occur near the center of the ridge. Earlier-formed rocks are broken and drifted, half to each side, in a symmetrical pattern.

Our model of ocean-floor spreading introduces the critical ingredient of geologic time. We noted that magnetic field reversals have been dated by the potassium-argon method (see Figure 10-35). According to the model, the anomaly strip at the very center of the ridge was formed during the past 700,000 years, the modern interval of "normal" magnetic directionality. Older strips can be identified and assigned to an age of formation by counting outward from the central axis of the pattern. It follows that the speed of ocean-floor spreading equals the distance (from the ridge crest) of a particular strip, divided by its age of formation (speed equals distance divided by time). An analysis confirms that various regions of the world ocean spread at different rates, as we might anticipate. Spreading half-rates* as fast as 6 or 7 centimeters per year in some places are known, but more typical values are 1 to 2 centimeters per year. As a rule of thumb, the Atlantic and

* Measured away from the ridge crest on one side only.

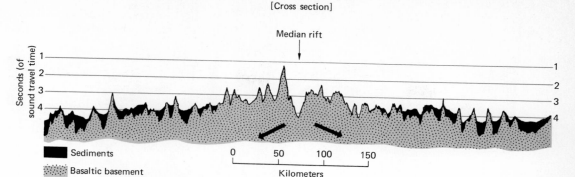

FIGURE 12-9 A sparker profile record across the Mid-Atlantic Ridge at latitude 44°N shows the increase in sediment thickness down the flanks of the ridge. Ocean-floor spreading (arrows) also explains the subsidence of guyots and atolls. As volcanic peaks drift laterally away from an elevated position at the ridge crest, they also plunge downward. [*After J. Ewing and M. Ewing, "I. Seismic Reflection," The Sea, vol. 4, part 1, Interscience Publishers, a division of John Wiley & Sons, Inc., New York, 1970.*]

Indian Oceans are widening, during a person's lifetime, a distance roughly equal to his body height. (The Pacific is shrinking, and will disappear in 200 million years unless the tectonic pattern changes.) About 2.6 square kilometers of new crust are generated throughout the length of the ocean ridge system each year.

Furthermore, since the oceanic crust becomes older away from a ridge, only thin layers or local pockets of recent sediments have accumulated at its very crest. In a direction perpendicular to the crest, the sedimentary carpet should become somewhat thicker, and older and older sediments should be located at the base of the strata (Figure 12-9). In general, data from sediment cores recovered from many parts of the world ocean are compatible with this prediction.

Subduction Zones

If the earth maintains a constant size (as evidence strongly suggests), then plates must be destroyed elsewhere at the same rate that they are created in the ocean ridge system. What happens where migrating plates converge? One possibility is a stupendous head-on collision: an interaction that seems exactly what we need to account for the horizontally directed forces that so evidently deformed the structures of fold mountains. India and Asia today are engaged in just such a confrontation along the Himalayan "battle front."

More commonly, one of the plates averts a direct collision by plunging abruptly into the mantle. Where the sinking plate noses down, the surface of the crust sags into an oceanic trench (Figure 12-10). Dipping planes of earthquake foci trace the sustained creep of some

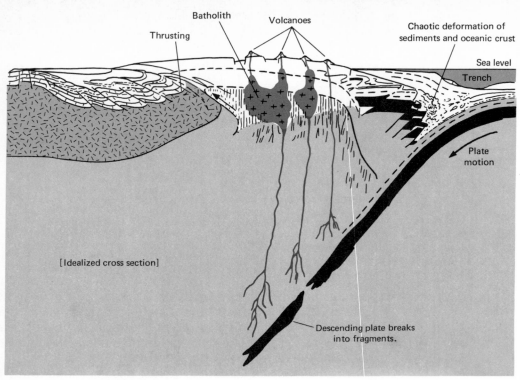

FIGURE 12-10 Where an active subduction zone dips beneath a continent, we find thrusting (and eventual disappearance) of a plate at great depth, generation of batholiths (some feeding upward to volcanoes), and fold mountains in the making. None of these geologic processes is very well understood. [*After J. F. Dewey and J. M. Bird, "Mountain Belts and the New Global Tectonics," Journal of Geophysical Research, vol. 75, no. 14, p. 2638, 1970. Copyright by American Geophysical Union.*]

TABLE 12-1
Birth and Demise of Oceans*

Stage	Example	Remarks
Embryonic	East African rift valleys	Domelike uplift, down-dropped fault blocks, deep lakes
Youthful	Red Sea	Trough-shaped rift occupied by ocean water
Mature	Atlantic Ocean	Well-developed mid-ocean ridge
Declining	Pacific Ocean	Trenches, island arcs, bordering fold mountains
Terminal case	Mediterranean Sea	Greatly diminished area, fold mountains
Deceased	Indus line, Himalayas	Ocean floor entirely subducted from view

* Every stage in the life cycle of an ocean basin is illustrated by modern examples.
SOURCE: J. T. Wilson, "2. Continental Drift, Transcurrent, and Transform Faulting," *The Sea*, vol. 4, part II, Interscience Publishers, a division of John Wiley & Sons, Inc., New York, 1970.

plates to depths of hundreds of kilometers. Oceanic trenches, then, are far more than mere depressions. They are the surface expression of *subduction zones* into which a plate disappears relentlessly into the depths of the earth.

Now we may understand the youth of oceanic crust and the extraordinary thinness of its sedimentary blanket in a new light. Oceans are a *dynamic* system that is continually devoured and replenished (Table 12-1). Continents, too, are constantly renewed because of activity in subduction zones. Eroded continental debris dumped into the trenches is carried downward, under the continent. However, sediment, because of its low density, tends not to go down very far. As it is rammed against the continent, it accumulates on the underside—it becomes the thick-

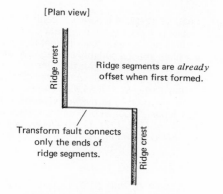

[Plan view]

Initial state

Ridge crest

Ridge segments are *already* offset when first formed.

Transform fault connects only the ends of ridge segments.

Ridge crest

A rupture has formed, but no movement of rocks on the two sides of the rupture has occurred.

FIGURE 12-11 A transform fault accompanies the growth of new crust as ocean floor spreads away from the ridge crest. With passing time, a particular strip of ocean basalt migrates, but the offset ends of ridge segments maintain a constant relative position.

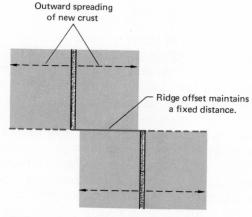

Outward spreading of new crust

Ridge offset maintains a fixed distance.

At a later time

Motion on the fault has created a displacement.

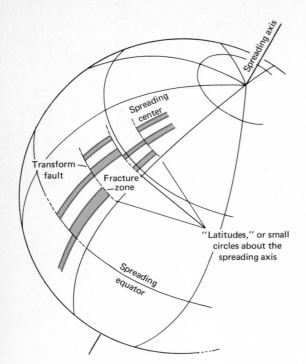

Spreading axis

Spreading center

Transform fault

Fracture zone

"Latitudes," or small circles about the spreading axis

Spreading equator

FIGURE 12-13 First-motion data from earthquakes have given a direct confirmation of the sense of movement of the plates. Earthquake-slip vectors (small arrows) indicate the stately drift of plates from their birthplace in the ocean ridge system to their burial ground in the trenches. Bold arrows summarize these motions. [*After B. Isacks, J. Oliver, and L. R. Sykes, "Seismology and the New Global Tectonics,"* Journal of Geophysical Research, *vol. 73, no. 18, p. 5861, 1968. Copyright by American Geophysical Union.*]

FIGURE 12-12 A theorem of geometry states that motion of a rigid plate in any direction across the surface of a sphere is equivalent to a simple rotation of the plate about a suitably chosen "spreading axis." Plate tectonic theory thus explains why fracture zones in the Atlantic follow concentric small circles (see Figure 11-8b). [*After K. S. Rodolfo, "Contrasting Geometric Adjustment Styles of Drifting Continents and Spreading Sea Floors,"* Journal of Geophysical Research, *vol. 76, no. 14, p. 3273, 1971. Copyright by American Geophysical Union.*]

ened "root zone" beneath a system of fold mountains. Friction between the subducted plate and the enclosing mantle material creates sufficient heat to melt pockets of granitic magma. This material, also of relatively low density, penetrates upward through the (metamorphosed) sediments, forming batholiths. A dense residue of the plate continues to plunge deeper into the mantle.

Transform Faults

A third component of plate tectonics is the shear zone that accommodates slippage of a migrating plate past its neighbors. Such zones are called *transform faults* because they connect various combinations of ridges and trenches; the fault transforms into something else at its two ends. The discovery of these faults was a triumph of logical reasoning. When the Canadian geophysicist J. T. Wilson postulated them in 1965, he knew they "had" to exist because the geometry of plate motion demands their existence. Today, these faults are identified with the abundant fracture zones in the ocean floor, and with a few land faults such as the San Andreas in California. Transform faults have some special properties that can explain the peculiar seismic data mentioned in the preceding chapter (Figure 12-11).

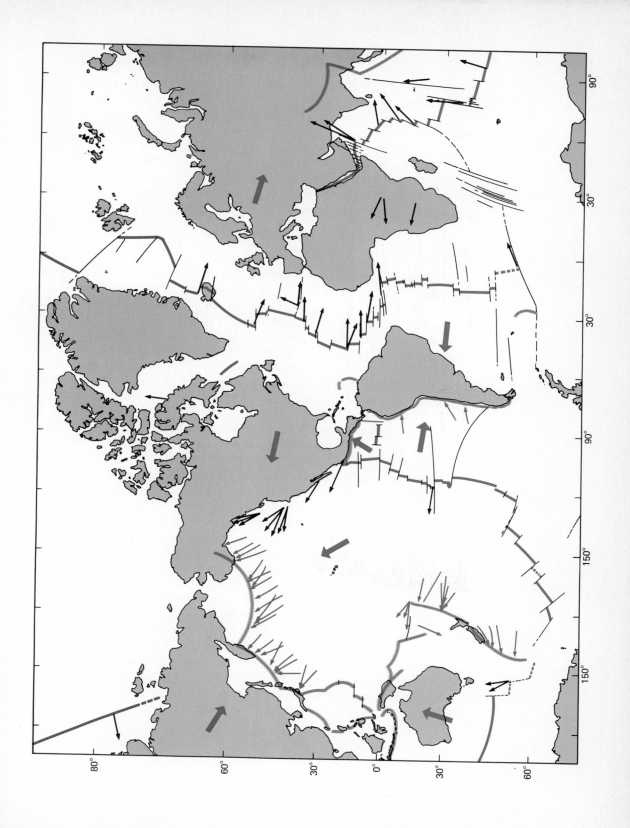

THE GREAT HEAT ENGINE

The development of our theme has been like peeling off layer after layer of explanation. First we examined surficial details (deep-ocean topography, arrangement of continents and mountain ranges, etc.), then we used these data to support the continental drift hypothesis. Continents and ocean floor, in turn, are carried along on the creeping plates. But how do we visualize the return flow of mantle material beneath the plates? How is the substance of the earth recycled back to the ocean ridge crest? What propels the tectonic machinery, and how? Unfortunately, our understanding here is far from complete.

Heat Flow

Certainly the primary energy is the earth's quota of *internal heat*. The earth is a great "heat engine," drawing upon this energy resource for power to do various things. At first sight, the heat released from within appears to be trivial; it is only 1/4,000 of the heat received from the sun. In spite of this disparity between the inward and outward sources of energy, the surface radiates heat away so effectively that the solar input is not able to penetrate deeper than 100 meters or so. Only that insignificant trickle of heat from the interior (sustained over millions of years) is available to be converted into the mechanical energy of plate movement.

Heat-flow values are obtained through a measurement of the temperature gradient in deep holes on land (or by a special probe thrust into the ocean sediments), coupled with data on the heat conductivity of the material. Geophysicists naturally assumed that heat flow through the continents would greatly exceed the flux emerging through the ocean floor. They knew that continental crust is highly enriched in radioactive elements (a potent heat source) relative to oceanic crust. How unpleasantly surprised everyone was to discover that, on a world-wide average, there is scarcely a difference in heat flow between the two—about 1.8 microcalories* escaping per second through each square centimeter of either type of crust. This unexpected result is a puzzle for which we yet have no fully satisfactory solution.

More importantly, perhaps, the heat flow diminishes from high values at the crest of an oceanic ridge to progressively lower values on the flanks. Lowest of all is the flux of heat through a trench floor. (Do these observations lend support to the plate tectonic theory?) According to a recent calculation, as much as 45 percent of the earth's heat budget may be expended in creating and consuming the plates. If so, the earth is a remarkably efficient heat engine indeed.

---

* Input of one calorie to a gram of water raises its temperature one degree Celsius. A typical square centimeter of the earth conveys a year's accumulation of about 57 calories. That quantity is not sufficient to warm even a cup of tea.

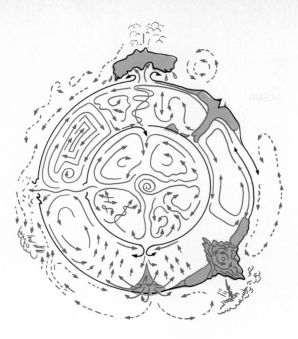

FIGURE 12-14 This geological cartoon summarizes the confusion of current ideas about the nature of deep internal motions in the mantle. Which flow pattern, if any of them, is correct? This fundamental question is under intensive study today. [*After R. S. Dietz and J. C. Holden, "Continents Adrift: New Orthodoxy or Persuasive Joker?" in* NATO Symposium on Continental Drift, *vol. 2, Academic Press, Inc., New York, 1974.*]

Convection

How can heat be transformed into mechanical energy? If heat were distributed unevenly within the deep earth, it seems certain that an unstable situation would result. A relatively warm portion of the mantle would expand, and because its density has decreased, it would buoy up toward the surface somewhat like an ascending plume of smoke. (The thought of "solid" rock flowing should cause us no difficulty at this point.) Model calculations suggest that a density change of only 1 part in 10,000 is sufficient to start the ascending motion. After the material has discharged its excess heat to the surface, it will descend again to complete an overturning, or convective, motion (Figure 12-14). By this interpretation, an ocean ridge is the region situated over the ascending limb of a convection cell, and a subduction zone occupies the descending limb.

Whatever the pattern of convection in the mantle, there can be no simple one-for-one relationship of surface flow of the plates to the return flow of material at depth (Figure 12-15). That is because there are more regions where new crust is being created than where crust is being consumed. For example, ocean ridges encircle Antarctica, and no subduction zone intervenes between the ridge and the continent. Since the ridge is a site of ocean-floor spreading, the geometry of motion demands that the ridge crest be pushing away from Antarctica in all directions. So not only do continents drift—everything is in motion!

More accurately, one of the plates could stay in a fixed position

373 THE GREAT HEAT ENGINE

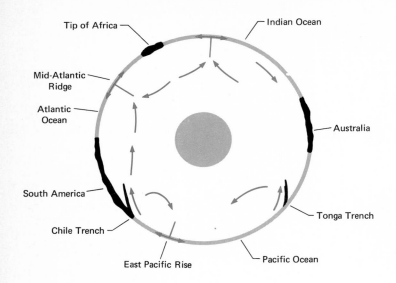

[Cross section]

Tip of Africa

Indian Ocean

Mid-Atlantic Ridge

Atlantic Ocean

Australia

South America

Chile Trench

Tonga Trench

East Pacific Rise

Pacific Ocean

FIGURE 12-15 A slice through the earth approximately at latitude 30°S intersects three ridge systems, but only two subduction zones. This inequality demands a flow pattern that is *at least* as complex as that shown by the arrows.

while all the others are moving. Perhaps the pattern of eruption that gave rise to the Hawaiian Islands can provide information on these motions. These islands are situated at the end of a long chain of volcanic islands and submerged seamounts stretching 5000 kilometers to the edge of the Aleutian Trench (Figure 12-16). Potassium-argon ages from dredge-haul material are oldest near the trench, becoming younger toward the Hawaiian chain. Kilauea, at the very end, is active today. This pattern would result if the Pacific plate were slowly drifting over a *hot spot*, a small area through which basaltic lava is fed upwards from the deep mantle. If the hot spot remained deep and fixed, the plate would carry the chain of volcanoes "downwind" from the hot spot, like smoke trailing from a chimney.

In this interpretation, Africa is situated upon the one plate that remains stationary. Volcanism above the African hot spots has created gigantic localized piles at the surface, not long strings of volcanoes as in the Pacific.

LOOKING BACKWARD AND FORWARD IN TIME

Our vision of the changing face of the earth gives us a certain fascinating insight into its appearance in both the past and the future. However, we would be ill advised to extrapolate too far into the distant recesses of time. Magnetic anomaly patterns suggest that fully half the

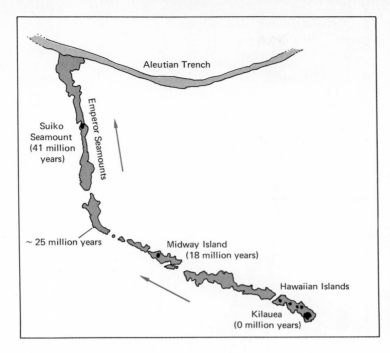

FIGURE 12-16 Earlier than about 25 million years ago, while the Emperor Seamount chain was being built, the Pacific plate drifted nearly due north. Since that time, the plate has drifted in a northwesterly direction. This abrupt change in ocean-floor spreading direction is signaled by a sharp "elbow" in the chain of seamounts.

ocean crust has been created during the Cenozoic Era, the latest 1.5 percent of earth history (Figure 12-17). We have no secure information about world geography earlier than the assembly of the great supercontinent (or perhaps, two supercontinents) toward the end of the Paleozoic Era. The Mesozoic and Cenozoic Eras have witnessed the breakup and scattering of the supercontinent, though not yet to the point that a "land hemisphere" is unrecognizable. Among the more spectacular events during the next 50 million years will be a noticeable widening of the Atlantic Ocean and the Red Sea. India will continue to be rammed beneath the Asian plate, and a slice of California west of the San Andreas Fault will become detached from the mainland.

Several inadequacies of plate tectonic theory, as presently understood, are apparent at this point. For example, metamorphosed slices of ocean floor are incorporated into many fold mountain systems. How can dense crust, seemingly poised for a long descent into a subduction zone, be carried upward to high elevations? How are descending plates reincorporated into the mantle? Why do sediments in the bottom of all active trenches appear not to be compressionally deformed? How can such enormous, but comparatively thin, plates travel such vast distances without shattering or buckling? Geologists will attempt to answer these and many other questions in the years to come.

375 LOOKING BACKWARD AND FORWARD IN TIME

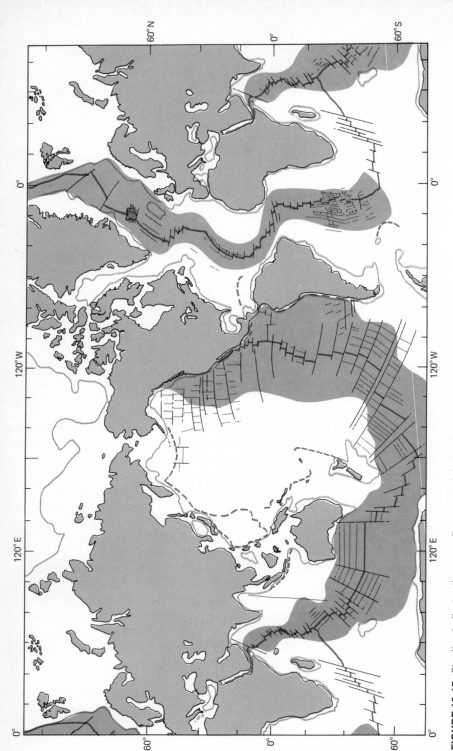

FIGURE 12-17 Shading indicates the ocean floor created during the past 65 million years. Ocean-floor spreading is about 3 times as rapid in the South Pacific as in the Atlantic. A segment of the ridge system south of Africa appears to be dead or dormant. Note that off the west coast of North America, less than half the magnetic anomaly pattern is preserved. Presumably, North America has overridden not only the nearby ridge, but also a portion of the ocean floor that lay originally on the far side of the ridge. [*After F. J. Vine, "The Geophysical Year," Nature, vol. 227, no. 5262, 1970.*]

FURTHER QUESTIONS

1 Refer to Figure 12-1. Iceland (between Greenland and Europe) and the Bahama Islands (east of Florida) are not shown. (They do not fit well into the continental reconstruction.) Research the geology of these islands, and determine their origin. Can you reconcile these data with an origin *after* continental drift had opened the Atlantic?

2 Estimate the width of the South Atlantic. If the spreading half-rate has stayed constant at 2 centimeters per year, how long ago did Africa and South America rift apart? In what geologic period did the breakup take place (refer to Figure 8-22). Does your answer agree with fossil evidence that *Lystrosaurus* roamed throughout the Southern Hemisphere in Early Triassic times?

3 Refer to Figure 12-10. The composition of basalt that erupted in Japan and other island arcs systematically changes along a direction away from the trench. Can you postulate how this observation might be connected with the increasing depth of the subducting Pacific plate underneath Japan?

4 The age of the earth, believed to be 4.6 billion years, has never been determined from any terrestrial rock. Is the absence of the most ancient rocks compatible with plate tectonic theory? Explain how.

5 Why is there a land hemisphere and a water hemisphere?

6 Compare the observation of Figure 11-9 with the explanation of Figure 12-11. Are these two compatible?

7 Compare the observation of Figure 11-8*b* with the explanation of Figure 12-12. Are the two compatible?

8 Refer to Figure 12-16. Why does the Emperor Seamount chain appear to end at the Aleutian Trench? Why does it not continue across the trench farther north into the Bering Sea?

9 Refer to Figure 12-16. Suiko Seamount, roughly 40 million years old, is 5000 kilometers from Kilauea, in the Hawaiian Islands. If both volcanoes formed during the drift of the Pacific plate over a fixed hot spot in the mantle, what has been the average speed of plate motion?

READINGS

Colbert, E. H., 1973: *Wandering Lands and Animals*, E. P. Dutton & Co., Inc., New York, 323 pp.

Marvin, U. B., 1973: *Continental Drift*, The Smithsonian Institution, Washington, 239 pp.

* Various authors, 1972: *Continents Adrift* (Readings from *Scientific American*), W. H. Freeman and Company, San Francisco, 172 pp.

* Available in paperback.

chapter 13

Depositional Systems

Point bars on the meandering Colorado River, Texas. [*University of Texas at Austin, Bureau of Economic Geology photograph.*]

The remainder of this book will focus mostly upon geologic processes at work upon the earth's surface. Our outlook on geologic time will sharpen also. Although these processes may have acted during most of earth history, the record they have created was best preserved over the last few million years. The nearness of the perspective naturally brings us to the role of geology in human affairs. Most of our fuel and mineral resources are located in the sedimentary rocks that overlie 70 percent of the continents. Some day soon, perhaps, we shall mine such resources from the sedimentary blanket that veneers an even greater proportion of the ocean floor. Modern exploration, spurred by a growing urgency to find more oil and gas, has revolutionized our understanding of the origin of sedimentary rocks. In some ways, the rapid progress of the science of sediments is like the exciting developments that led to plate tectonic theory. After slowly maturing for many years, suddenly both subjects were revitalized in the 1950s, and both have been organized into a Grand Synthesis with important economic overtones. This chapter is devoted to the *terrigenous clastics*—sediment made of particles derived by weathering and erosion of the continents. To draw this subject together, we shall trace the progress of clastic sediments down the Mississippi River to its mouth, and out into the Gulf of Mexico.

DEPOSITIONAL PROCESSES

Deposition can take place under diverse circumstances in a variety of environments. Examples are the *fluvial* (river) environment, which merges into the *delta* environment, where the stream dumps its sediment load (Figure 13-1). These two environments make up a "dip-fed" depositional system—that is, one where gravity is dominant in moving the sediment downslope. Upon reaching the coast, sediment may be distributed along the shore in a "strike-fed" depositional system. Strike is defined as the orientation of a horizontal line drawn on the surface of

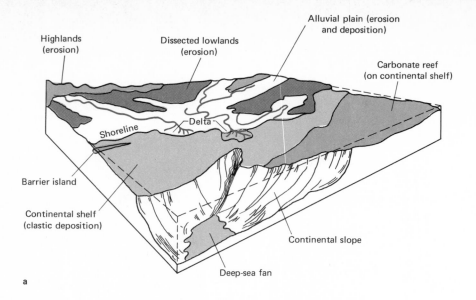

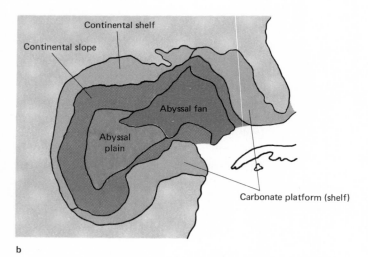

FIGURE 13-1 (*a*) Local environments of deposition or erosion (a few of which are shown in this schematic block diagram) each consist of a mosaic of related subenvironments. Recently completed maps of sedimentary geology along the Texas coast recognize 135 distinct kinds of environment. [*Courtesy Howard Gould and Esso Production Research Company.*] (*b*) Other distinctive environments, not very well studied as yet, are characteristic of the ocean. This map identifies some of the major environments in the Gulf of Mexico. [*University of Texas at Austin, Bureau of Economic Geology.*]

an inclined plane. The dip direction is down the plane, perpendicular to strike. Sediment drifted along laterally in a strike-fed environment has accumulated as beaches, dunes, and other shoreline features along the Gulf Coast.

The open Gulf contains yet other depositional environments. Terrigenous clastics are deposited from the muddy waters along the north and west coasts, whereas on the east and south are regions of clear water where limestone is accumulating. Other depositional systems, known elsewhere, are not important in the Gulf of Mexico. Among these are the glacial, the lacustrine (lake deposits), and the eolian (wind-blown sediment prominent in deserts). We should keep in mind that uplift and erosion may lay bare any kind of sediment of any age on the surface of the land. A task of the geologist is to interpret the environment in which ancient sedimentary rocks originated.

What clues are present to guide his interpretation? Was a particular sandstone body deposited in a stream channel, in an advancing delta front, as a wind-blown dune, on a beach? Finding the answer demands careful study not only of the sandstone, but of all the surrounding rocks. Even more important, the ancient record can make sense only if we understand modern streams, deltas, dunes, and beaches. The big breakthrough in the 1950s came when geologists began to take seriously Hutton's famous dictum, "The present is the key to the past." Today, certain of these modern environments have been mapped in vivid three-dimensional detail. Surface observations are supplemented by drilling of 60,000 new wells that, all told, penetrate 75,000 kilometers of the subsurface each year in the United States alone!

As noted in Chapter 6, the central interior of North America is a plain (or more accurately, a wide trough) gently sloping toward the Gulf. Drainage in this interior lowland is funneled mostly into tributaries of the Mississippi, which alone carries about 40 percent of the runoff from the continental United States.* A case study of this terrain is profitable for several reasons. Through drilling and surface mapping, it has come to be known better than any other area of sedimentary deposition. Continuous observations of the Mississippi River have been carried out for more than 200 years. Moreover, the Gulf of Mexico is one of the oldest oceanic basins still surviving. Because the Gulf region is so stable, deposition has gone on ever since Late Mesozoic times without any disturbance except for adjustments due to piling on of sediment. About half the volume of the original Gulf cavity has been filled in with sediments. Thus our theme is to be the story of sand and mud!†

---

* By far the largest drainage area of any river belongs to the Amazon in South America. The Amazon's discharge (volume of water delivered per unit time) exceeds the combined discharge of the eight next largest rivers. Area of the Mississippi's drainage basin is third in size, but in average discharge the Mississippi ranks a poor seventh.

† Why just sand and mud? As Figure 5-4 shows, sedimentary grains come in any size from submicroscopic particles to giant boulders, but sand- and mud-sized particles are the most abundant.

A grain of sand on its downward path to the sea travels first through a zone of erosion, later to reach an area of deposition. Normally, the soil cover is not part of a depositional system, for soil and other loose debris that mantle a hill slope are in a transient state of flow toward a stream channel. On the other hand, neither are sedimentary rocks stored permanently as such, for they may be uplifted, only to participate in the cycle of erosion and deposition all over again. Fluvial environments are the transitional region, where erosion and deposition are in constant interplay. Sediment transport is greater and sediment storage is less in the fluvial system than in any of the other environments of deposition. In general, we may distinguish three kinds of time scale in these processes. A particle may take thousands of years to creep or wash downhill to a channel. Once in a major river system, it may need millions to tens of millions of years to reach the ocean. Deltaic and marine sediments may stay hundreds of millions of years before uplift and erosion have exposed them on the dry surface once again.

Large river systems consist of two parts. Streams join together to form ever larger streams in the *contributive net*, which occupies the upper reaches of the basin (Figure 13-2). Tributaries are united together

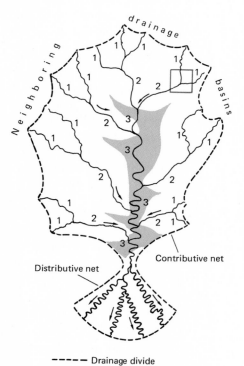

FIGURE 13-2 On this map of a contributive and distributive net, first- and second-order streams merge into the master channel, a third-order stream. Tributaries form a dendritic pattern that resembles the trunk and branches of a tree (the Greek *dendron* means "tree"). A shaded area outlines that part of the basin in which precipitation runs directly into the master stream. It raises the question of effective flood control: Should we impound flood waters behind one very large dam far downstream, or retard them with many small dams placed higher up on the tributaries? The latter solution is definitely helpful, but in view of the large area that feeds directly into the master stream, it appears that *both* kinds of dam are needed. [*After J. R. L. Allen, "A Review of the Origin and Characteristics of Recent Alluvial Sediments,"* Sedimentology, *vol. 5, no. 2, 1965.*]

Distributive net

Contributive net

– – – – Drainage divide

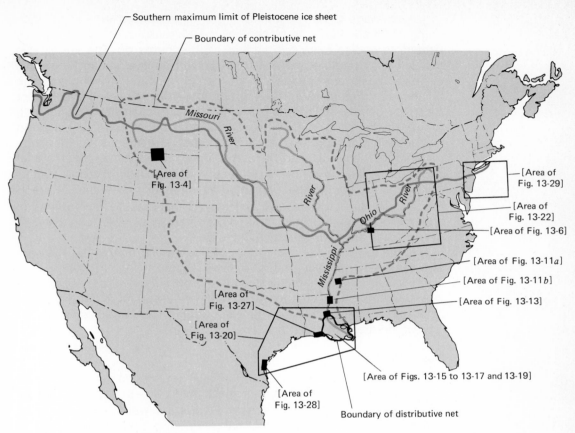

FIGURE 13-3 An index map locates the Mississippi River, its two largest tributaries, and the maximum southward encroachment of an ice sheet that repeatedly occupied the upper Mississippi Valley. The close fit of the ice limit to the modern Ohio and Missouri Rivers suggests that these valleys were eroded when meltwater surged along the margin of the ice.

like the branches and main trunk of an immense, two-dimensional tree. Suppose we designate all the streams that have no tributaries as being of first-order rank. Second-order streams are those that have only first-order tributaries. Third-order streams receive only first- and second-order tributaries, and so forth to higher orders. Since three or four smaller streams typically drain into each larger stream, the number of streams rapidly increases as we go to lesser orders. On this basis, the Mississippi is a tenth- or eleventh-order stream (depending upon the detail of the estimate) whose principal tributaries are the Missouri and Ohio Rivers (Figure 13-3). The entire Mississippi network includes approximately 300,000 streams, of which first-order channels comprise about two-thirds of the total. These smallest of tributaries average only a kilometer or two in length.

A drainage basin becomes larger by headward erosion (lengthening of tributaries) at its fringe. For various reasons, one tributary may extend its course more rapidly than its neighbors do. The aggressive tributary may capture and divert the flow of another stream; in fact, a drainage basin can invade a large portion of a neighboring basin in this manner (Figure 13-4). Thus a river is not the same age everywhere along its course. As a tree grows in both directions by putting forth branches above and roots below, the Mississippi system has extended its outreach headward ever since the shallow Cretaceous sea that once covered the central United States withdrew. At the same time, material deposited at the mouth has built the land oceanward (a process called *progradation*), forcing the river to wander farther and farther to reach the sea. Thus the Mississippi system is a respectable several tens of millions of years old, but some other rivers that have long drained shield areas are undoubtedly much older.

At the lower end of many large rivers is a *distributive net*, the second major component of the system (Figure 13-2). In this region, smaller *distributary* channels split off and fan out over the delta, repeatedly dividing the discharge. A distributive network is invariably much smaller than the contributive network; in the Mississippi system, the ratio of the areas is about 1:50.

Fluvial Regimes

Although every river, great and small, experiences periods of flood and of low water, the pattern of these intervals—the regime* of the river—varies from stream to stream. Floods are well-known agents of geologic change, but events that happen during the less vigorous, but longer, intervals of normal flow are also known to be effective. Even in a basin where climate and the geology of the bedrock are uniform, the rise and fall of water level in small streams tend to be more abrupt than in larger streams. A local summer thundershower over a creek drainage in the Mississippi Valley can create a severe flash flood that subsides just as quickly as it rose. When the flood pulse reaches the Mississippi, the effect is negligible, for the hydrologic inertia of the master stream is tied to thousands of similar small basins in which other things were happening on that same summer day. Even if a single storm were to deluge the entire Mississippi basin, the runoff pattern would be smoothed out because one of the fallen raindrops may take only a minute, but others as much as a number of weeks, to reach the Gulf. In short, a small stream is sensitive to day-by-day variations of the *weather*, whereas the regime of a giant river reflects the slower rhythms of changing *climate*.

---

* *Regime:* a systematic procedure of a natural process. Here, regime refers both to the fluvial environment and the processes that shape it.

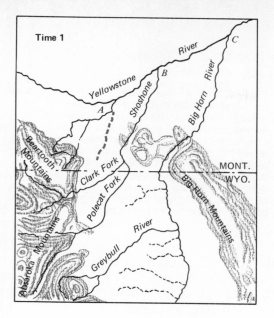

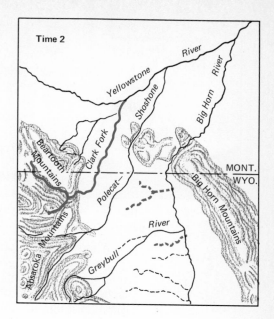

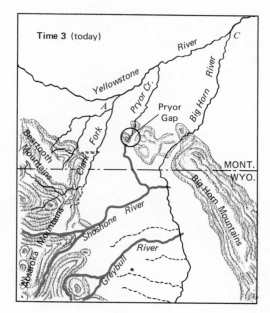

– – – – – Stream about to make a capture

——— Captured stream

FIGURE 13-4 In a rugged terrain, especially during early stages of erosion, the streams may not yet be adjusted to stable drainage basins. Mountain streams in parts of Montana and Wyoming sometimes flow through canyons cut into massive crystalline rocks, but elsewhere they cross low basins floored by weakly coherent shale and sandstone. Minor tributaries (in color) draining these easily eroded sediments rapidly extended headward, capturing and diverting much larger streams. The net result was an integration (merger) of drainages of three major streams (A, B, C) into two streams (A, C). Pryor Creek today is a ridiculously small stream where it trickles across the huge valley of Pryor Gap. [*After J. H. Mackin, "Erosional History of the Big Horn Basin, Wyoming,"* Geological Society of America Bulletin, *vol. 48, 1937.*]

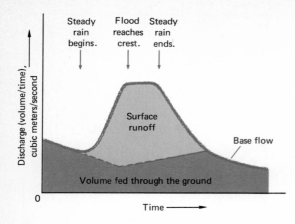

FIGURE 13-5 A simple hydrograph shows how stream discharge responds to a storm that begins and ends abruptly. The area under the curve corresponds to volume per unit time, times the total time, or total volume of water. It can be "decomposed" into a subarea that represents both surface runoff and water that enters the stream via percolation through the ground. A decision on how the boundary should be drawn may be difficult, but it is legally important because in many places the groundwater belongs to the private citizen, but the runoff belongs to the state.

A stream's response to presence and absence of input is usefully pictured by a *hydrograph:* a plot of stream discharge at a given point within a specified time (Figure 13-5). Once a rainstorm has started, the discharge begins to climb, but slowly at first because some of the precipitation infiltrates the soil, and time is needed for the flood crest to have traveled downstream to the gauging station. Eventually a steady state (flat top) is reached when the soil is saturated and all the precipitation runs off. After the storm, the hydrograph returns to the base flow, or low-water stage, during which the stream is fed only by water infiltrating through the ground.

Differences between the behavior of large and small streams are nicely illustrated by hydrographs of the Ohio River at Louisville, Kentucky, and Beargrass Creek, one of its small tributaries. The extremely "spiky" hydrograph for Beargrass Creek during selected periods of 1950 reflects the small area and limited storage capacity of the basin (Figure 13-6a). A corresponding hydrograph for the Ohio River (Figure 13-6b) is notably more broad and less abrupt, primarily because the

FIGURE 13-6 (a) Beargrass Creek drains a 48-square-kilometer basin at ▶ Louisville, Kentucky. Surface runoff depends upon a complex relationship among timing of the rainfall, size and shape of the basin, vegetation, saturation of the soil, and many other variables. And yet, predictions of the characteristics of the hydrograph of a particular stream are vital to the proper designing of bridges, flood-control channels, road elevations, and the like. In an attempt to understand the controlling factors, Stanford University hydrologists generated a computer simulation (broken curve) that fits the observed discharge moderately well. [*After N. H. Crawford and R. K. Linsley, "Digital Simulation in Hydrology: Stanford Watershed Model IV,"* Technical Report No. 39, *1966.*] (b) A gauging station on the Ohio River 8 kilometers downstream from the mouth of Beargrass Creek provided data for this hydrograph during the same periods of 1950. On May 10, a local storm created a huge spike on the Beargrass hydrograph, but hardly a ripple on the Ohio River hydrograph. [*Data from* U.S. Geological Survey Water-Supply Paper 1173, part 3, Ohio River Basin, *1953.*]

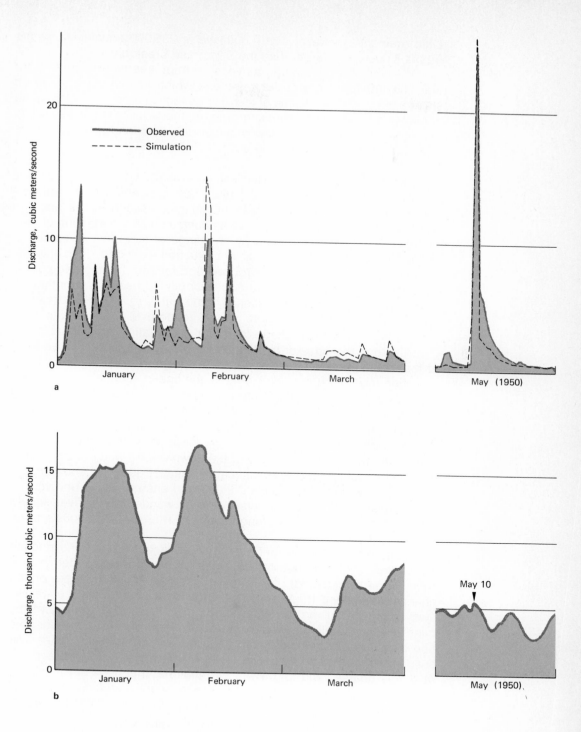

Ohio drainage upstream from Louisville is 236,000 square kilometers (about 5,000 times larger than the Beargrass Creek basin).

The flow of water through a river system proceeds very much faster than the transport rate of the sediment, which travels via a series of starts and stops. The regime of the river governs the configuration of the sediment en route and how it is moved. In this regard the Mississippi is an interesting case, for its deposits record a drastic change in regime that took place a few thousand years ago.

The Pleistocene braided Mississippi At that time, an ice sheet* had overrun the northern part of the United States, pushing south approximately to the position of the present Ohio and Missouri Rivers (see Figure 13-3). Storage of such massive amounts of water in the form of continental ice was accompanied by shrinkage of that ultimate source of water, the ocean. The effects of this situation extended deep into the continent, for as the shoreline dropped and retreated to the edge of the continental shelf, the ancestral Mississippi's course was lengthened. This in turn induced an adjustment in the *long profile* of the stream. A cross section down the length of the modern Mississippi (its long profile) is typical of the shape found for nearly all rivers (Figure 13-7).

Although the base of the Mississippi's channel is below sea level at the mouth, the river system cannot excavate the land surface to an elevation lower than sea level. That is, the ocean surface is a *baselevel* that establishes the approximate position of the lower end of the long profile. Just why the long profile should be concave upward is not well understood, as the shape depends on at least eight variables, such as stream depth, velocity, slope, sediment load, etc. One suggestion takes note of the rate at which energy of the downflowing water is dissipated in creating turbulence and, ultimately, heat. Equations show that a concave-upward profile results if the rate of energy loss is uniform per unit of river length. But of course we may wonder why a river should spend its energy uniformly along its length. There is no obvious answer to this question.

Lowered sea level during glacial advances in the Pleistocene Epoch was a disturbance to which the Mississippi responded by *entrenching* (downcutting) its valley. Consequently, the slope of the channel steepened (Figure 13-7). During the summers the ice wasted away rapidly, causing torrential floods of meltwater to spill down the river system. So violent was the turbulence during flood periods that fine clay and silt particles were kept in *suspension* in the water. The suspended sediment load was swept on to the Gulf, whereas the *bedload*—coarse sand to gravel-sized material that hops or rolls along the channel bottom—lagged behind. Continually shifting islands of coarse debris choked the river channel, forcing the water to wander

* Glaciers and the Ice Ages are discussed in the following chapter.

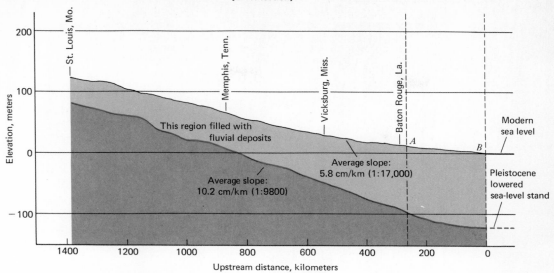

[Cross section]

FIGURE 13-7 Lowered sea level during the Pleistocene permitted the Mississippi to cut a deeper and steeper long profile (lower curve). The Mississippi responded to today's elevated sea level by depositing sediment until the upper profile was established. Volume of sediments placed in storage between the two curves exceeds 4000 cubic kilometers. An amazingly low average gradient (1: 17,000) of the modern stream is sufficient to enable the river to carry to its mouth more than a *million* metric tons of sediment per day! Within the past 5000 years, the front of the Mississippi delta has prograded approximately 200 kilometers seaward from an ancestral shoreline near Baton Rouge (vertical line *A*) to the modern shoreline (vertical line *B*). [*After H. N. Fisk*, Geological Investigation of the Alluvial Valley of the Lower Mississippi River, *Mississippi River Commission, Vicksburg, Miss., 1944.*]

among them in paths that split and rejoined (Figure 13-8). During the wintertime, when the ice ceased melting, the river flow was reduced to an ineffectual trickle. Steep gradients, intermittent periods of high discharge, and an overload of coarse sediment characterized this network of channels in the *braided* Mississippi River.

The modern meandering Mississippi As the ice margin continued to melt back, other drainage basins were uncovered through which melt-water was diverted elsewhere. Zones of climatic extremes also followed the northward retreat of the ice, to be replaced in the south by a more moderate precipitation pattern. Sea level rose, and with it came another major readjustment of the long profile, this time an elevation of the curve (Figures 13-7 and 13-9). The Mississippi built its bed higher by depositing a broad plain of alluvium over the older braided stream deposits. These braided stream sediments are now destined to remain

389 RIVERS

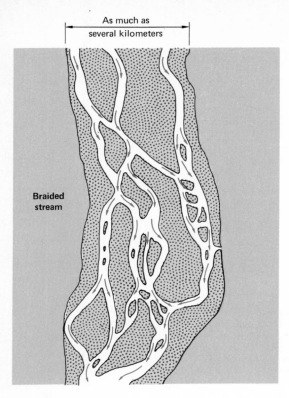

As much as
several kilometers

Braided
stream

FIGURE 13-8 During a low-water stage, shown on this map of a segment of a braided river, the channels are threaded among shifting gravel and sand bars. During flood stage the entire channel is submerged. [*After J. R. L. Allen, "A Review of the Origin and Characteristics of Recent Alluvial Sediments,"* Sedimentology, *vol. 5, no. 2, 1965.*]

in storage for millions of years, unless widescale refrigeration should once again cause the sea level to drop, initiating a new cycle of entrenchment.

Slowly the Mississippi evolved into its modern regime in which *meanders* dominate the fluvial pattern. These smooth sinuous loops, named after a stream in Turkey once known as the Maiandros, are another feature whose origin is little understood. Obviously the Mississippi is not constrained by solid rock walls to follow a meandering path, for its floodplain is composed of easily dislodged, loose sediment. Since the river is freely able to erode the unconsolidated material in one place and deposit it in another, why doesn't the channel simply adjust into a line aimed straight to the sea? Surely a straight-line path would meet the least resistance to flow, yet we find that a stream rarely continues in a straight line for a distance greater than 10 times the width of its channel.

An eminent student of land forms, W. M. Davis, perfectly summarized the popular notion about meanders when he said that meandering is the aimless wandering of a stream too sluggish to accomplish any work of erosion. At least in this surmise, however, he was wrong. Meanders are not "aimless"; they are quite regular curves whose size

[Block diagrams, very large vertical exaggeration]

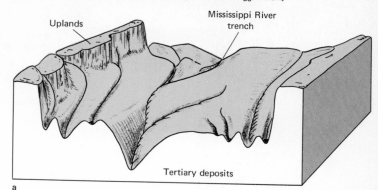

Uplands

Mississippi River
trench

Net erosion

Tertiary deposits

a

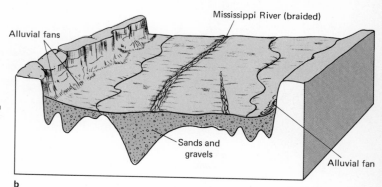

Mississippi River (braided)

Alluvial fans

Net deposition

Sands and
gravels

Alluvial fan

b

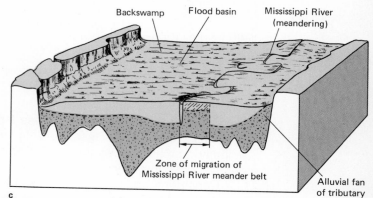

Backswamp   Flood basin   Mississippi River
(meandering)

Net deposition

Zone of migration of
Mississippi River meander belt

Alluvial fan
of tributary

c

FIGURE 13-9 (*a*) When sea level was about 130 meters below the present position, the Mississippi River deeply entrenched its valley. (*b*) While sea level rose to within 30 meters of the modern level, vast quantities of glacial meltwater poured down the Mississippi, creating a braided stream choked with sand and gravel. (*c*) Deposits of today's meandering Mississippi, established at a yet higher sea-level position, have buried the older braided stream deposits. [*After S. A. Schumm, "Fluvial Geomorphology: Channel Adjustment and River Metamorphosis," River Mechanics, vol. 1, H. W. Shen, Fort Collins, Colo., 1971.*]

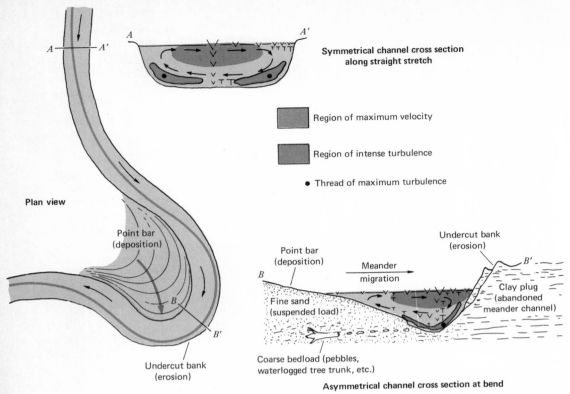

Symmetrical channel cross section
along straight stretch

Region of maximum velocity

Region of intense turbulence

• Thread of maximum turbulence

Plan view

Point bar
(deposition)

Undercut bank
(erosion)

Point bar
(deposition)

Meander
migration

Undercut bank
(erosion)

Clay plug
(abandoned
meander channel)

Fine sand
(suspended load)

Coarse bedload (pebbles,
waterlogged tree trunk, etc.)

Asymmetrical channel cross section at bend

FIGURE 13-10 Forward velocity of water within the channel is shown by relative sizes of V in the cross sections through a stream. Turbulence, denoted by various sizes of T, is the chief controlling factor of sediment deposition. As the channel migrates laterally to the right at cross section BB', the point bar keeps pace with erosion by building to the right also. "Lag" gravels, waterlogged tree trunks, and other bedload debris are incorporated into the base of the point bar deposit. (Giant trees of the famous Petrified Forest, Arizona, were buried in a fluvial environment in Triassic times.)

grows larger as the stream becomes larger. Nor do rivers necessarily flow lazily through the meanders. In many rivers, the water velocity actually increases downstream, the apparent sluggishness being deceptive when the flow is in a large channel.

Consider the processes at work in a meander (Figure 13-10). Force of moving water is directed against the outer bank at a bend (BB'), creating turbulence and erosion of the bank material. At that point, scour deepens the channel and occasionally undermines the soft bank materials, allowing them to slump into the stream. Because the channel is asymmetrical at bends, its deepest part wanders from side to side— a matter of great importance to river boatmen during low-water stages.

Where turbulence is less, along the inner side of the bend, sand is deposited as a broad, gently sloping *point bar*. Sediment carved from an outer bank tends to be carried cross-river by the spiraling motion of the water (arrows), to be deposited on the next point bar downstream. As this happens, a meander migrates laterally in the direction indicated by the bold arrow. A meander loop may grow more and more accentuated, eventually forming almost a complete circle that may even be breached across at the neck. Stranded segments of meander loops, called *oxbow lakes*, are dotted about on the face of the Mississippi floodplain (Figure 13-11).

Catastrophic processes in an alluvial valley eventually destroy the abandoned meanders. As floodwaters overflow the bank, their turbulence is abruptly reduced, and so too is their power to hold sediment in suspension. Coarser particles (silt-sized) drop along the river's edge where they build up an embankment, the *natural levee* (Figure 13-13). Clay-sized particles spill over into the broad floodplain where they settle out from quiet waters. An oxbow lake is filled simply by the overbank deposits of hundreds of floods.

Perhaps the interplay of the fluvial processes can help us to understand meanders in a new light. Why *doesn't* a river head straight to the sea by the shortest path? If it did, the slope of the bed would be the steepest possible. Any excess energy of the moving water would be used to erode and transport sediment, thereby reducing the slope of the long profile. Meandering is a more efficient way to consume this excess energy, for then not only does the river scour sediment off its bed; it also erodes *laterally* by undercutting its banks. In the economy of nature, a process will always follow the least energy-consuming route available to it. What *can* happen most easily, *will* happen.

Can the changing regime of the Mississippi afford some useful clues to fluvial processes of the distant geologic past? When North America was still locked in the Pleistocene's icy grip, rivers of meltwater shifted an immense load of coarse sediment through the wide, shallow channels of the braided stream regime. Much earlier, before "higher" (terrestrial) vegetation evolved in the Silurian Period, none of the land surface was protected from the catastrophic effects of flash flooding. Perhaps most of the rivers in those days were of the braided type. Today, fine sand is the coarsest bedload that can be moved along through the comparatively narrow and deep channel of the meandering Mississippi. Sporadic floods succeed only in laying down a mask of fine clay in the extensive backswamps bordering the modern stream.

Whether a fluvial system was braided or meandering may be important economically. Deposits of a meandering stream contain the essential ingredients for storage of petroleum and natural gas (Figure 13-14). Porous sands of buried point bars and channel fills—the reservoir—are capped by impermeable backswamp clay which assures that the oil

a

FIGURE 13-11 (*a*) Viewed from high altitude, a tributary stream in the Mississippi floodplain appears as a fantastic tangled maze of abandoned meander scars. Each lateral swing of a meander loop has erased an older deposit, replacing it with a new point-bar deposit. Depositional patterns so obvious from above are too subtle to be noticed by a casual passerby on the ground. [*Tobin Research, Inc., photograph.*] *Opposite page:* (*b*) A highway map shows that Mississippi meanders have changed significantly on a time scale of decades. Solid color lines identify segments of the river that were active when Louisiana and Mississippi became states (1812 and 1817). Today these abandoned boundary loops project absurdly into the state on the opposite side of the river. [*Map base*, © *Rand McNally & Company*, R. L. 73Y138.]

will not leak out of the system. A braided stream, on the other hand, fills its valley almost entirely with coarse sand and gravel: hardly a secure trap for oil and gas.

DELTAS

About 70 percent of all terrigenous clastic sediment accumulates in *deltas*, where streams discharge into an open body of water. The term was coined nearly 25 centuries ago by Herodotus, who was struck by the similarity between the Greek letter Δ and the triangular shape of the

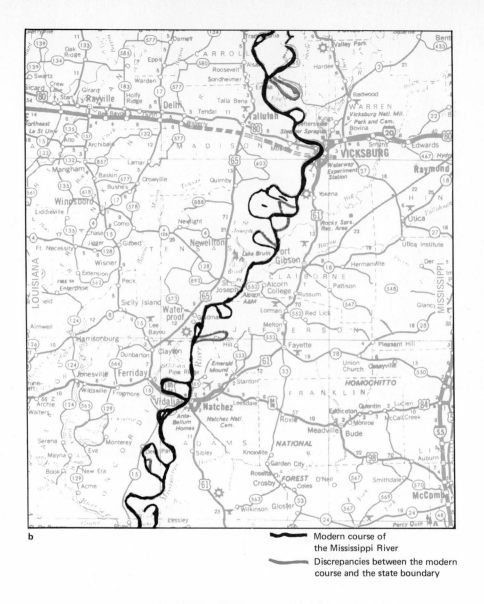

b

──── Modern course of
the Mississippi River

──── Discrepancies between the modern
course and the state boundary

Nile delta. Shapes of other deltas are very different, however, for they respond sensitively to processes that act in continual opposition to one another. A delta is an uneasy place, an arena claimed by both land and sea. Although the net outcome is progradation of the land, the architecture of a delta depends upon whether sediment is simply brought in by the river in one instance, or whether it is reworked by waves, tides, and longshore currents in another. Internally, every large delta contains a variety of depositional environments. Some 20 have been identified in the Mississippi delta alone, which will serve as a case study of deltaic processes in general.

395

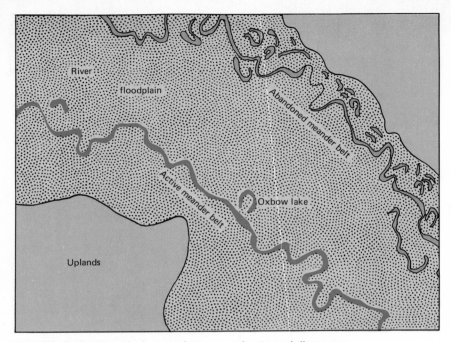

FIGURE 13-12 Not only do meanders grow, migrate, and disappear locally, but on a longer time scale (every thousand years or so), a river may jump abruptly from one meander belt to pursue its course down a totally separate belt. In the meantime, abandoned loops, sealed off with a filling of fine clay, are important in confining the river to a restricted belt. When an active meander encounters one of these tight clay plugs, its lateral migration is curbed. [*After H. A. Bernard and others,"Recent Sediments of Soutneast Texas (A Field Guide to the Brazos Alluvial and Deltaic Plains and the Galveston Barrier Island Complex),"* University of Texas Bureau of Economic Geology Guidebook II, *1970.*]

Setting the Scene

Of course, the river environment continues across the part of a delta that lies above sea level (the *delta plain*). Compared to the immensity of this plain, though, even the Mississippi appears to be scarcely more than a winding thread. How was deposition accomplished in regions that are as much as 250 kilometers from the modern river? A clue to the explanation is seen in the inset of Figure 13-15. Left to itself, the lower Mississippi would be captured by its main distributary, the Atchafalaya River. This is because the path of the Atchafalaya, a shortcut to the sea, is about 3 times steeper than the course of the main stream. Even with careful control of the diversion by a system of locks, the Atchafalaya intercepts 30 percent of the total flow.

The figure makes clear what has happened again and again during the 10,000 years before man was present to interfere. Wherever the Mississippi enters the Gulf, it builds forward a *delta lobe.** Core material

* *Lobe:* a rounded projection.

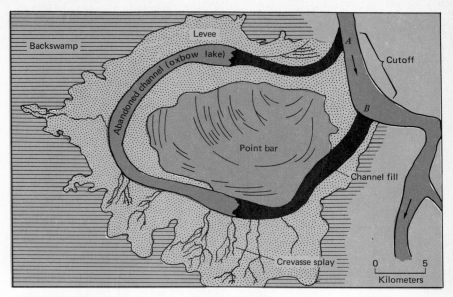

FIGURE 13-13   A short distance northwest of Baton Rouge, Louisiana is a neck cutoff of a large Mississippi meander. Since the river falls more steeply by the direct route from *A* to *B* than it does via the meandering path, the latter is abandoned and quickly sealed off by a clay plug (black). Sometimes a river breaks through its natural levee during a flood. A miniature delta deposit builds out from the breach, or *crevasse*, into the backswamp. Reestablishment of the natural levee tends to heal a crevasse after a few years. This cutoff was established over a period of about a century, around the year 1600. [*After H. N. Fisk*, Fine-grained Alluvial Deposits and Their Effects on Mississippi River Activity, *Mississippi River Commission, Vicksburg, Miss., 1947.*]

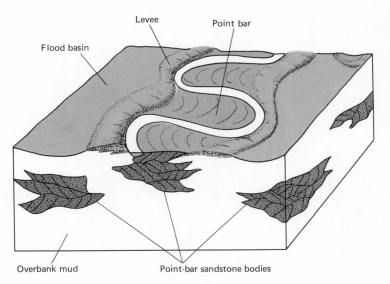

[Block diagram]

FIGURE 13-14   Drilling for petroleum in fluvial deposits is an important, if tricky, business. The problem is to strike the highly productive, but localized, sandstone reservoirs. Not even a direct penetration guarantees success, because the buried point-bar deposits may have acted like a huge spirit level. If the oil and gas have migrated for some distance, finally to be trapped as a "bubble" beneath some discontinuity in the reservoir sands, a misplaced exploratory well may encounter only sand saturated with salt water. [*University of Texas at Austin, Bureau of Economic Geology.*]

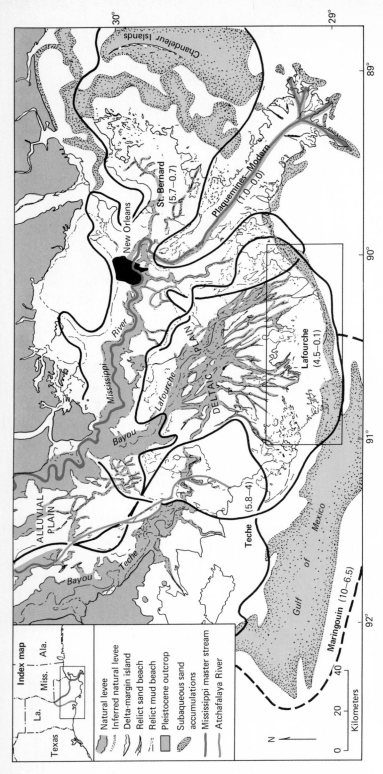

FIGURE 13-15 A map displays some of the environments of deposition in the five principal delta-lobe complexes (each consisting of smaller sediment lobes) that make up the Recent Mississippi delta. Periods of active progradation of each lobe are given in thousands of years before the present. At certain times, two lobes were being fed simultaneously. Each lobe is like a person's arm and hand. The "arm" is the upstream trunk of the Mississippi. At the "wrist," the river splits into distributaries outlined by a delicate tracery of natural levees. Sand and silt of the channel-levee system are like radiating bones of the hand, whereas the "flesh" between them is made of peat, other organic-rich marsh deposits, and fine overbank mud. Partially submerged natural levees project as "fingers" into the Gulf. Waves battering the open shore have winnowed away the mud, leaving broad areas of reworked delta-front sheet sand. (The Chandeleur Islands are the only delta-front sand remaining above sea level.) All the Maringouin lobe and much of the Teche and St. Bernard lobes have subsided beneath the sea. Rectangular box: area of Figure 13-16. [*After D. E. Frazier, "Recent Deltaic Deposits of the Mississippi River: Their Development and Chronology," Transactions, Gulf Coast Association of Geological Societies, vol. 17, 1967.*]

from numerous borings has revealed at least 17 such lobes, which are grouped into 5 major delta-lobe complexes. Eventually the building of a delta lobe is self-defeating, for as the river travels farther across previously deposited sediment, its slope becomes less. With reduced gradient, the ability of the river to transport sediment also declines. Meanwhile, neighboring delta lobes are slowly sinking beneath the waves, typically by about 10 centimeters per century. Part of the subsidence is an isostatic adjustment to sediment loading, but most is due to a slow squeezing of water out of the mud, which may have been as much as 80 percent water by volume when deposited. Settling of an abandoned delta lobe also steepens the local gradient to the sea, thus encouraging the master stream all the more to switch its course into some remote corner of the delta. It is no coincidence that the Atchafalaya is threatening to divert the master stream into the region of the Teche and Maringouin lobes, the two oldest and most sunken of the abandoned lobe complexes.

Dating of organic debris by the carbon-14 method (described in the following chapter) shows that these abrupt diversions have taken place every few hundred years. An entire lobe complex containing three or more individual lobes may have been a few thousand years abuilding. And so the "struggle" continues—progradation, abandonment, subsidence, and reinvasion by marine waters—to complete the chain of events in a *delta cycle*. In the Mississippi delta, ground is being gained today only at the tip of the master stream and in a lake being infilled by the Atchafalaya River. Elsewhere, an average of 40 square kilometers per year of the land surface is disappearing (Figure 13-16). At that rate, Louisiana is losing an area the size of New Orleans every 12 years!

The Birdfoot

Around 600 years ago, the modern delta lobe was established. Often referred to as a "birdfoot" because just a few persistent distributaries radiate almost from a single point, this lobe is typical of the first landform to appear in a new delta-lobe complex. What processes operate in this depositional environment? Density contrasts between different water and sediment types are the controlling factor here. A natural sorting of the clastic particles takes place at the distributary mouth. In this strange world that is neither securely land nor water, only the paired natural levees project above sea level (Figure 13-17). Steered to the mouth by the levees, the sandy bedload drops immediately, forming a channel-mouth bar. The river water itself (even though mud-laden) is less dense than seawater. A freshwater plume spreads out over the Gulf, eventually to drop its burden of mud when the fresh water is mixed into the ocean (Figure 13-18). (Salt water neutralizes the mutually repul-

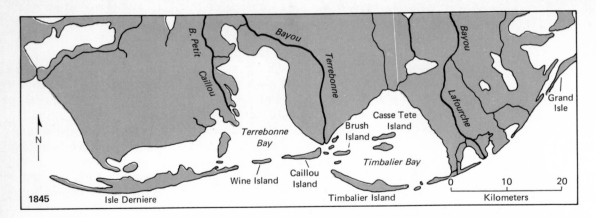

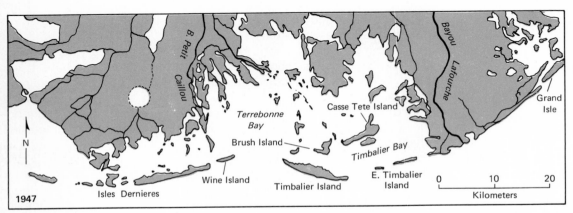

FIGURE 13-16 It took only a century for subsidence to bestow a distinctly moth-eaten appearance to the exposed shore of the abandoned Lafourche lobe. [*After H. J. Kwon, "Barrier Islands of the Northern Gulf of Mexico Coast: Sediment Source and Development,"* Coastal Studies Series 25, *Louisiana State University Press, Baton Rouge, 1969.*]

sive electrostatic charges on the clay particles, permitting them to clump together.) This offshore deposit of *prodelta clay* makes up about 95 percent of the bulk of the Mississippi delta (Figure 13-19). (You recall that shale, or lithified mud, is the most abundant sedimentary rock type.)

Deltas and Man

The Nile delta symbolizes how important delta environments have always been to civilization. Satellite photographs of the fertile delta plain set in the midst of the bleak North African desert are dramatic witness to the most time-honored usage: agriculture. Ironically, the entrapment of sediment behind the Aswan Dam some 800 kilometers up-

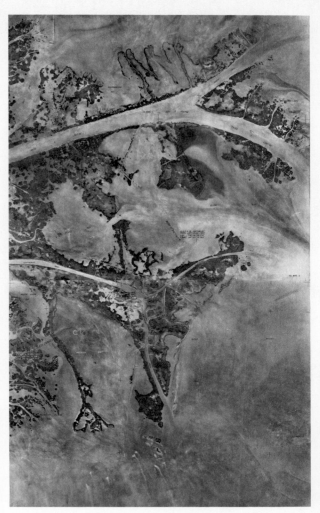

FIGURE 13-17 Near each of the Mississippi's several mouths, natural levees project a meter or two above water. Water in the broad areas between distributaries is no more than a meter or two deep. Submerged levee ridges extend long distances beyond the farthest point of land. [*Tobin Research, Inc., photograph.*]

stream has placed this green oasis in jeopardy. No longer will floods deliver enough sand and overbank mud to build the sediment surface higher as it continues to subside. Over the next few decades the Egyptian coast will undergo serious wave destruction, coupled with invasion of salt water into the land behind the coastal barrier dunes. Like some other well-intentioned engineering projects, the Aswan Dam will produce disastrous side-effects.

Demand for fossil fuel underscores a more recently exploited characteristic of deltas. About 30 percent of the world's production of petroleum and natural gas comes from ancient deltaic sediments. Within these deposits are located the fabulously productive oil fields of the Texas and Louisiana Gulf Coast, the North Slope of Alaska, and many

401 DELTAS

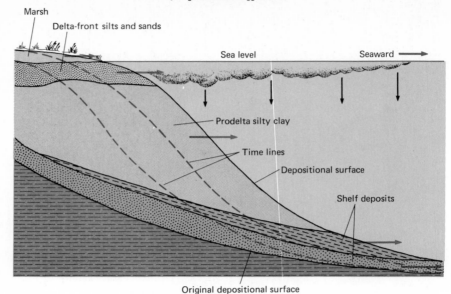

[Large vertical exaggeration]

FIGURE 13-18 A plume of muddy fresh water, delivered from a distributary mouth located just behind the plane of this cross section, deposits the prodelta clay and silt. As time goes on, delta-front sand, followed in turn by marsh deposits, are laid on top of the prodelta. [*After P. C. Scruton, "Delta Building and the Deltaic Sequence," in* Recent Sediments, Northwest Gulf of Mexico, *American Association of Petroleum Geologists, Tulsa, 1960.*]

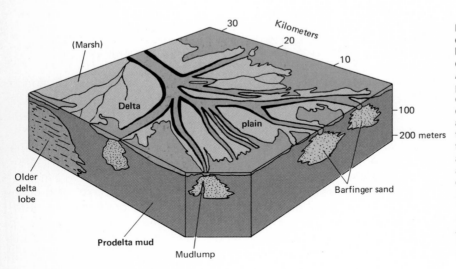

FIGURE 13-19 A block diagram of the active Mississippi delta lobe explains its birdfoot shape. As thick, water-saturated prodelta mud continues to compact (subside), the distributary channels become all the better able to conduct the flow of water. Hence, thick "barfinger sands" accumulate beneath just a few well-stabilized distributary channels. [*After H. N. Fisk and others, "Sedimentary Framework of the Modern Mississippi Delta,"* Journal of Sedimentary Petrology, *vol. 24, no. 2, 1954.*]

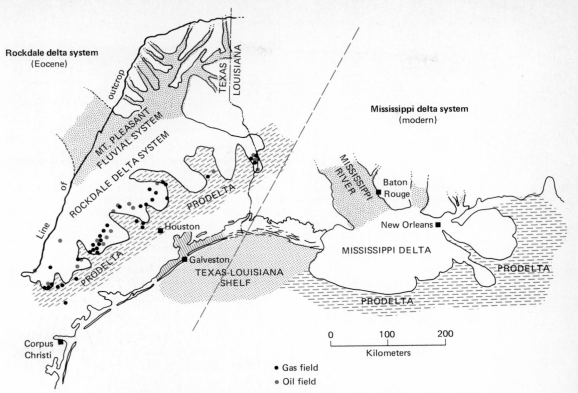

Rockdale delta system
(Eocene)

Mississippi delta system
(modern)

TEXAS
LOUISIANA

MT. PLEASANT
FLUVIAL SYSTEM

ROCKDALE DELTA SYSTEM

Line of outcrop

PRODELTA

PRODELTA

MISSISSIPPI RIVER

Baton Rouge

New Orleans

MISSISSIPPI DELTA

PRODELTA

PRODELTA

■ Houston

■ Galveston
TEXAS-LOUISIANA
SHELF

Corpus
Christi

0   100   200
Kilometers

● Gas field
● Oil field

FIGURE 13-20 Early in the Eocene Epoch a delta system, much like the Mississippi delta in size and shape, developed along the Texas Gulf Coast (but somewhat inland from the modern coast). The map shows a reconstruction of the Eocene deltas as they were, not as they are. Today these gently coastward-dipping rocks are exposed only along a narrow arcuate outcrop. In the vicinity of Houston, they lie buried thousands of meters beneath younger sediment. Numerous oil and gas fields are peppered about in the delta-front sands. Beneath the Mississippi delta is a thick sequence of oil-rich Miocene deltas. [*After W. L. Fisher and J. H. McGowen, "Depositional Systems in Wilcox Group (Eocene) of Texas and Their Relationship to Occurrence of Oil and Gas,"* Bulletin of the American Association of Petroleum Geologists, *vol. 53, no. 1, 1969.*]

others (Figure 13-20). Delta sediments far excel fluvial sediments as a source of hydrocarbons.* In both environments the mud and sand have been sorted into intermingled but sharply distinct "packages." Hydrocarbons in deltas are found almost entirely in sediments laid down in the submarine part of the depositional system. Organic debris from the delta plain and the remains of floating marine organisms (plankton) rain down on the prodelta mud. Protected from the oxidizing power of the atmosphere, the buried mud "matures"; the organic soup turns into

* Hydrocarbons include the mixture of hundreds of carbon- and hydrogen-containing substances in petroleum, and methane (CH_4), the chief ingredient of natural gas.

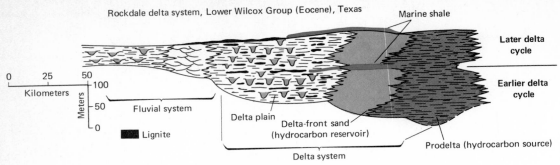

FIGURE 13-21 A schematic cross section through an Eocene delta exhibits the vertical stacking of rocks of two delta cycles. Deposition of impermeable marine shale atop the oil sands terminates each cycle. Deposition rates of these two sediment types are enormously different. A given thickness of the sand is estimated to accumulate about 6000 times faster than the same thickness of shale. [*After W. L. Fisher and J. H. McGowen, "Depositional Systems in Wilcox Group (Eocene) of Texas and Their Relationship to Occurrence of Oil and Gas,"* Bulletin of the American Association of Petroleum Geologists, *vol. 53, no. 1, 1969.*]

petroleum, which is slowly flushed out of the compacting mud along with abundant water. The migrating fluids are conveniently trapped as a series of layers—water beneath the oil, gas above it—in the porous delta-front sands nearby. Even the delta cycle works to the advantage of the oil seeker. Subsidence followed by progradation of later delta lobes creates a many-storied stack of reservoirs (Figure 13-21).

Nearly all the earth's coal originated in delta marshlands. Some of these were great saltwater marshes that grew above an abandoned, subsiding delta lobe. More significant were the freshwater tracts set between distributaries during the progradational stage. Both types of marsh were deposition sites of peat, the initial substance from which lignite, then bituminous coal, and finally anthracite coal can be formed by progressive burial (Figure 13-22).

Not all aspects of the delta environment are favorable to the works of man. Subsidence threatens New Orleans, Venice, and other cities built upon delta plains. Parts of New Orleans have sunk below sea level, protected from flooding only by a system of artificial levees. Moreover, because the ground settles unevenly, sewer pipes are subjected to bending stress that has broken them in countless places.

COASTAL SYSTEMS

What happens to clastic sediment once it has arrived at the ocean? Most of it is trapped in deltas; the small volume released farther into the open ocean either plunges straight on to great depth, or is distributed

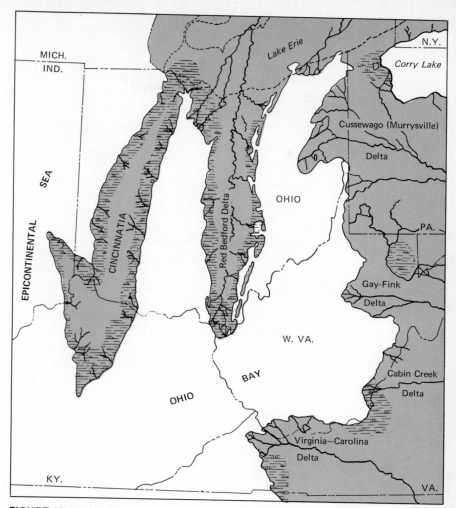

FIGURE 13-22 In early Mississippian time, rivers fed a system of deltas shown in this reconstruction of the famous coal-producing region of Ohio, Pennsylvania, and West Virginia. European geologists use the term "Carboniferous" for the Mississippian plus Pennsylvanian Periods of North American terminology. Apparently the Carboniferous was well named! [*After J. F. Pepper and others, "Geology of the Bedford Shale and Berea Sandstone in the Appalachian Basin," U.S. Geological Survey Professional Paper 259, 1954.*]

along the shore. Surprisingly, only a trivial amount of sediment finally comes to rest near the brink of the continental shelf.

Attempts to classify landforms along coasts are in a state of confusion. Is the shore a zone of high energy (high waves or tides) or low energy? Is it composed of weakly cemented particles or strong, resistant rock? Is it stable, or is it sinking or rising with respect to sea level?

No classification scheme based on these variables (where they are known) seems able to explain adequately the appearance of a particular coast. Perhaps our difficulty is that we are living in unusual times (geologically speaking) because of the recent retreat of the polar ice sheets. Sea level, having risen rapidly for many thousands of years, has stabilized only since about 5000 years ago. No wonder the world coastline appears so unsettled. In some places the active geologic processes may have already brought the coast into a state of balance, while elsewhere, equilibrium will not yet be restored for thousands of years.

Shoreline Processes

Why are coastal processes not more rapid agents of change? Who has not marveled at the awesome power of the surf crashing on a beach? Typically, about 8000 waves will hammer that beach in a single day. According to studies on the New Jersey coast, a net quantity of roughly 250,000 metric tons of sand per year may be conveyed past a given point on the shore. The Mississippi dumps that much sediment into its delta in less than 6 hours! We may calculate how rapidly energy is "consumed" (transformed into heat) in the process of moving sediment down the Mississippi River, compared with the rate that energy is dissipated as waves move the same quantity of sediment along the shore. There are so many variables that the estimate is only approximate, but nevertheless instructive. As a sediment carrier, the Mississippi is roughly a thousand times more efficient than the waves. Our amazement at the power of breaking waves is well founded, but it only deepens the paradox. All that energy, and such a trivial net effect!*

Directedness of the energy spells the difference between the dip-fed and strike-fed depositional systems. In rivers and deltas the water flows essentially in one direction, whereas waves and tides are forever tediously moving particles up on the shore, then out to sea again (Figure 13-23). Water waves come in many sizes, from ripples a few centimeters across to the two tidal bulges whose wavelength is half the circumference of the earth. We saw in Chapter 2 that the earth's gravitational interaction with the sun and moon causes the tides. Another, infrequent source of water waves is a submarine earthquake. It can initiate the dreaded *tsunami*† that races 200 meters per second unnoticed beneath a ship, only to devastate entire communities upon reaching shore.

By far the greatest proportion of wave activity comes from the more persistent urging of the wind. As a wave enters shallower water, the

---

* Sea cliffs have been eroded back a number of meters by a single great storm. Such devastating storms are exceedingly rare—the sort of storm that happens once in several centuries—and even they do not do much to solid rock.
† Often incorrectly called a tidal wave because of its long wavelength.

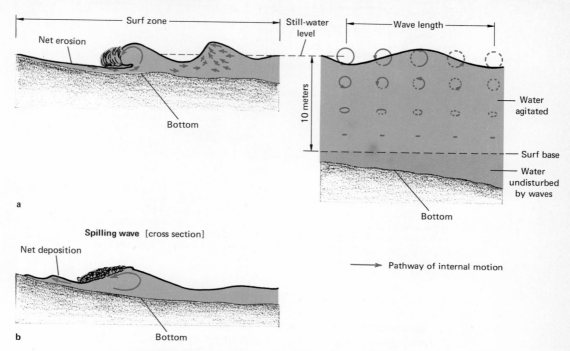

Plunging wave [cross section]

Surf zone — Still-water level — Wave length

Net erosion

10 meters

Bottom

Water agitated

Surf base

Water undisturbed by waves

Bottom

a

Spilling wave [cross section]

Net deposition

Pathway of internal motion

b   Bottom

FIGURE 13-23 Breaking waves are a geologic process whose time scale is but a few seconds' duration. Oil droplets of neutral buoyancy introduced into water in a wave tank serve as visible tracers of internal motion. The droplets show that a particle at the surface describes a circular orbit equal to the height of the wave. Particles at increasing depths trace out a smaller circle, then a yet smaller flattened ellipse, and finally, a tiny straight line back and forth. All the particles in a vertical column of water move to the right and to the left in unison (that is, in phase). Surf base lies at about 10 meters depth, below which only extraordinary storm waves have the ability to entrain loose sediment. When waves 3 meters high pound a beach, their power is equivalent to a solid line of automobiles, each with 270 horsepower, advancing side by side at full throttle.

bottom begins to interfere with its organized motion. Friction retards the tempo of motion in the deeper part of the wave, while momentum of the wave crest causes it first to pitch forward steeply and then to collapse. Plunging waves (Figure 13-23a) topple violently, whereas, in spilling waves (Figure 13-23b), the crest gradually dissipates by tumbling down into the trough ahead. In general, plunging waves erode; spilling waves accomplish deposition. With the arrival of each new season, wave intensity changes in most parts of the world (Figure 13-24).

When the wind blows obliquely at the shore, some of the energy of the breaking waves goes to support *longshore drift*. A swash of water

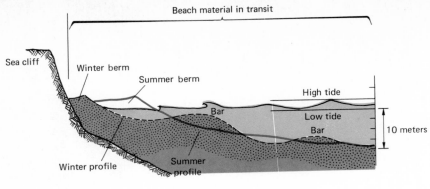

[Cross section]

FIGURE 13-24 On a time scale of months, the profile of a beach is self-regulating. Gentler summer waves lift the sand up onto a terrace, or *berm*, whose position is relatively low and wide. As higher waves in the winter cut away at the summer berm, they build a narrow but more elevated winter berm and stow away the remaining sand as a series of offshore bars. The next year, the cycle repeats. [*After Willard Bascom*, Waves and Beaches, *Doubleday & Company, Inc., New York, 1964.*]

rushes diagonally up on the beach, but it retreats straight downslope. Consequently, as a sand grain entrained in the water is dragged along shore, it traces out a ragged, saw-toothed pattern (Figure 13-25). In the surf zone a few meters offshore, the water itself may slowly drift parallel to the beach. If the new season brings a reversal of the wind direction, much of the sand will drift right back where it started. Only about 15 percent of the sand motion along the beach in the New Jersey study represents a *net* transport.

Permanent changes do take place, of course, and they will continue wherever there is **any** movement of the ocean with respect to the shore. The recent rise of sea level is an example of such change; so are tectonic movements of the land and the progradation of a delta. A competition exists in all depositional environments between processes that bring sediment in and those that move it out. Deltas are "point sources" from which sand is introduced into a longshore system. As the waves keep washing sand grains up on the beach and out to sea again, some of the particles finally escape from the system. Exposed sand on the beach can be blown inland. Sediment pours down into submarine canyons or is carried by powerful storms out into water too deep for waves to take possession of it again. As long as the sea was rapidly advancing across the land, abundant sand lay in its path, available to be incorporated into beaches. But with today's sea-level standstill, the world's beaches appear to be losing sand—a gloomy but irreversible fact.

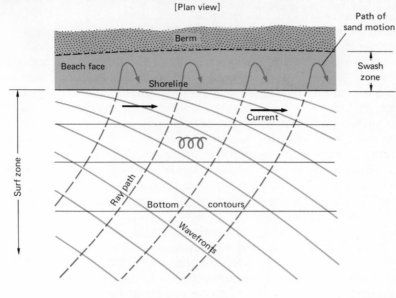

[Plan view]

Path of sand motion

Berm

Beach face

Shoreline

Swash zone

Surf zone

Current

Ray path

Bottom   contours

Wavefronts

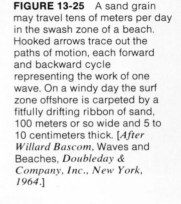

FIGURE 13-25 A sand grain may travel tens of meters per day in the swash zone of a beach. Hooked arrows trace out the paths of motion, each forward and backward cycle representing the work of one wave. On a windy day the surf zone offshore is carpeted by a fitfully drifting ribbon of sand, 100 meters or so wide and 5 to 10 centimeters thick. [*After Willard Bascom*, Waves and Beaches, *Doubleday & Company, Inc., New York, 1964.*]

FIGURE 13-26 On a time scale of centuries, longshore drift coupled with erosion can straighten out an irregular coast. As the advancing waves "feel" an uneven bottom, the effect is to refract (bend) the distribution of wave energy. Ray paths, showing the directions of energy flow, are focused on the headlands, and spread farther apart in the bay. Longshore drift builds a spit (an elongated sand beach) across the bay mouth. [*After Willard Bascom*, Waves and Beaches, *Doubleday & Company, Inc., New York, 1964.*]

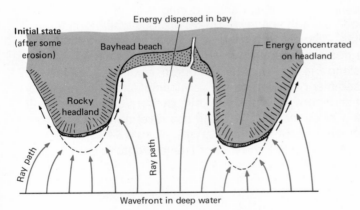

[Plan view]

Energy dispersed in bay

Initial state (after some erosion)

Bayhead beach

Energy concentrated on headland

Rocky headland

Ray path

Ray path

Wavefront in deep water

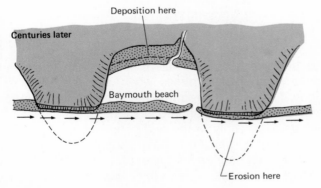

Deposition here

Centuries later

Baymouth beach

Erosion here

Strandplains and Barrier Islands

Included in our case study of depositional systems is an interesting exception to the picture of declining beaches. Prevailing winds maintain westward-drifting longshore currents in southwestern Louisiana (Figure 13-27). During episodes when the Mississippi was feeding into a lobe on the west side of the nearby delta, the longshore currents deposited large mud flats that were quickly overgrown by a topping of marsh deposits. When the Mississippi was diverted to an easterly, more distant lobe, the sediment supply to the mud flat was reduced. Wave attack outdistanced the influx of sediment. Clay was winnowed out while the waves erected a sandy, shelly beach. Then the Mississippi again broke through to the west, to deliver a new supply of abundant sediment. Shifting balance between the processes of erosion and resupply in this environment has left a *strandplain:* old beach ridges stranded ashore among scattered remnants of mud flats and marshes. Strandplain deposits may be widespread, but generally they are quite thin because subsidence does not accompany deposition.

Farther west, other distinctive environments of deposition front the Texas coast and continue southward to the vicinity of Tampico, Mexico (Figure 13-28). Here, the coast is double. Indenting the inner coastline are *estuaries* where river valleys, carved during time of low sea level, are now "drowned" by high sea level. Rivers are busily filling in the estuaries with deltas. Offshore is a shallow lagoon, and still farther seaward is an outer coastline made up of *barrier islands*. Although the origin of barrier islands is still disputed, modern evidence points to a local supply of sediment scoured off the continental shelf, not primarily to lateral feed of sediment via longshore drift. Barriers and lagoons will disappear in a few more thousand years if sea level remains constant. These features occupy about 1 percent of the world's 440,000 kilometers of coastline. More than half the barrier island length is found along the East and Gulf Coasts of the United States.

FROM CONTINENTAL SHELVES INTO THE ABYSS

Our case study concludes with a brief account of depositional systems on the continental shelf and in the deeper waters of an ocean basin. These locales revealed a number of surprises when intensive study began in the 1950s. Soon it was discovered that the sediments do not fit the expected pattern. According to standard doctrine of the times, coarse sediment would be found only in the surf zone because only there is the energy high enough to transport heavy particles. Proceeding offshore, we should encounter finer and yet finer grain sizes. A continental shelf (it was affirmed) makes an excellent sediment trap. It

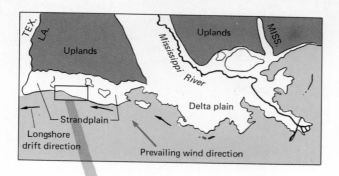

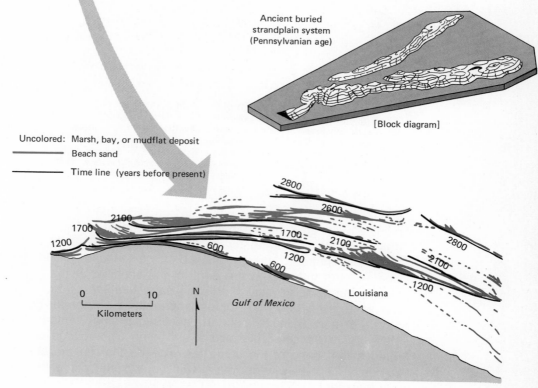

Uncolored: Marsh, bay, or mudflat deposit
—— Beach sand
—— Time line (years before present)

Ancient buried
strandplain system
(Pennsylvanian age)

[Block diagram]

FIGURE 13-27 Wood and shell material dated by the carbon-14 method has provided a time scale for the accretion of this strandplain in southern Louisiana. Some 2800 years ago, the active beach lay about 15 kilometers inland from the modern coast. Successively younger deposits are found nearer the coast. Each episode of deposition can be identified with a major shift of the outlet of the nearby Mississippi River, the source of the sediment. Buried "shoestring sands" in Kansas (block diagram) are a petroleum reservoir of probable strandplain origin. [*After H. R. Gould and E. McFarlan, Jr., "Geologic History of the Chenier Plain, Southwestern Louisiana," Transactions, Gulf Coast Association of Geological Societies, vol. 9, 1959.*]

411 FROM CONTINENTAL SHELVES INTO THE ABYSS

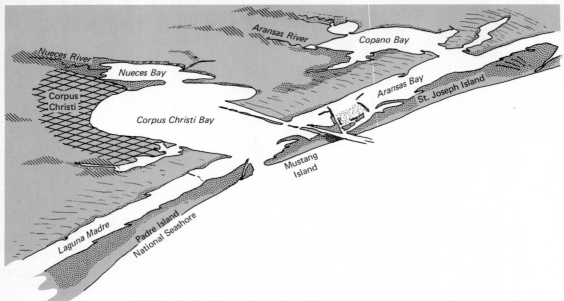

FIGURE 13-28 This oblique view of the south Texas coast near Corpus Christi includes four major components: barrier islands, lagoons behind the barriers, deeply indented estuaries, and deltas that are filling in these estuaries. [*After J. H. McGowen and others, "Effects of Hurricane Celia: A Focus on Environmental Geologic Problems of the Texas Coastal Zone,"* Bureau of Economic Geology Circular 70-3, *University of Texas at Austin, 1970.*]

should be "filled up," say, nearly to sea level before permitting sediment to overstep it into deep water. The continental slope was presumably a smooth surface lightly mantled by a sheet of descending debris.

These notions, though plausible, were found to be incorrect. Coarse sand may lie on the brink of the shelf, or at the base of the slope, or upon an abyssal plain a thousand kilometers from land. Deepsea sand is heaped into ripples, apparently by vigorous bottom currents. Far from being smooth, the continental slope is scarred by numerous furrows, some the size of a canyon. The volume of sediment piled at the base of the slope far exceeds that on the shelf above. Evidently Neptune's underwater realm is a busier place than anyone had foreseen!

Continental Shelf Processes

An imaginary journey along the North American shelf helps us to understand this odd situation. Off the New England coast the shelf bottom is highly irregular. Piles of boulders strewn about, sandy beds, and patches of naked bedrock can be discerned. Off New Jersey the bottom is a series of low ridges and shallow depressions. Dredge hauls bring up marsh deposits and teeth of Pleistocene mammoths. A broad western Florida shelf is underlain by thick limestone. Off Texas and northern Mexico we find a gently sloping underwater plain composed of terrigenous clastic sediments. The California shelf is narrow, rocky, broken by deep canyons and closed depressions. In brief, the continental shelf is rather similar to the land immediately onshore.

An estimated 70 percent of the worldwide continental shelf is floored by *relict** sediments. These sediments were deposited, not in today's marine environment, but in other environments during low sea level when the shelf was exposed. Naturally these environments would be similar to those still presently on land near the modern shoreline. It is as though the continental shelf were a Pleistocene museum. The outer shelf appears abandoned, insulated from the effects of all but the most violent processes. Plastic "bottom drifters" released at many points on the shelf tend to be stranded ashore, further emphasizing the sense of this isolation (Figure 13-29). Even so, the relict sediments are subject to a peculiar sort of modification. Hurricanes can agitate the entire water mass overlying the shelf. Every few years the bottom sediment gets stirred up and moved a short distance, first in one random direction, then another.

* *Relict:* something that has survived a destructive process.

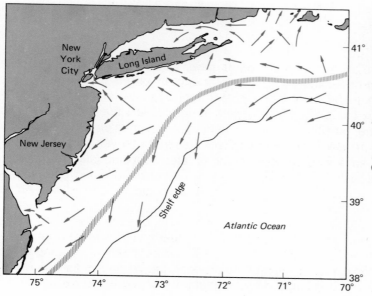

FIGURE 13-29 Arrows show the travel paths of simple plastic devices that were permitted to drift along the bottom across the continental shelf. The pattern suggests that estuaries are rapidly being choked by river sediment plus the sediment drifted in by bottom currents. The hachured line marks the boundary between regions of inward and outward drift. [*After D. F. Bumpus, "Residual Drift along the Bottom on the Continental Shelf in the Middle Atlantic Bight Area,"* Limnology and Oceanography, *Supplement to vol. 10, 1965.*]

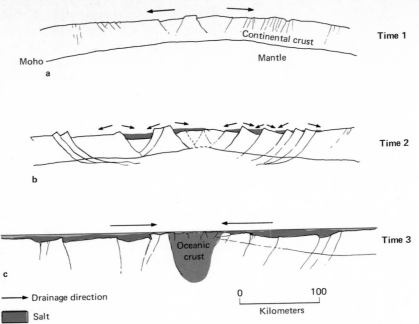

Time 1

Continental crust

Moho

Mantle

a

Time 2

b

Time 3

Oceanic crust

c

⟶ Drainage direction

▮ Salt

0   100
Kilometers

FIGURE 13-30 (*a*) Regions where continental crust was rifted apart (for example, the Atlantic Ocean, the Red Sea, the Gulf of Mexico) probably evolved according to this sequence of cross sections. First, the crust bulged upward, stretched, and broke. (*b*) Then a partial collapse occurred, creating fault-block mountains and intervening troughs where salt was deposited. As the crust foundered, the entire region later received a layer of salt. (*c*) Finally, basaltic crust was injected into the central axis: a new ocean was born. Marginal fault blocks rotated backward toward the coast.

Geology of Salt

Before examining some of the deep-water environments, let us return to our case study, the Gulf of Mexico. Its origin, long a mystery, is beginning to be understood in the light of plate tectonic theory described in the preceding chapter. In particular, the theory helps to explain the presence of vast quantities of common salt (NaCl), an evaporite mineral. The salt has given rise to some very peculiar structures that have modified the course of later deposition. (More recently, similar deposits of thick salt have been discovered along both coasts of the Atlantic Ocean and along the Canadian border of the Arctic Ocean.)

Probably the Gulf began as a narrow rift, or zone of splitting of the continental crust. We imagine that in its early stages it resembled the modern Red Sea and the Dead Sea of the Middle East. According to the plate tectonic model, rifting began when the mantle bulged upward, stretching and thinning the continental crust above it (Figure 13-30). As the crust began to separate, it broke into numerous fault blocks tilted away from the central rift axis. Between the fault-block mountain chains lay valleys with interior drainage, like the Dead Sea valley. Tectonic movements repeatedly connected some of the valleys to the sea, then slowly isolated them once again. As each new influx of seawater evaporated, it left behind a deposit of salt. (Modern examples help us to visualize this process. For example, a thickness of salt and gypsum exceeding 4 kilometers has concentrated in the Dead Sea rift within the past 3 million years.) Finally, the rift widened enough to permit free circulation of normal seawater, terminating the epoch of salt deposition. Thus, by mid-Mesozoic times the newly forming Gulf of Mexico had come to be floored by an expanse of salt thousands of meters thick, over which other sediments began to accumulate.

Now, salt is an extraordinary sediment. Unlike sand and mud, it is a water-free, rather incompressible substance from the moment of deposition. Its density, 2.2 grams per cubic centimeter, is greater than that of water-saturated clastic sediments, but less than the density of sediment from which deep burial has expelled the water. Inevitably, an unstable situation developed once a thickness of a kilometer or two of clastic sediment had piled up, for then *less* dense salt lay beneath *more* dense

FIGURE 13-31 The initial salt layer, fondly known as the mother bed, is depleted around the base of a rising salt structure. A progression of forms, ranging from pillows to continuous walls, correlates with increasing original thickness of the mother salt. Note that a mass of salt detached from the mother bed may even continue to rise as a "bubble." [*After F. Trusheim, "Mechanism of Salt Migration in Northern Germany,"* Bulletin of the American Association of Petroleum Geologists, *vol. 44, no. 9, 1960.*]

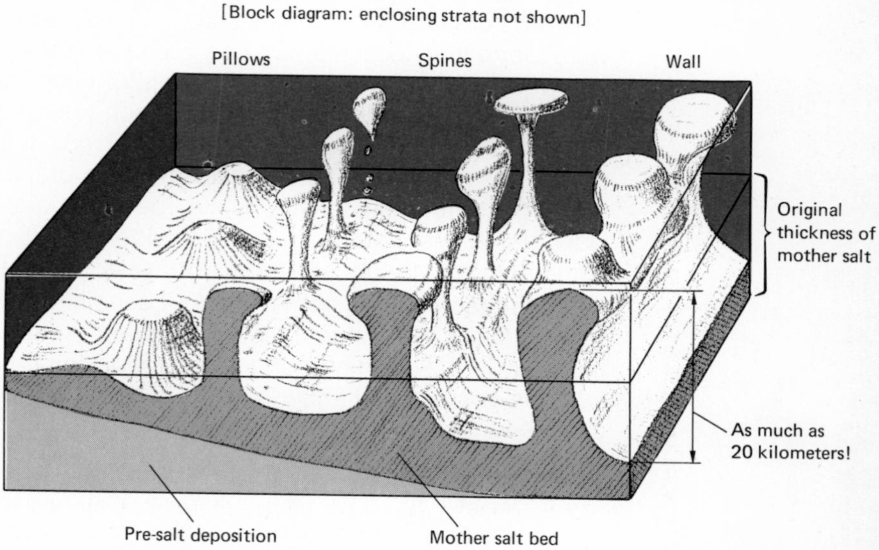

[Block diagram: enclosing strata not shown]

Pillows   Spines   Wall

Original thickness of mother salt

As much as 20 kilometers!

Pre-salt deposition   Mother salt bed

overburden. The salt responded to the density inversion by penetrating through the overlying sediment. Where the salt pierced upward it assumed different shapes—pillows, tall spines, mushroomlike overhangs, or even a continuous wall (Figure 13-31). In the process, the salt flowed plastically; the enclosing sediments were profoundly disturbed by faulting and folding.

Occasionally the top of a salt intrusion emerged briefly to the surface (modern examples are Avery Island, Louisiana, and High Island, Texas). Surface water flushed away the easily dissolved salt, leaving a sheath, or *caprock*, of less soluble impurities, notably anhydrite ($CaSO_4$). Caprock environments may contain water, calcium sulfate, and oil and gas—all the ingredients necessary for sulfate-reducing bacteria to flourish. One of the bacterial waste products is H_2S, which has been converted into large quantities of valuable native sulfur (S) in many caprocks (Figure 13-32). More than 400 salt structures in the Gulf region have already been drilled. Doubtless, many more await discovery under the continental shelf and slope.

Deep Water

Textures of clastic sediment on the ocean floor are distinctively different from those in the other environments. In many places we find a monotonously repetitious series of beds, each a few centimeters thick. A single continuous layer may extend over a distance of hundreds of kilometers. In each stratum the particle size gradually decreases from coarse sand at the base to fine clay at the top. Above the clay is a sharp boundary, then the next bed with its upward gradation from coarse to fine. This texture, called *graded bedding*, is exactly what would be created if a violently churned mixture of water, sand, and mud were permitted to settle. Coarse material would precipitate at once, followed by the more leisurely descent of ever finer particles, perhaps over a period of weeks, depending upon how high the sediment was stirred up into the water.* There are other puzzling facets of the situation. Remains of nearshore organisms, tree branches, and all kinds of exotic trash from the continents may be incorporated into the sediment. How does this catastrophic and widespread depositional process come about? Why is not a remote abyssal plain even more a place of solitude than the continental shelf? It seems so out of character.

Consider the status of sediment piled on the brink of the shelf, for instance in the vicinity of the Mississippi delta. Its stability is precarious, for any unusual disturbance—a storm, an earthquake, or simply oversteepening of the sediment interface—can cause a large mass

---

* Calculations suggest that a clay-sized particle takes about 30 years to settle from the surface to the bottom of the ocean.

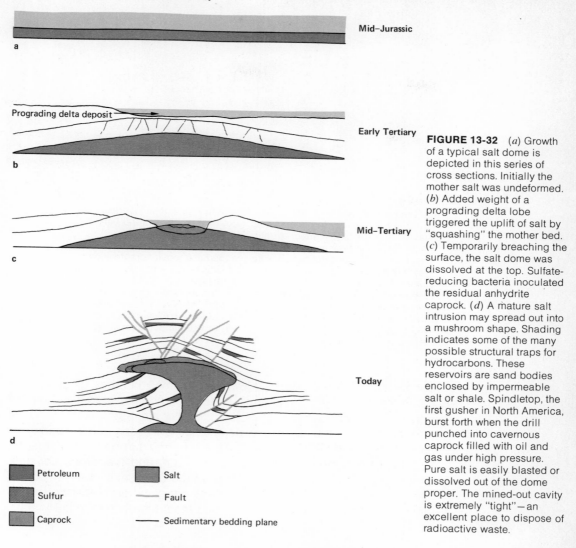

[Cross sections]

Mid–Jurassic

a

Prograding delta deposit

Early Tertiary

b

Mid–Tertiary

c

Today

d

Petroleum

Sulfur

Caprock

Salt

—— Fault

—— Sedimentary bedding plane

FIGURE 13-32 (*a*) Growth of a typical salt dome is depicted in this series of cross sections. Initially the mother salt was undeformed. (*b*) Added weight of a prograding delta lobe triggered the uplift of salt by "squashing" the mother bed. (*c*) Temporarily breaching the surface, the salt dome was dissolved at the top. Sulfate-reducing bacteria inoculated the residual anhydrite caprock. (*d*) A mature salt intrusion may spread out into a mushroom shape. Shading indicates some of the many possible structural traps for hydrocarbons. These reservoirs are sand bodies enclosed by impermeable salt or shale. Spindletop, the first gusher in North America, burst forth when the drill punched into cavernous caprock filled with oil and gas under high pressure. Pure salt is easily blasted or dissolved out of the dome proper. The mined-out cavity is extremely "tight"—an excellent place to dispose of radioactive waste.

suddenly to slump into the depths. As it slides along, an amazing change of physical properties takes place. The slurry of water and suspended material becomes a dense, coherent fluid. As the *turbidity current* gains momentum down the continental slope, it can scour (erode) the bottom. Farther out, where the ocean floor is nearly level, the current slackens, finally releasing its load (Figure 13-33). A *turbidite* has been deposited.

Thus it would appear that turbidity currents can nicely explain both the presence of terrigenous clastics in the deep ocean and the origin of

417 FROM CONTINENTAL SHELVES INTO THE ABYSS

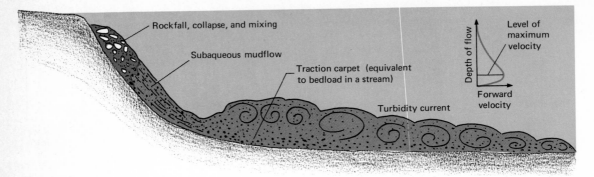

[Cross section, very large vertical exaggeration]

FIGURE 13-33 Deposits of a turbidity current starting high up on the continental shelf or slope may be laid down with catastrophic suddenness. In 1929 an earthquake triggered a submarine slump on the Grand Banks, near Newfoundland. A number of trans-Atlantic cables were stretched across the ocean floor immediately downslope from the earthquake epicenter. One by one the cables were snapped in a regular progression. Timing of the cable breakages shows that the disturbance rushed downslope initially at a speed of 100 kilometers per hour. It was still moving 20 kilometers per hour when the most distant cable broke, some 750 kilometers from the region of origin. [*After R. C. Morris, "Classification and Interpretation of Disturbed Bedding Types in Jackfork Flysch Rocks (Upper Mississippian), Ouachita Mountains, Arkansas:* Journal of Sedimentary Petrology, *vol. 41, no. 2, 1971.*]

certain submarine canyons. However, a small number of geologists claim that turbidity currents are unimportant compared to the "normal" persistent bottom currents. They point out that although we can produce turbidity flows in the laboratory, no one has ever seen one in action in a natural setting in the deep ocean. This is hardly surprising, considering that the environment is uninhabited by man, and the event, consummated in a matter of days, may take place only once in a century, or perhaps once every 10,000 years in any one location. Better documentation of the origin of this abundant sediment type is needed.

FUTURE STUDIES

We have seen the path of a sand grain from mountain top to ocean bottom to be an immense journey in space and time. Along the way, the particle occupies an interesting succession of depositional environments. A point bar, a delta-front sheet, a beach or turbidite—all these owe their existence to the force of gravity. More directly, their size and shape, their origin (and destruction) are governed by a variety of complex local processes. Our understanding of these environments has

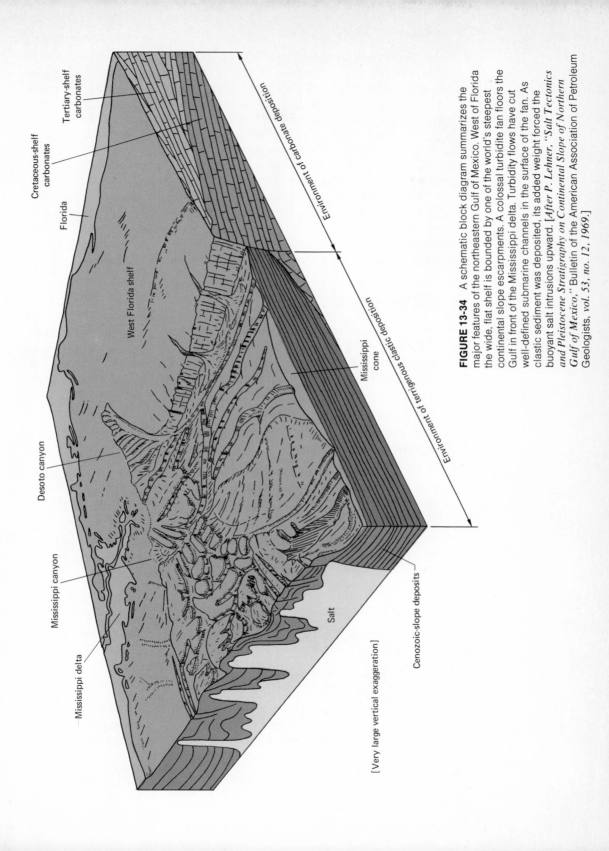

FIGURE 13-34 A schematic block diagram summarizes the major features of the northeastern Gulf of Mexico. West of Florida the wide, flat shelf is bounded by one of the world's steepest continental slope escarpments. A colossal turbidite fan floors the Gulf in front of the Mississippi delta. Turbidity flows have cut well-defined submarine channels in the surface of the fan. As clastic sediment was deposited, its added weight forced the buoyant salt intrusions upward. [*After P. Lehner, "Salt Tectonics and Pleistocene Stratigraphy on Continental Slope of Northern Gulf of Mexico," Bulletin of the American Association of Petroleum Geologists, vol. 53, no. 12, 1969.*]

Cretaceous-shelf carbonates

Tertiary-shelf carbonates

Florida

West Florida shelf

Desoto canyon

Mississippi canyon

Mississippi delta

Mississippi cone

Salt

Cenozoic-slope deposits

Environment of carbonate deposition

Environment of terrigenous clastic deposition

[Very large vertical exaggeration]

been based quite properly on field studies. If we only knew enough about the physics of motion of fluids and particles, could we have predicted all this without seeing it? Even if we could, of what value are theoretical arguments when the deposition can be quite simply observed in progress? Yet there are several reasons why future sedimentary research must become a more theoretical, mathematically exact field of study. Some environments are not easy to probe, especially in the deep ocean. A better theory would help us to interpret ancient sediments that formed in a manner for which no modern analogue is known.

Sedimentary deposition also takes place on the planet Earth's near neighbors in space. How does the turbulent, rarefied atmosphere of Mars or the turbulent, hot, superdense atmosphere of Venus govern sediment behavior? Chances are, even with the best of theories, these planets will afford plenty of surprises when examined firsthand.

FURTHER QUESTIONS

1   Refer to Figure 13-2. Note the relationships where streams join. Suppose you knew that an aerial photograph of the area of the small rectangle was located in the contributive net. How could you tell that the downstream direction is to the southwest, not to the north or east?

2   Refer to Figure 13-4. What stream capture is about to take place? How would it alter the drainage pattern? Would it be possible to integrate the entire drainage into one river system?

3   Refer to Figure 13-6a. The Beargrass drainage basin was a case study in the effects of urbanization. How would the construction of numerous streets and rooftops modify the hydrograph? Is the change a desirable effect?

4   What would happen to the port of New Orleans if the Mississippi River were permitted to establish its primary course into the Atchafalaya River, which is at present its main distributary?

5   Subsidence threatens much of New Orleans. What other natural process could flood the city permanently?

6   Refer to Figure 13-19, and note the mudlump intruded up into the barfinger sand. Explain this intrusion.

7   Refer to Figure 13-22. Although deltas formed in abundance in earlier times, prominent coal deposits did not appear before the mid-Paleozoic. Why?

8   Refer to Figure 13-32d. If more sediment is deposited over a rising salt structure, it responds by rising yet higher. Why?

READINGS

* Bascom, Willard, 1964: *Waves and Beaches*, Anchor Books, Doubleday & Company, Inc., Garden City, N.Y., 267 pp.

  Bird, Eric C. F., 1969: *Coasts*, The M.I.T. Press, Cambridge, Mass., 246 pp.

* Bloom, Arthur L., 1969: *The Surface of the Earth*, Prentice-Hall, Inc., Englewood Cliffs, N.J., 152 pp.

* Gagliano, S. M., and J. L. van Beek, 1970: *Hydrologic and Geologic Studies of Coastal Louisiana*, vol. 1, Louisiana State University, Coastal Studies Institute and Department of Marine Sciences, Baton Rouge, 140 pp.

  Leopold, Luna B., 1962: "Rivers," *American Scientist*, vol. 50, no. 4, pp. 511–537.

  Morisawa, Marie, 1968: *Streams*, McGraw-Hill Book Company, New York, 175 pp.

* Available in paperback.

chapter 14

The Great Ice Ages

An icecap on Ellesmere Island, Canadian
Northwest Territories, spills down into valley
glaciers that occupy a fjord. [*National Photo
Library, Surveys and Mapping Branch,
Department of Energy, Mines, and Resources,
Ottawa, Canada.*]

Some of the strangest circumstances on this remarkable planet are due to the extraordinary properties of water. We have noted the role of water in a great variety of natural processes — in the origin of rocks, in shaping the landscape, in sustaining life. The very existence of an ancient fossil record suggests that a delicate temperature balance, rather nearer the freezing point than the boiling point of water, has been maintained for billions of years. (If the earth's surface had been only slightly warmer, life today would be very different. Life would have vanished altogether at a temperature exceeding the boiling point.) But it is equally clear that in the midst of a generally moderate climatic history, there were episodes of harsh contrasts between the polar and tropical regions. The present happens to be one of these interesting and unusual times. Though nearly 98 percent of the surface H_2O is in liquid form, one continent (Antarctica) and the largest island (Greenland) are mantled by ice sheets. Smaller glaciers by the thousands are scattered about, especially in the northern continents. Ice covers about 10 percent of the land; truly we are living in an "ice age," though it is not now at its height. Over the past million or so years, the ice sheets have expanded and retreated many times. At their maximum, they overspread an estimated 30 percent of the continents and locked up about 8 percent of the water (Figure 14-1).

Except for the special properties of water, these climatic extremes would have been far more severe. One stabilizing factor is the large amount of heat required to melt ice. Approximately 80 calories must be added to one gram of ice at 0°C just to convert it into water at the same temperature. This heat energy is used to break chemical bonds in converting the systematic atomic structure in an ice crystal to the disordered arrangement in a liquid. Similarly, 80 calories per gram must be removed in the transition from water to ice at 0°C. Approximately 600 calories of heat must be added to a gram of water at 0°C to convert it to water vapor. Or to put it another way, water contains more than twice

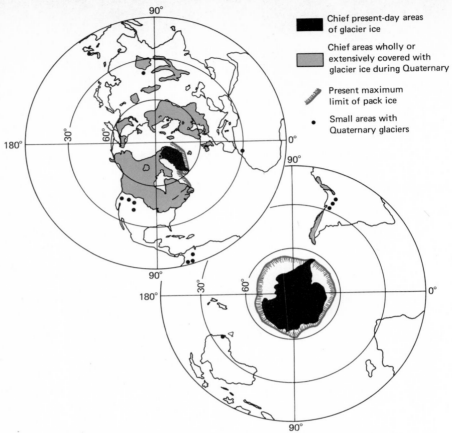

FIGURE 14-1 Quaternary ice sheets have repeatedly occupied much of North America and Europe but only isolated parts of Siberia. Africa and Australia supported no more than minor mountain glaciers. An overspreading ice sheet makes Antarctica the most elevated of the continents today. Glaciers in Antarctica and Greenland discharge icebergs into the ocean; consequently these ice sheets were never much larger than they are now. Icebergs simply broke off the fringe more frequently. [*After J. R. L. Allen*, Physical Processes of Sedimentation, *American Elsevier Publishing Company, Inc., New York, 1970.*]

the internal energy of ice, and water vapor contains 5 times the internal energy of water at a given temperature. Vast quantities of heat can be moved during transformations from solid to liquid to gas, all at roughly constant temperature. Water is an excellent thermal "buffer."

Another peculiarity of H_2O is the "open," rather loose crystal structure of ice. Consequently, ice is about 9 percent less dense than water—an exception to the usual density relation between the solid and liquid states of matter. Since compressing a material makes it more dense, ice may melt solely because it has been placed under pressure (Figure 14-2).

424 THE GREAT ICE AGES

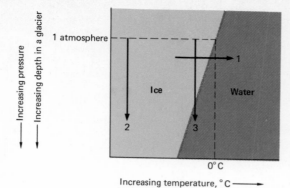

Increasing pressure

Increasing depth in a glacier

1 atmosphere

Ice

Water

1

2

3

0°C

Increasing temperature, °C ⟶

FIGURE 14-2 A phase diagram shows how temperature and pressure affect the stability of water and ice: two phases of H_2O. Each point located in the field of the diagram represents a unique combination of pressure and temperature. Arrow 1 depicts the melting of ice by a temperature increase at constant pressure. If the ice is quite cold, it remains solid when placed under higher pressure (arrow 2). But if the temperature is initially just below the melting point, raising the pressure will melt the ice even at constant temperature (arrow 3).

GLACIER ICE: A METAMORPHIC ROCK

In some ways, a glacier is like a mechanically weak metamorphic rock kept near its melting point. Nourished by falling snow (a sediment), the substance of a glacier continually recrystallizes while in the solid state (metamorphism), finally to lose identity by melting (becoming "igneous"). Unlike ordinary metamorphic rock, glacier ice is plainly visible while all these processes are at work, though sampling the interior of a glacier still presents technical difficulties. Gravity is the controlling force, causing the ice to compact and flow under its own weight. The physics of glacier flow is complex and incompletely understood. Elegant (and partly contradictory) equations to describe this flow have been developed by mathematicians, some of whom probably have rarely set foot on the frozen objects of their study.

Warm and Cold Glaciers

Speaking of cold and warm (or temperate) glaciers may seem a little absurd. Who has seen ice that is not "cold"? And yet, just as conditions of metamorphism deep in the earth may vary over a range of several hundred degrees, the temperature of glacial metamorphism varies also. In glaciers the ice temperature is never higher than 0°C (the maximum possible), nor lower than about −50°C. Even this modest contrast makes for considerable differences between glaciers in polar and in temperate climatic zones.

How does snow become ice? Melting and refreezing may seem the most obvious way to make ice, but in many glaciers, as in central Antarctica and Greenland, melting never occurs. Instead, the texture of snow passes through a series of changes with time and increasing depth of burial. The delicate lacework of a fresh snowflake (Figure 14-3a) is quickly destroyed not only because the fragile points are easily broken off, but also because the surface area of the crystal is so

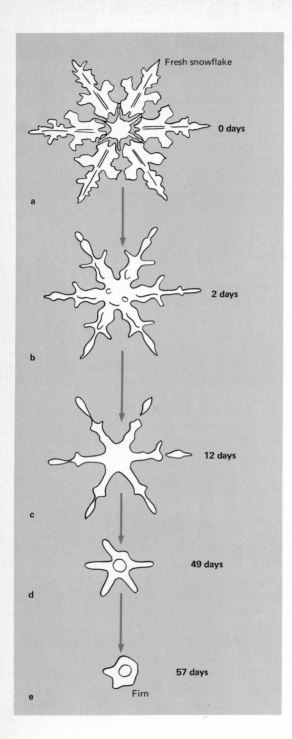

Fresh snowflake

0 days

a

2 days

b

12 days

c

49 days

d

57 days

Firn

e

FIGURE 14-3 Transformation of fresh snow into firn takes but a few days or weeks in all but the coldest glaciers. These changes are brought about primarily by *sublimation*, a process in which H_2O molecules are transferred from an ice crystal into the vapor state, and directly back to the solid state again. (The liquid state is bypassed.) Corners and edges are especially prone to loss by sublimation, whereas broad surfaces are good sites for redeposition. Consequently the firn particle (e) may contain the same volume as the original snowflake, but its surface area was greatly reduced as material was redistributed. [*After H. Bader and others,* Snow and Its Metamorphism, *U.S. Army, Corps of Engineers, Snow, Ice, and Permafrost Establishment, Transl. 14, 1954.*]

large relative to its size. Always the tendency during compaction is to reduce the surface area and to consolidate small particles into larger ones. Soon the dry, powdery snow has converted into rounded grains of "old snow," or *firn* (Figure 14-3e). With further squeezing the firn becomes less porous, and finally, when the air spaces between grains are sealed off, glacier ice is the result.

In cold glaciers the transformation from fresh snow, to firn, to ice is accomplished entirely in the solid state. The situation is quite different in temperate glaciers, where some of the surface melts each summer. As the meltwater seeps down into the body of the glacier and refreezes, it contributes heat to the ice (80 calories per gram). This highly effective means of heat transfer soon brings the entire glacier exactly to the melting point (Figure 14-4). Does the temperature of a warm glacier increase or decrease at greater depth? Is this the usual direction of temperature change downward in the earth? Solid ice is first encountered at a shallow depth in temperate glaciers, but is much deeper in cold glaciers (Figure 14-5).

FIGURE 14-4 Temperatures are below the freezing point everywhere throughout a cold ice sheet, but the deep ice is actually warmest because geothermal heat flows into the glacier from the bedrock. Temperature throughout the body of a warm glacier is exactly at the freezing point of water (melting point of ice). [*After W. S. B. Paterson*, The Physics of Glaciers, *1969. Reprinted with the author's permission. Copyright 1969, Pergamon Press Ltd., Oxford.*]

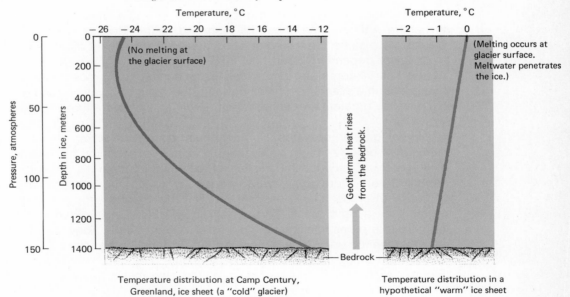

Temperature distribution at Camp Century, Greenland, ice sheet (a "cold" glacier)

Temperature distribution in a hypothetical "warm" ice sheet

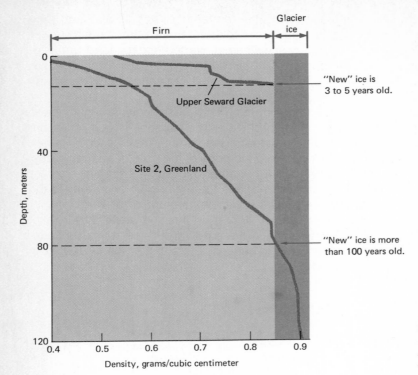

Firn | Glacier ice

"New" ice is 3 to 5 years old.

Upper Seward Glacier

Site 2, Greenland

"New" ice is more than 100 years old.

Depth, meters

Density, grams/cubic centimeter

FIGURE 14-5 Firn has compacted into solid ice at a depth of about 13 meters on Upper Seward Glacier, Yukon (a temperate glacier), but not above a depth of about 80 meters in the Greenland ice sheet, a cold glacier. Snow accumulation rates differ greatly, as suggested by the age of the most recently formed ice in each glacier. Polar ice sheets are deceptive, for though they consist of nothing but H_2O, they are among the most arid regions on earth. Annual precipitation (liquid equivalent) at Site 2, Greenland, is about 35 centimeters per year. Over the Upper Seward Glacier the annual precipitation averages about 10 times greater than at Site 2, and 50 times greater than at the South Pole. [*After W. S. B. Paterson*, The Physics of Glaciers, *1969. Reprinted with the author's permission. Copyright 1969, Pergamon Press Ltd., Oxford.*]

Glacier Dynamics

Whether a glacier is warm or cold, whether it is a small tongue or the entire Antarctic ice sheet, its response to the force of gravity is basically the same. On the other hand, there are significant differences in behavior depending upon how thick the ice is and how much rock debris it can incorporate. Let us look first at some of the processes that take place in a glacier confined between the walls of a valley. (Sometimes valley glaciers are called Alpine because they were first studied in the Alps.)

Every glacier operates on an annual budget in which the *accumulation* of ice at higher elevations results in a continual migration down-glacier to balance the *wastage* of ice and snow near the terminus (Figure 14-6). There are several means by which the ice can accomplish this flow. One possibility is that the glacier simply slides downhill as a rigid block. To test this idea, glaciologists have sunk long, flexible pipes into Alpine glaciers at various points. They discovered that the ice does in fact carry the pipes along bodily, as required by the rigid block model, but in the meantime they become distorted (Figure 14-7). That is, *internal* deformation also makes an important contribution to the forward motion.

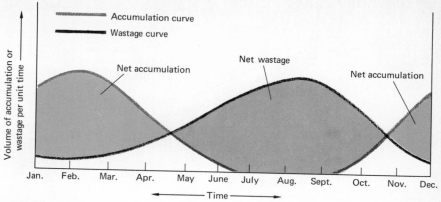

FIGURE 14-6 An annual budget for a Northern Hemisphere temperate glacier shows that accumulation ceases altogether during the summer. Although wastage is continual, it greatly speeds up in the summer months. In a sense, the budget resembles the hydrograph of a stream (see Figure 13-5). The area between a curve and the horizontal axis equals the volume of *total* accumulation or wastage (or volume/time, times time). Shaded areas represent *net* accumulation or wastage. If a glacier is in a steady-state condition (a rare situation), the two types of shaded area are equal. [*After R. P. Sharp*, Glaciers, *University of Oregon Books, Eugene, 1960.*]

When placed under stress, glacier ice will flow plastically: the deformation is permanent even when the stress is released. Presence of deep open cracks, or crevasses, in many glaciers proves that ice within 100 meters of the surface may act in a brittle manner while plastic flow goes on at depth (Figure 14-8, plan view). The flow velocity of some Al-

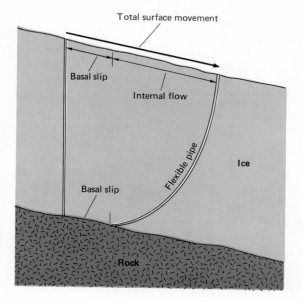

FIGURE 14-7 Basal sliding, lubricated by a film of water at the bottom of a temperate glacier, may account for as little as 10 percent to as much as 90 percent of the total forward motion. Cold glaciers are frozen to the bedrock; their motion is accommodated entirely through internal deformation. [*After R. P. Sharp*, Glaciers, *University of Oregon Books, Eugene, 1960.*]

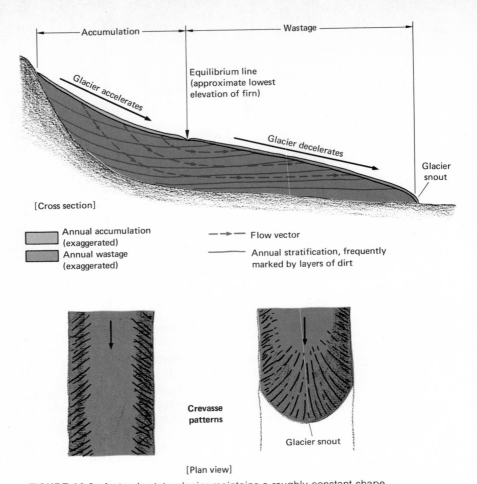

FIGURE 14-8 A steady-state glacier maintains a roughly constant shape and size. Since the rate of accumulation or wastage varies from point to point, the material balance is preserved only if velocity increases as the ice approaches the zone of wastage, then decreases as the ice nears the terminus. In addition, the flow vectors must first plunge *into* the ice; then *out* again farther down-glacier. Unlucky Alpine climbers, buried by an avalanche high up on a glacier, have unwittingly confirmed this flow pattern. Years later their frozen bodies reappeared at the surface at the snout, as predicted by theory. [*After R. P. Sharp*, Glaciers, *University of Oregon Books, Eugene, 1960.*]

pine glaciers may approach zero (stagnation), but normally it lies between 10 and 200 meters per year.

Flowage within a polar ice sheet is not confined by rock walls as it is in a valley glacier. A schematic cross section through the Greenland ice sheet emphasizes the domelike contour of the surface of a continental glacier (Figure 14-9). As an ice layer, deposited initially at the crest, subsides into the interior, it becomes enormously thinned and stretched.

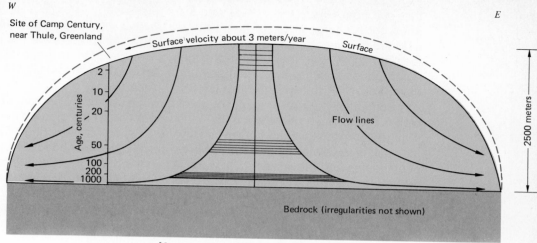

W

E

Site of Camp Century,
near Thule, Greenland

Surface velocity about 3 meters/year

Surface

Age, centuries

2
10
20
50
100
200
1000

Flow lines

2500 meters

Bedrock (irregularities not shown)

[Cross section, vertical exaggeration: 130 times]

FIGURE 14-9 Flow lines trace the paths taken by individual water molecules in the Greenland ice sheet. At Camp Century, data from a continuous core drilled to the base of the ice were analyzed according to a mathematical model. The model nicely illustrates the effect of thinning. Ice about midway down the length of the core is believed to be about 2800 years old, whereas ice at the very base is vastly older—about 100,000 years. A broken line locates the glacier profile that would be established if annual snowfall were to double. Calculations suggest that the ice sheet would grow only slightly thicker, but its flow velocity would increase dramatically. [*After W. Dansgaard and others, "One Thousand Centuries of Climatic Record from Camp Century on the Greenland Ice Sheet," Science, vol. 166, no. 3903, 1969. Copyright 1969 by the American Association for the Advancement of Science.*]

Small-scale internal adjustments in the moving ice include various possibilities. For example, a tiny volume of ice could repeatedly thaw and freeze again after the liquid or vapor had moved a short distance. This process cannot function at all in a cold glacier, and even in most temperate glaciers it is considered not to be important. Perhaps the ice crystals could roll past one another like beans in a beanbag. If this were the case, we would expect the crystals to become spherical, and to be pulverized into smaller grain sizes down-glacier. Instead, just the opposite is seen. The crystals grow larger* and delicate projections extend outward, locking them together. This observation is better explained if the flow is accommodated not so much *between* crystals as *within* them. Crystals weld together, and they deform by slipping along planes of atoms while in the solid state.

Most valley glaciers waste away by melting. In contrast, melting is minor in the great polar ice sheets which advance relentlessly until they enter the sea. Nearly all the loss of mass from Antarctica and about half

* Single crystals a meter or so in diameter are known at the snout of an old, far-traveled glacier such as the Malaspina Glacier, Alaska.

the loss from Greenland comes about by release of icebergs. Source areas for the largest icebergs are the *ice shelves* that fringe the polar land areas. These gigantic sheets are attached to land where the ice is fed into them, but their leading edges are afloat. On at least two recorded occasions, individual icebergs larger than Massachusetts have broken loose from the Ross Ice Shelf (Antarctica).

Sometimes a rather inactive glacier will suddenly become unstable. A thickened bulge appears in the upper part of the glacier, then it migrates in a wavelike manner down the axis, much faster than the ice itself is traveling. As the wave sweeps by, it leaves a chaotic jumble of pinnacles and crevasses in its wake. When the bulge reaches the far end, the terminus may "snap" forward; for 2 or 3 years the flow velocity jumps to as much as 100 times the normal rate (Figure 14-10). Velocities of 20 meters per day are common. Surveys show that a *glacier surge* is purely an internal transaction—a rapid transfer of ice from higher to lower elevation.

None of the many theories put forward to explain glacier surges appears to be very convincing. Surges are very infrequent, and in fact only a handful of glaciers are known to have surged more than once. Although surging glaciers are found in abnormal abundance along the south coast of Alaska, they occur elsewhere too. The surging phenomenon does not correlate with altitude, local climate, glacier size or shape, or whether the glacier is of the warm or cold variety. Occasional earthquakes or large avalanches of snow on a glacier do not seem to trigger a surge.

GLACIAL EROSION AND DEPOSITION

Although small Alpine glaciers had long attracted notice, geologists did not begin to realize until the early 1800s that an ice sheet once overspread northern Europe and North America. An acquaintance with modern ice sheets would have helped to break down opposition to this radical idea, but at that time Antarctica was unknown and Greenland unexplored. Evidence for former continental glaciation was all indirect; it centered in the effects of erosion and deposition by the vanished ice sheet. Even today, little has been done to organize this information into a glacial "depositional system" like those mentioned in the preceding chapter.

Pure ice is probably an ineffectual engine of erosion (it deforms too easily), but ice that is armed with rock debris is quite another matter. The embedded fragments scour the bedrock, themselves becoming worn smooth (faceted) on one face (Figure 14-11). Other facets develop if the abrasion tool is occasionally forced to rotate in its icy matrix.

Debris left by a melted glacier is thus a highly distinctive clastic sediment, rich in fragments from boulder-sized to submicroscopic. If

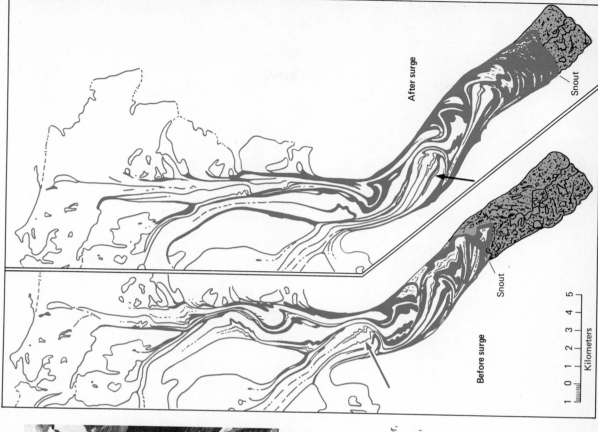

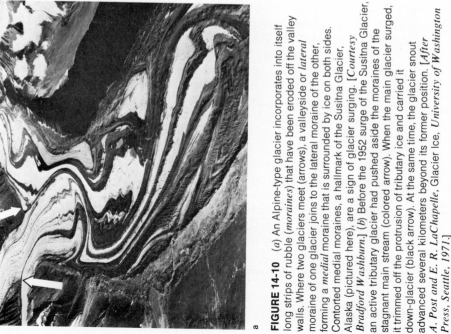

FIGURE 14-10 (a) An Alpine-type glacier incorporates into itself long strips of rubble (*moraines*) that have been eroded off the valley walls. Where two glaciers meet (arrows), a valleyside or *lateral* moraine of one glacier joins to the lateral moraine of the other, forming a *medial* moraine that is surrounded by ice on both sides. Contorted medial moraines, a hallmark of the Susitna Glacier, Alaska (pictured here), are a sign of glacier surging. [*Courtesy Bradford Washburn.*] (b) Before the 1952 surge of the Susitna Glacier, an active tributary glacier had pushed aside the moraines of the stagnant main stream (colored arrow). When the main glacier surged, it trimmed off the protrusion of tributary ice and carried it down-glacier (black arrow). At the same time, the glacier snout advanced several kilometers beyond its former position. [*After A. Post and E. R. LaChapelle, Glacier Ice, University of Washington Press, Seattle, 1971.*]

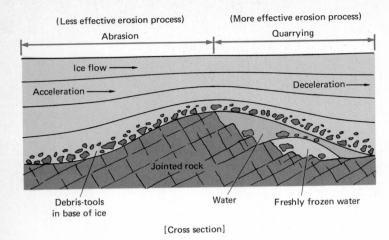

(Less effective erosion process)   (More effective erosion process)
Abrasion   Quarrying

Ice flow →

Acceleration →   Deceleration →

Jointed rock

Debris-tools   Water   Freshly frozen water
in base of ice

[Cross section]

FIGURE 14-11 Debris embedded in the sole of a glacier can abrade the bedrock. If the bedrock contains open cracks (joints), meltwater can penetrate them and refreeze, prying apart the rock so that it is easily dislodged. This "plucking" action requires an alternation between freezing and thawing. Freezing at night and thawing during the day are common in the summer at the head of a temperate Alpine glacier. [*After J. R. L. Allen, Physical Processes of Sedimentation, American Elsevier Publishing Company, Inc., New York, 1970.*]

the material is an unsorted "boulder clay," it is called *till*—a name familiar to the old Scottish farmers who must have cursed the ground because it is so difficult to till. This material contains a fascinating assortment of cobbles and boulders (some of them enormous) that have been quarried from distant regions. Studies of distinctive rock fragments that can be matched with the original outcrop suggest that most of the pebbles have traveled only a few kilometers at most. In a few instances, transportation of more than a thousand kilometers has been documented. Mapping indicates that the pieces are "sprayed out" over a fan-shaped area immediately down-glacier from the outcrop. What is now the largest copper mine in Finland was discovered when ore-containing glacial boulders were traced back to their source. Scores of excellent pebble-sized diamonds have been found in glacial deposits in Pennsylvania and New York, and in the Midwestern states near the Great Lakes. No one has yet discovered the source(s), which is assumed to lie in the Canadian Shield.

Near the border of a continental glacier there is a transition between thick ice that is still moving, and thin, brittle ice that has stagnated at the very edge (Figure 14-12). The flowing ice lifts to the surface a mass of debris as the ice shears across and rides up over the stagnant fringe. So effectively has the coating of sediment insulated the stagnant ice of the Malaspina Glacier that a dense forest of century-old spruce trees became established on top. In some places, stagnant ice has taken as long as 3000 years to melt. Finally, a thin but extensive till sheet is let down on the ground beneath. Much of the till is reworked by the meltwater that pours out of the front of a wasting glacier. The glacial streams promptly convert the unstratified debris into stratified deposits by means of the now familiar fluvial processes described in the preceding chapter. In addition, there are some peculiar environments of deposition found within the stagnant ice just before its disappearance (Figure 14-13).

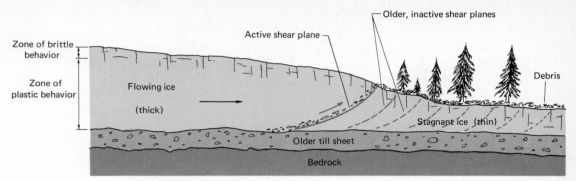

FIGURE 14-12 A shear plane marks the boundary between the region of plastic flow and the zone of brittle stagnation. Wastage is causing the glacier to retreat, even though ice is continually moving forward. Ice is moving up along only one shear plane at any one time. Plastic flow is possible only if the ice is at least approximately 100 meters thick. [*After Dwight E. Deal*, Geology of Rolette County, North Dakota, *North Dakota Geological Survey, Grand Forks, Bulletin 58, 1971.*]

FIGURE 14-13 Meltwater cuts channels both on and below the surface of disintegrating ice. A *kame* terrace is a stratified deposit formed along the glacier edge. Melted blocks of stranded ice leave behind pockmarks, or *kettles*. An *esker* is a stratified, ridge-shaped deposit that may snake across the countryside for hundreds of kilometers. It may climb over low divides and creep down across valleys in complete disregard of present topography. This peculiarity suggests that the meltwater was confined to deep channels or to tunnels under the ice. Stratified glacial deposits are important sources of sand and gravel for the construction industry. [*After R. F. Flint*, Glacial and Quaternary Geology. *Copyright © 1970. By permission of John Wiley & Sons, Inc., New York.*]

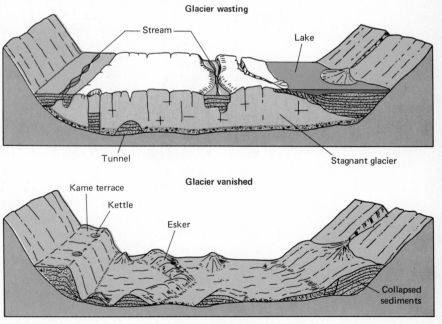

[Block diagrams]

Glaciated Landscapes

Most of the low-lying shield areas of the world apparently have been glaciated at one time or another. Do these terrains owe their flat profile to glacial erosion, or did the glaciers merely skin off a weathered soil zone? The volume of material scraped off the Canadian Shield is quite modest, suggesting that erosion there was insignificant once the ice had scoured down to a cleaned and polished bedrock surface. On the other hand, the estimate of sediment volume may fall far short if much of the glacial scrapings have gone down the Mississippi River system. In any case, glacial erosion is spectacular where an ice sheet can take advantage of a preexisting valley. A series of large freshwater lakes are arrayed along the boundary between the crystalline shield in Canada and its overlapping Paleozoic sedimentary cover (Figure 14-14). Glacier ice has scooped out the bottoms of some of these lakes well below sea level.

Continental ice sheets are responsible for an interesting application of the principle of isostasy. Such great thicknesses of ice represent a considerable load pressing down upon the crust. (The floor of Hudson Bay [Figure 14-14] and much of the rock surface in Greenland and Antarctica have been depressed below sea level by the weight of ice.) After the Canadian ice sheet disappeared, the land surface began to rebound upward as rapidly as several centimeters per year at first, then more and more slowly but continuing until the present. In a few more thousand years with continued isostatic recovery, Hudson Bay will have shrunk to a small inlet.

Glaciation of a highland may create a variety of landforms. If a continental ice sheet is thick enough to overrun the high country (as in parts of Antarctica today), the contours of the landscape exposed after ice retreat are rounded and subdued (Figure 14-15). The Adirondack Mountains, New York, are a good example of ice-smoothed terrain (Figure 14-16*a*).

Other mountainous regions, eroded by Alpine glaciers, contain some of the most magnificent scenery known to man (see Figure 14-16*b* and lead photograph of this chapter). In these localities the ice was confined to preexisting valleys. At the valley heads, the glaciers have carved huge, steep-walled depressions called *cirques* (serks). In some places, glaciers on opposite sides of a ridge have excavated to the point that only a towering, knife-sharp partition separates the cirques on either side (Figure 14-15). Glaciers flowing down several sides of a mountain (for example, the Matterhorn in the Alps) have chiseled it into a pyramid. At lower elevations, the ice became loaded with debris avalanched down from the valley walls. The glaciers effectively used these abrasion tools to straighten out originally crooked valleys and to reshape the V-shaped profile of the valley walls into the U-shaped profile characteristic of glaciated terrains. Where a tributary flowed into a master ice stream, the upper surface of the ice joined

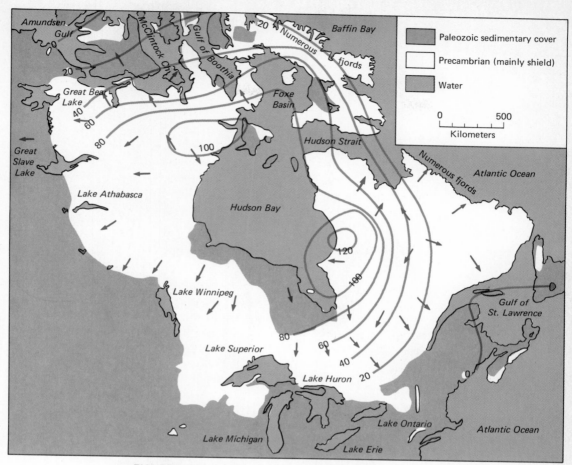

FIGURE 14-14 Erosion, deposition, and vertical movements of the earth's crust have operated on a grand scale in the glaciated expanses of Canada. Arrows point the directions of ice movement, as indicated by striations (grooves), etc., in the bedrock. Broadly speaking, they radiate from two focal areas, one east and the other northwest of Hudson Bay, suggesting that the most recent ice sheet was thickest in these two places. Till deposits and their stream-reworked equivalents are concentrated in the fringing one-third of the glaciated area. Fertility of much of the soil in the Corn Belt is due to the materials transported and deposited by glaciers. Along the eastern shores of Baffin Island and Labrador, the ice gouged numerous fjords. Contours indicate the amount of isostatic uplift (in meters) during the past 8000 years.

smoothly between the two glaciers. Moreover, the cross sections of each glacier were similar in shape; consequently the master glacier eroded a channel much deeper and wider than that of the tributary. When the ice melted away, the feeder valleys were left hanging high above the sheer walls of the main valley. Yosemite National Park, California, contains some outstanding examples of this type of glacial modification.

437 GLACIAL EROSION AND DEPOSITION

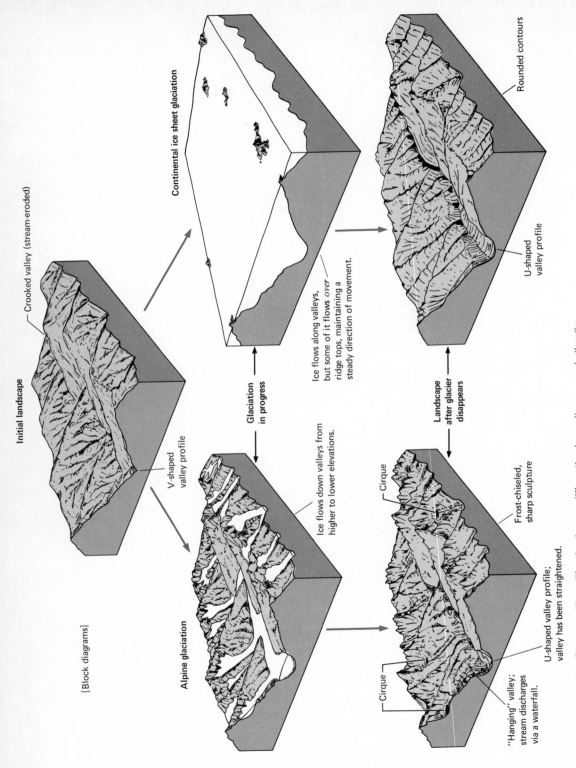

FIGURE 14-15 Glaciers will modify a landscape differently, depending upon whether the ice was restricted (Alpine type) or whether it blanketed the region. Unlike streams, an ice sheet attacks the entire landscape simultaneously. [*After R. F. Flint, Glacial and Quaternary Geology, Copyright © 1970. By permission of John Wiley & Sons, Inc., New York.*]

Initial landscape

Crooked valley (stream-eroded)

[Block diagrams]

V-shaped valley profile

Continental ice sheet glaciation

Glaciation in progress

Ice flows along valleys, but some of it flows *over* ridge tops, maintaining a steady direction of movement.

Landscape after glacier disappears

Rounded contours

U-shaped valley profile

Alpine glaciation

Ice flows down valleys from higher to lower elevations.

U-shaped valley profile; valley has been straightened.

"Hanging" valley; stream discharges via a waterfall.

Cirque

Cirque

Frost-chiseled, sharp sculpture

a

b

FIGURE 14-16 (*a*) Contours of the landscape in this part of the Adirondack Mountains, New York, have become smoothed and rounded by an ice sheet that once completely buried it. [*Courtesy Jerome Wyckoff.*] (*b*) When valley glaciers retreated in the Sierra Nevada, California, they left behind a rugged landscape. [*U.S. Forest Service photograph.*]

439

ICE AND ISOTOPES*

Most of us live where vegetation and industry are continual sources of haze in the atmosphere. Only the occasional traveler has witnessed the extreme clarity of the polar air, which usually enables one to see perfectly to the horizon no matter how distant it is. Polar ice sheets also are exceedingly pure,† so it is no surprise that annual layering in the ice may appear as no more than subtle variations of texture—variations that are often so subtle that they are invisible! How can we learn anything about an apparently featureless pile of ice? What can serve as a "tracer" of its internal structure? Fortunately, it turns out that the frozen H_2O is its own best tracer. Hydrogen and oxygen that make up the very substance of the ice contain information about its history. Let us see how this is so.

In nature, the element hydrogen consists of two stable isotopes, ^1H and ^2H (sometimes called protium and deuterium). Oxygen has three stable isotopes, ^{16}O, ^{17}O, and ^{18}O.‡ You recall that different isotopes of a given element contain different numbers of neutrons in the nucleus, but that the chemical bonds they form with other atoms all behave the same. More accurately, they behave *almost* the same. Geochemists have discovered that the stable isotopes of hydrogen and oxygen are selectively "partitioned" into different environments. For example, if water and water vapor are in contact, the oxygen-16 atoms tend to concentrate ever so slightly in the vapor. Similarly, oxygen 18 slightly prefers the liquid phase. Oxygen in the liquid is "heavier" than oxygen in the vapor.

Suppose we designate oxygen in ocean water as a standard of reference. (It is an ideal choice because the ocean is so large, because all other forms of surface H_2O eventually return to it, and because its isotope composition is uniform.) As clouds form, they selectively take up ^{16}O, but there is 100,000 times more water in the ocean than in the earth's cloud cover at any one moment. Hence the isotopic composition of the ocean, an "infinite" reservoir, is scarcely changed. A cloud is a much smaller reservoir, however. Falling rain or snow preferentially drains the heavy isotope (^{18}O) from the cloud, leaving the vapor enriched in oxygen 16 (Figure 14-17). As the precipitating cloud sweeps inland over an ice sheet, oxygen in the remaining "dregs" of water vapor becomes lighter and lighter. (At the South Pole the value of

---

* See Chapter 1 for a review of isotopes.
† For example, the abundance of insoluble dust particles in central Antarctica ice is only about 2 parts per billion. A fair proportion of the dust consists of micrometeorites (particles of cosmic origin).
‡ The proportion of atoms of ^1H to ^2H is roughly 6500:1. The proportion $^{16}O/^{17}O/^{18}O$ is approximately 2500:1:5.

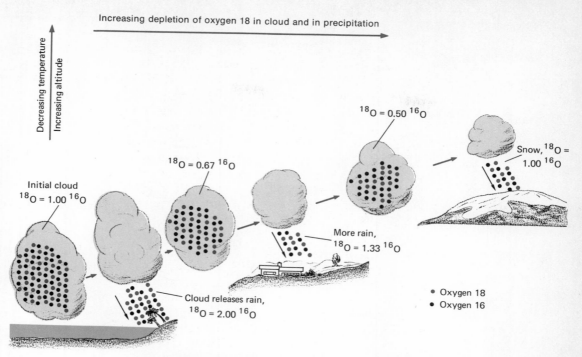

Decreasing temperature

Increasing altitude

Initial cloud
$^{18}O = 1.00\ ^{16}O$

$^{18}O = 0.67\ ^{16}O$

$^{18}O = 0.50\ ^{16}O$

Snow, $^{18}O = 1.00\ ^{16}O$

More rain,
$^{18}O = 1.33\ ^{16}O$

Cloud releases rain,
$^{18}O = 2.00\ ^{16}O$

● Oxygen 18
• Oxygen 16

FIGURE 14-17 If oxygen 18 is more abundant in precipitation than in the source cloud, both the cloud and continuing precipitation will become more depleted in oxygen 18. In this schematic drawing, the ^{18}O atoms are precipitated twice as readily as ^{16}O atoms. The degree of fractionation and the abundance of ^{18}O are highly exaggerated for emphasis.

$^{18}O/^{16}O$ is about 5 percent less than it is in ocean water.) Moreover, lowered temperatures enhance this partitioning effect, so that oxygen in winter snow falling *at a given spot* on the ice sheet is lighter than in summer snow.

Oxygen isotope analyses of the ice core from Camp Century, Greenland (Figure 14-9), have revealed the local climatic history in great detail. Near the surface where the ice is not so compacted, isotopic variations trace the summer and winter snow deposition year by year (Figure 14-18a). A thickness of ice representing but a few years' deposition high in the ice sheet corresponds to a duration of many thousands of years near the base, where the ice is stretched and thinned. Isotopic data from the lower part of the core are useful to detect the broad variations of climate recorded over the past 1200 centuries. Small wiggles in the curve, representing annual layers, are superimposed upon large swings that suggest three major episodes (Figure 14-18b). From 120,000 to 70,000 years ago, and from 12,000

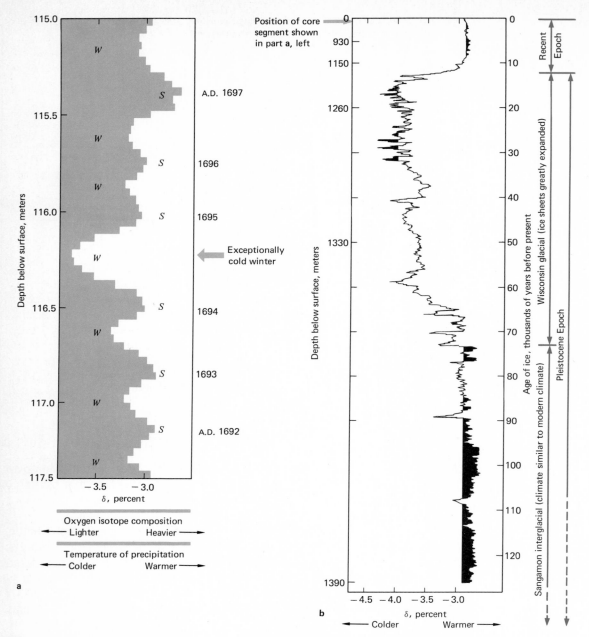

FIGURE 14-18 (*a*) Oxygen isotope compositions of some 7500 samples cut from the Camp Century ice core were painstakingly analyzed. A short section about 115 meters beneath the surface contains the isotopic record of the frigid weather of the "little ice age." Historical documents mention that the winter of A.D. 1694–1695 in nearby Iceland was extremely cold. Values of "delta" (δ) are the percent difference between $^{18}O/^{16}O$ in the sample and in ocean water, the reference standard. Negative values simply mean that oxygen in the sample is lighter (has less ^{18}O) than the standard. *W* and *S* indicate winter and summer respectively. (*b*) Data from the entire core are plotted against a linear time scale (right-hand margin) but a foreshortened scale of ice thickness (left-hand margin). Delta values that are exceptionally high or low are set off in black. Even in the presence of minor fluctuations, three climatic periods stand out clearly. They appear to be the Recent warm episode, the Wisconsin glacial episode, and the Sangamon interglacial episode, all of which were long ago identified from other evidence. [*After S. J. Johnsen and others, "Oxygen Isotope Profiles through the Antarctic and Greenland Ice Sheets," Nature, vol. 235, no. 5339, 1972.*]

years ago until the present, the climate (though cold enough to maintain the ice) was relatively warm. Between 70,000 and 12,000 years ago, the climate was colder. The latter was also a time when continental ice sheets were greatly expanded.

THE CARBON-14 CLOCK

Again and again we have seen that explanations of geologic phenomena draw heavily upon knowledge of their *timing*. Geologic time is important in any theory of the ice ages, for example, in considering why the continents bear extensive ice sheets only now and then. Geologists have long resorted to superimposed glacial deposits as evidence for glacial advances and retreats, but this is mainly to establish a *relative sequence* of past events, not their ages. Evidence from stable isotopes is intriguing, but we must keep in mind that the ages at Camp Century are calculated from a mathematical model that possibly is inaccurately calibrated. The uranium-lead, rubidium-strontium, and potassium-argon age methods are not applicable to glacial deposits (or to most other sediments). Nor are these methods suitable to measure ages spanning the past few thousand years—times for which the glacial record is best preserved. Is there another isotopic clock that fulfills these special requirements?

Groundwork for such a dating procedure was laid several decades ago by physicists who were studying *cosmic rays*. They learned that

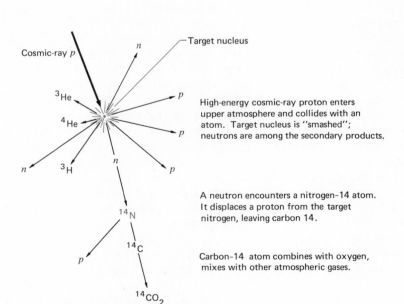

Cosmic-ray p

Target nucleus

High-energy cosmic-ray proton enters upper atmosphere and collides with an atom. Target nucleus is "smashed"; neutrons are among the secondary products.

A neutron encounters a nitrogen-14 atom. It displaces a proton from the target nitrogen, leaving carbon 14.

Carbon-14 atom combines with oxygen, mixes with other atmospheric gases.

FIGURE 14-19 A complicated chain of steps leads to a steady production of radioactive carbon dioxide in the air.

443

FIGURE 14-20 Annual layers in gnarled, stunted bristlecone pine trees living on the ragged edge of existence vary greatly in width, depending upon the whims of the weather prior to and during a particular growing season. Each growth ring is a reservoir of carbon 14 that is isolated from further exchange with atmospheric CO_2 when the wood dies. [*Courtesy H. C. Fritts and Laboratory of Tree-Ring Research, University of Arizona, Tucson.*]

these rays are extremely high-energy particles (mainly protons) that constantly bombard the earth from all directions. (There is evidence that supernovas [Chapter 1] are the source of most of these particles.) When a cosmic ray collides with an atom in the upper atmosphere, it splits the target nucleus into smaller particles (Figure 14-19). Among these secondary products are neutrons that in turn react with nitrogen in the air, producing carbon 14. Thus carbon 14 is continuously being created,* but because it is a radioactive isotope (its half-life is 5730 years), it is also constantly disappearing. Broadly speaking, the rate of creation and radioactive decay are in a steady-state balance such that the entire earth contains about 60 metric tons of ^{14}C at any one moment.

In 1949 the nuclear chemist W. F. Libby and his associates proposed a clever way to use carbon 14 (radiocarbon) as a geologic clock. They likened the earth to a complicated plumbing system of interconnected reservoirs. A newly created atom of radiocarbon promptly combines with oxygen, forming $^{14}CO_2$, which mixes among all the other gas molecules in the atmosphere reservoir. Of course, the atoms of stable carbon isotopes (^{12}C and ^{13}C) greatly outnumber the ^{14}C atoms. Before man had tampered with the atmosphere's balance, the ratio $^{14}C/^{12}C$ was maintained at about 1:800 billion, equivalent to 14 radioactive decays of ^{14}C per minute per gram of carbon.

Other reservoirs, such as the ocean and the biosphere, are in constant communication with the atmosphere. Suppose that a bristlecone pine tree were to use some of the CO_2 in the production of its wood.

* At a rate of two atoms per second per square centimeter of the earth's surface.

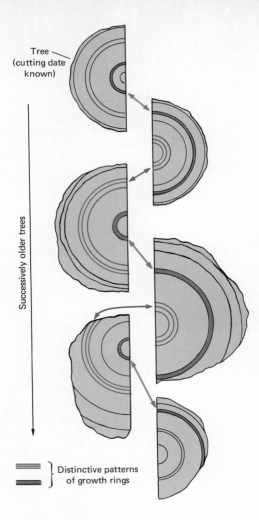

Tree
(cutting date
known)

Successively older trees

≡≡≡ } Distinctive patterns
▬▬▬ } of growth rings

FIGURE 14-21 Growth-ring patterns can be correlated from a modern tree back through a succession of more ancient trees whose lifetimes overlapped. Ancient wood can thus be dated by a means that is independent of the radiocarbon method. Since an occasional year's growth may be represented by a double ring, or no ring at all, the correlation must take into account the *entire* set of rings, not just a few as pictured here. Living bristlecone pine trees as old as 46 centuries are known, and by means of growth-ring correlation, the ages of tree rings going back more than 8000 years can be precisely identified. (However, the oldest living organism is the lowly lichen that coats the surfaces of rock. Slow-growing individuals in arctic regions are estimated to be 9000 years old!)

Once a growth ring has been deposited (Figure 14-20), it no longer shares in the life processes of the tree. Carbon 14 in this wood merely continues to dwindle away by radioactive decay. Old tree rings are "dead-end" reservoirs, unable to donate or accept carbon atoms from the environment (Figure 14-21). Similarly, radiocarbon in animal tissue is maintained at a steady concentration. Animals continually incorporate radiocarbon from their food (whether plants or other animals) and excrete carbon-containing substances in their waste products. When the animal dies, this steady-state exchange ceases.

Decay of radiocarbon provides the basis of the dating method. A tree ring that is 5730 years old (one half-life) contains only half the abundance of radiocarbon found in a modern (living) tree ring. Wood that is 11,460 years old (two half-lives) contains one-fourth as much ^{14}C,

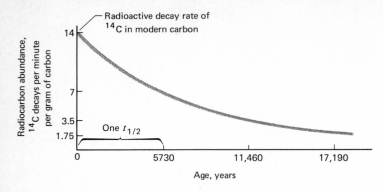

FIGURE 14-22 The radioactive decay rate is proportional to the amount of carbon 14 present in a sample. Carbon-14 abundance in an isolated reservoir (the sample) declines according to the familiar radioactivity curve (compare with Figure 6-5).

etc. (Figure 14-22). By the time the material is about 40,000 years old, so little carbon 14 remains that tiny amounts of contamination by "modern" carbon appear large by comparison. This factor sets a practical limit to the oldest age that can be measured. The radiocarbon method has been applied to a variety of materials—wood, charcoal, bone, peat, shells, even to air dissolved in glacier ice. It has become an indispensable tool to archaeologists as well as to glacial geologists.

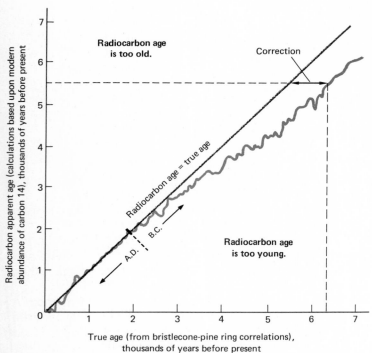

FIGURE 14-23 Ancient, naturally caused fluctuations in the rate of creation of radiocarbon pose a serious difficulty to the carbon-14 method. Analyses of bristlecone pine wood of known age (from tree-ring correlations) lead to a calibration curve (in color). For example, a correction (horizontal arrow) must be added to a radiocarbon apparent age of 5500 years to obtain the true age (6350 years).

CAUSES OF THE ICE AGES

The origin of the Ice Ages remains one of the most stubborn mysteries in geology. It is only partially solved in spite of the fact that more than 600 theories of continental glaciation have been proposed! By now it has become clear that the great Ice Ages do not have *a* cause—they have *many* causes that interact in complex ways. What are the essential conditions of an ice age? In simplest terms, if, over a period of years, more snow falls than is melted, a glacier is certain to appear. Mere cold is not enough; a plentiful supply of snow is also required. Pearyland (northernmost Greenland) is extremely cold, but it does not support an ice sheet today because it is too dry. At the same time, glaciers flourish in high-precipitation areas (for example, the Olympic Peninsula, Washington) even though the average temperature there may be several tens of degrees warmer than in Pearyland. Climate today hovers close to the brink of another full-blown glaciation. Calculations suggest that if the mean annual temperature in Canada were to drop about 7°C, another ice sheet would begin to grow. The balance is delicate indeed, and easily disturbed.

History of the Ice Ages

Any successful theory of the Ice Ages must answer the question of why continental glaciers have appeared so seldom and at such irregular intervals. Little detail is known of two *Cryptozoic* episodes that occurred around 2.2 billion and 600 million years ago. (Other Cryptozoic ice ages may await discovery.) During the *Paleozoic* Era a major ice sheet prevailed over the southern continents. Profound *Cenozoic* glaciation began in Antarctica around 13 million years ago, as inferred from the appearance of glacial debris in late Miocene ocean sediments in the vicinity. Later, the glaciation spread to the Northern Hemisphere (Figure 14-1). At times other than these four major episodes, world climate apparently was too warm to sustain glaciers.

A closer look at the Cenozoic record brings to light other complexities. Studies of till in the Midwestern states and in Europe have demonstrated that ice sheets pushed far to the south, then retreated a number of times. Till sheets, strongly weathered at the top, may be overlain by younger, unweathered deposits (Figure 14-24). Such deep weathering signifies that the readvancing glacier did not deposit the next till sheet until thousands of years had gone by. Spores and pollen in buried peat bogs indicate that plants living during these *interglacials* (warm periods) were much the same as those living in the same areas today.

The late Cenozoic *multiple glaciations* have even left an imprint in the tropical ocean, far removed from the realm of the ice. Deep-ocean sediment cores are informative from many viewpoints. For example, a

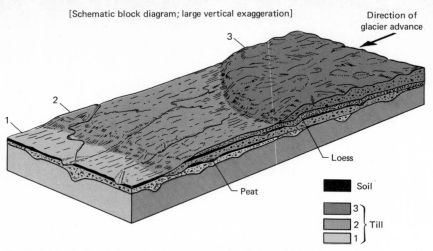

[Schematic block diagram; large vertical exaggeration]

Direction of glacier advance

Loess

Peat

Soil

3 ⎫
2 ⎬ Till
1 ⎭

FIGURE 14-24 For thousands of years, an early till sheet (1) lay exposed while weathering developed a thick soil cover (black). Large granite boulders "rotted" into masses of crumbly grains. Peat bogs filled depressions in a humpy surface upon which drainage was disoriented and incomplete. Another long weathering period followed a glacial readvance (till sheet 2) that stopped short of the first advance. A deposit of wind-blown silt (loess) was laid down in the arid terrain fronting the advance of the most recent ice sheet (till 3). Modern drainage is disrupted on the surface of the uppermost deposit, but it has had time to become organized into a dendritic pattern on the older till sheets. [*After C. O. Dunbar and K. M. Waage,* Historical Geology, *Copyright © 1969. By permission of John Wiley & Sons, Inc., New York.*]

record of magnetic field reversals is preserved as a succession of normally and reversely magnetized layers of sediment (Figure 14-25). Because the polarity reversals were sudden and worldwide, geophysicists have been able to establish a "magnetic stratigraphy" that can be correlated from one area to another even though sediments were not everywhere deposited at the same rate. We recall that the more recent magnetic reversals have been dated by the K-Ar method. This enables us to attach a time scale to the deposition of the sediments.

Paleontologists have studied the forams (tiny fossils) preserved in the same core samples (see Figure 5-9). Population abundance of several foram species is sensitive to small changes of the water temperature. Census counts of these fossils show that the balance between warm- and cold-water species has swung wildly back and forth 20 times or more during the past 2 million years (Figure 14-25). Geochemists have reached the same conclusion from evidence based upon oxygen isotopes in the foram skeletons ($CaCO_3$). A skeleton in contact with cold water selectively incorporates more oxygen 18 than it does when in contact with warmer water. The isotopic data suggest that tropical water temperature varied only approximately 5°C between glacial and interglacial periods.

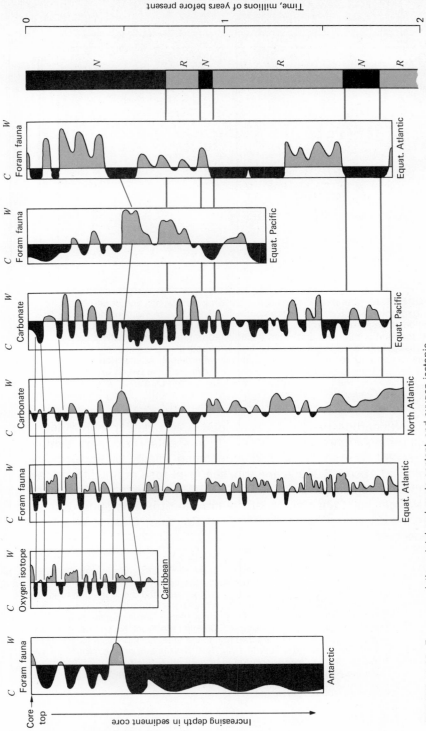

FIGURE 14-25 Foram populations, total carbonate content, and oxygen isotopic compositions have been examined in deep-ocean cores from a variety of places. All these factors indicate water temperature and, indirectly, whether the prevailing condition was glacial or interglacial. Pictured core lengths are adjusted so that magnetic reversals, which were recorded simultaneously everywhere, are aligned across the page. Correlations of warm (*W*) and cold (*C*) periods match rather well throughout the Atlantic Ocean, but less well between the Atlantic and Pacific where glacial influence was less pronounced. Antarctic waters have remained cold almost continually since at least 1.5 million years ago. The most recent, or Wisconsin, continental glaciation corresponds to no more than the uppermost wiggle or two of these curves. [*After W. F. Ruddiman, "Pleistocene Sedimentation in the Equatorial Atlantic: Stratigraphy and Faunal Paleoclimatology," Geological Society of America Bulletin, vol. 82, 1971.*]

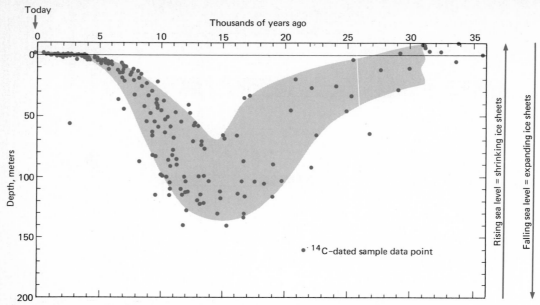

FIGURE 14-26 Samples of salt-marsh peat, drowned forests, and coral, dredged from various depths on the continental shelf, are an index of the position of the former shoreline. In general, the data agree as to when sea level reached a minimum and when it stabilized in the modern position, but they disagree as to *extent* of sea-level drop. Probably the continental shelf is slowly warping up in one place and down somewhere else, independently of sea level. [*After K. O. Emery and others, "Post-Pleistocene Levels of the East China Sea,"* The Late Cenozoic Glacial Ages, *Yale University Press, New Haven, Conn., 1971.*]

Radiocarbon dates have provided a fascinating insight into the waning days of the latest continental ice sheets. Rising sea level (Figure 14-26) was an index of the declining vitality of these glaciers. Beginning about 15,000 years ago, sea level rose rapidly; then it stabilized and remained so, since about 5000 years ago. Radiocarbon dates from the Canadian Shield portray the disintegration of the ice even more vividly (Figure 14-27). Final stages of disappearance were almost catastrophic.

FIGURE 14-27 Contours based upon radiocarbon dates, glacial moraine ▶ directions, and other evidence follow the headlong retreat of the Canadian ice sheet a few thousand years ago. Ice persisted the longest in precisely those areas where it originally was thickest (compare with the outspreading flow pattern, Figure 14-14). For a while, the ice sheet was melting back nearly a meter per day along a 7000-kilometer front. No wonder that flood stories were an important tradition to the ancient Hebrews and to many other cultures the world around! [*After R. A. Bryson and others, "Radiocarbon Isochrones on the Disintegration of the Laurentide Ice Sheet,"* Arctic and Alpine Research, *vol. 1, no. 1, 1969.*]

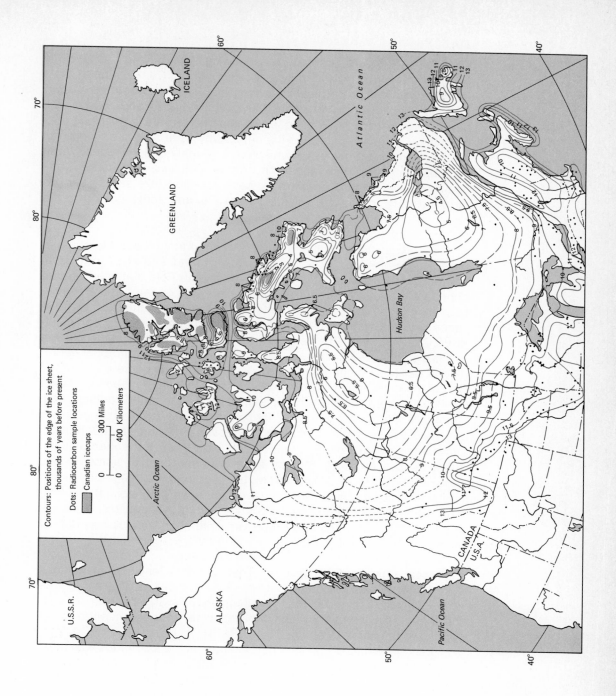

Contours: Positions of the edge of the ice sheet,
thousands of years before present

Dots: Radiocarbon sample locations

Canadian icecaps

0   300 Miles
├─────────┬─────────┤
0   400 Kilometers

ICELAND

GREENLAND

Atlantic Ocean

Hudson Bay

Arctic Ocean

CANADA
U.S.A.

ALASKA

U.S.S.R.

Pacific Ocean

Facts, Theories, and Fancies

It seems that the problem of explaining the Ice Ages has become very complicated. How can we account for both a *long* time scale (hundreds of millions of years between major glaciations) and a *short* time scale (thousands of years between advances and readvances during the Cenozoic*)? Why do ice sheets disintegrate in less than one-tenth the time needed to build them up? Why were glacial advances in the Northern and Southern Hemispheres roughly simultaneous? What initiates the cycle of advances and recessions, and what turns it off? Some of the proposed answers to these questions are solidly backed by good data; some are plausible hunches, and some belong in the realm of fantasy.

Tectonic plates and sheets of ice   An important clue to the origin of the Ice Ages is seen in the distribution of land and sea today. Both polar regions are thermally isolated—the South Pole stationed upon a continent, and the North Pole in a fairly restricted ocean basin. Because heat cannot be readily transported into the high latitudes, climatic zones tend to become sharply distinct. World climate would be far different if a pole were centered in the vastness of the Pacific Ocean.

If ice ages result from an accident of geography, what about earlier glacial periods? What if geography were different in some earlier day? One of the most satisfying by-products of plate tectonic theory (Chapter 12) is its power to explain the Paleozoic glaciation in the southern continents. At that time, you recall, these lands were massed together as one supercontinent. Moreover, in the Ordovician Period, North Africa lay athwart the South Pole (Figure 14-28). Slowly, as the supercontinent drifted across the pole, the ice sheet migrated with it across Africa and South America. A glacial record appears in many places as striations of the bedrock and as layers of *tillite* (lithified till). By Permian times, just before the breakup of the supercontinent, ice sheets lay over Antarctica and Australia. (However, Antarctica was ice-free during the Mesozoic and early Cenozoic.)

Cenozoic multiple glaciations   The multiple Cenozoic glaciations are difficult to explain, partially because a number of different factors were involved. It may be that the climate will swing between glacial and interglacial conditions only when several of these happen to be working in the same direction.

One conjecture is that the energy output of the sun varies with time. Not all the solar energy falling upon the earth is retained at the surface. The earth's *albedo* (al-BEE-doe)—the fraction of solar energy reflected

* Limited evidence suggests a number of glacial "pulsations" in the Paleozoic episode as well.

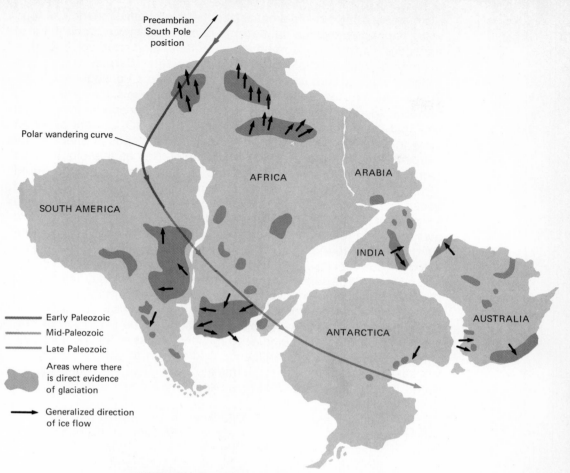

Precambrian
South Pole
position

Polar wandering curve

AFRICA

ARABIA

SOUTH AMERICA

INDIA

AUSTRALIA

ANTARCTICA

Early Paleozoic

Mid-Paleozoic

Late Paleozoic

Areas where there
is direct evidence
of glaciation

Generalized direction
of ice flow

FIGURE 14-28 A curve traces the Paleozoic wanderings of the South Pole. Its path, obtained from paleomagnetic measurements (Chapter 10), is equivalent to a drift of the supercontinent over a fixed pole position, or to a combination of continental drift and polar wandering. Glaciation may have continued unrelentingly for as long as 400 million years. In fact, this episode may be a continuation of the latest Precambrian ice age evidenced in Britain, Scandinavia, and Greenland. If so, it implies that in a period of 600 million years, Greenland has drifted nearly from pole to pole. Note that ice flowed *inland* onto what is now southern Australia. This pattern makes no sense unless Australia was attached to another land mass at the time.

directly back into space—is about 33 percent. Energy absorbed in the atmosphere and on the surface is partly converted into heat used to drive winds and ocean currents, make clouds, etc. Model calculations suggest that a 1 percent change in solar input could lead to a highly significant 8°C change in average surface temperature. Unfortunately, we have no way to test the idea that the sun is a slowly "flickering" star.

453 CAUSES OF THE ICE AGES

A related hypothesis emphasizes the role of atmospheric CO_2 and water vapor. Some of the light energy passing downward through these transparent gases is absorbed at the surface, then returned upward in the form of longer wavelength (infrared) radiation. Carbon dioxide and water strongly absorb and reemit the infrared. As a consequence, even though these gases are present only in trace amounts, they are able to trap considerable solar energy in the lower atmosphere. We might anticipate that climate would grow warmer if the CO_2 content of the air were to increase. On the other hand, just the opposite effect might happen if more rapid evaporation prompted the earth's cloud cover (currently about 50 percent) to expand. In that case, the albedo would also increase, causing more of the sunlight to be lost by direct reflection. Since 1900, the carbon dioxide content of the air has risen some 15 percent through burning of fossil fuels.* During the same period, average world temperature increased at first, but turned sharply downward following 1940 (Figure 14-29). Even the most sophisticated theory is too simple to predict with confidence how changing CO_2 content will affect climate. The same is true for studies of the effect of dust particles injected by volcanic eruptions. On balance, the absorption and reflection of solar energy by dust particles are dominated by their reflection factor. Several exceptionally cold years followed the explosion of Krakatoa (Figure 4-5), which emplaced millions of tons of fine dust into the stratosphere. Any climatic influence would have to persist for several thousand years to initiate (or terminate) growth of a continental ice sheet. It is unlikely that intense explosive volcanism would be sustained for so long.

A Yugoslav astronomer, M. Milankovitch, developed another interesting theory based upon the changes of direction of the earth's spin axis and of the earth's orbit about the sun (Figure 14-30a). He noted that the *tilt* of the axis with respect to the plane of the orbit varies between about 22° and 24°. This tilt is the well-known cause of the seasons. The larger the angle of tilt, the greater becomes the contrast between summer and winter temperatures at any given latitude. In addition, the axis accomplishes a *precession*; its orientation slowly traces out a circle in the sky. Thirdly, the earth's orbit about the sun is an ellipse whose *eccentricity*, or degree of elongation, varies between 0.0† and 0.06. Since the orbital motions continually change but at different rates, sometimes their effects on climate tend to cancel out, while at other times they reinforce one another. That is, the *total amount* of solar energy received over the course of one year does not vary, but its *distribution* does vary. For example, at a given time the middle latitudes

---

* Over the next century, CO_2 in the air will probably increase 400 percent or more.
† An eccentricity of 0.0 corresponds to a perfect circle, which is a special kind of ellipse.

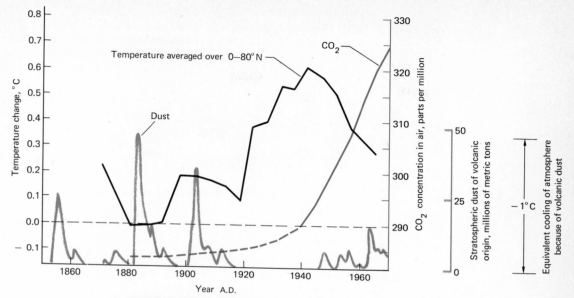

FIGURE 14-29 Fluctuations in average temperature do not correspond to the steady rise of CO_2 concentration in the air during the past century, or to injections of fine dust by occasional volcanic outbursts. Seemingly, the effects of CO_2 and dust concentration are minor compared with the influence of other factors, all of which remain poorly understood. [*After J. M. Mitchell, "A Preliminary Evaluation of Atmospheric Pollution as a Cause of the Global Temperature Fluctuation of the Past Century,"* Global Effects of Environmental Pollution, *D. Reidel Publishing Company, Dordrecht, Holland, 1970.*]

may experience long cold winters and short hot summers.* Thousands of years later the summers and winters may be about equal in length and relatively mild.

Calculations suggest that ice sheets are rapidly destroyed by the hot summers when seasonal contrasts are sharp. Ice sheets expand during periods of weak seasonal contrast (Figure 14-30b). Temperature indications from oxygen isotopes and fossils in deep-ocean sediment correspond quite well (though imperfectly) with the timing of glaciations predicted by the Milankovitch curves. However, these astronomical factors obviously cannot be the sole cause of glaciation, for they were equally significant when the earth was not in the grip of an ice age. Perhaps they are a "trigger" of glacial advances and retreats only when other circumstances are just right.

Yet another proposal, by the American geophysicists M. Ewing and W. Donn, stresses the need for abundant moisture in building an ice

---

* At present, the summer half-year in the Northern Hemisphere is 7 days and 14 hours longer than the winter half-year.

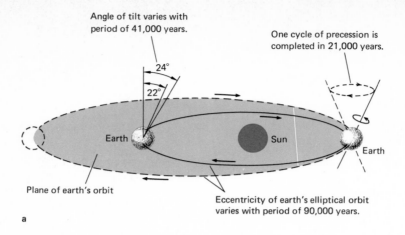

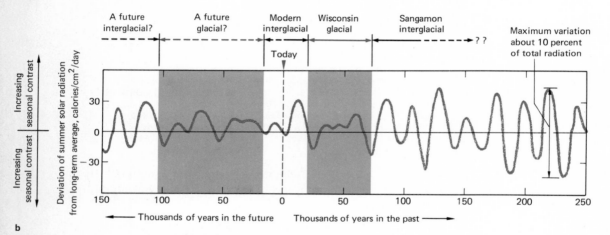

FIGURE 14-30 (*a*) Tilt and precession of the earth's spin axis and the eccentricity of its orbit vary with periods of approximately 41,000, 21,000, and 90,000 years. These periods in turn vary somewhat in response to the gravitational attraction of neighboring planets. Since the periods are unequal, their superimposed effect on climate at any one latitude is highly irregular with passing time. The effect also varies between latitudes and between the Northern and Southern Hemispheres. (*b*) A Milankovitch curve for latitude 45°N traces the variations of radiation received during summer over the past quarter-million years, and projects these variations 150,000 years into the future. Marked seasonal contrasts tend to accompany times when solar radiation is unusually high or low. Growth of an ice sheet is equated with times of low seasonal contrast; terminations correspond to times of high contrast. If the Milankovitch mechanism indeed governs the Cenozoic multiple glaciations, we appear due for another major advance over the next 50,000 or so years. [(*b*) *after W. S. Broecker and J. van Donk, "Insolation Changes, Ice Volumes, and the O¹⁸ Record in Deep-Sea Cores,"* Reviews of Geophysics and Space Physics, *vol. 8, no. 1, pp. 190, 195, 1970. Copyright by American Geophysical Union.*]

sheet. They recognized that only a trivial warming would suffice to melt the 3-meter layer of pack ice coating the Arctic Ocean. Evaporation from an open water surface is likely to be a factor of 10 to 50 times greater than the rate of evaporation from ice. Arctic lands fringing the ocean would transform from the cold desert that they are now to a cold, high-snowfall region. Ice sheets would appear and expand southward where they would be further nurtured by moist air brought from the North Atlantic and the Pacific. By this time, the ice is no longer simply a *product* of changing climate; it has become a *creator* of further changes which prove to be its undoing. The cold temperature and high albedo of ice (greater than 90 percent) cause the surrounding air and water to cool. Again the Arctic Ocean surface freezes over, and evaporation from the neighboring open oceans declines. Deprived of nourishment, the ice sheet starves away. Of course, while the retreat is in progress, the regions of moisture supply begin to warm up again, but only slowly. By then it is too late; the ice sheet has faded away except for the isolated remnants such as we see today.

Solar radiation, reflectivity of the earth, severity and contrast of seasons, availability of moisture—which of these and other possible variables controls the rise and fall of the mighty ice sheets? A more accurate time scale of glaciation in the Southern Hemisphere will help to resolve this question. If glacial pulsations always occur at the same time in the two hemispheres, the "flickering sun" hypothesis is favored. According to the Ewing-Donn proposal, in the southern latitudes glaciation need not be tied to events far to the north. The Milankovitch theory predicts that sometimes northern and southern glaciations coincide, sometimes they alternate in time.

Changes in the glacial regime may be rapid enough to have serious consequences for mankind. For example, some glaciologists contend that the Antarctic ice sheet is poised for a sudden surge. Were that to happen, sea level could rise nearly 20 meters in less than a century! This is but one of many hazards to the delicate balance between the welfare of man and of his environment—a subject we shall explore in the concluding chapter.

FURTHER QUESTIONS

1 How would the earth appear if ice were more dense than water? Would life be possible? (Use your imagination.)

2 An ice skater actually slides along on a thin film of water. Refer to Figure 14-2, and explain why this is so.

3 Refer to Figure 14-4. As geothermal heat is added to the base of the temperate glacier, does its temperature rise? If not, why not?

4 Refer to the caption beside Figure 14-5. What is the annual precipitation at the South Pole? Why is it so low?

5 Refer to Figure 14-7. Suppose that stakes were implanted in a straight line across the width of a valley glacier. How would the pattern appear after the ice had traveled a short distance? Justify your answer.

6 Refer to Figure 14-11. A continental ice sheet is too thick for daily temperature variations to reach the base of the ice. And yet, even under a thick glacier, the ice can melt on the up-glacier side of an obstruction, and refreeze on the down-glacier side. Explain how this happens.

7 Refer to Figure 14-22. Since the late 1800s, immense quantities of "dead" CO_2 (devoid of radiocarbon) have poured into the atmosphere from burning of fossil fuels. Beginning about 1950, this dilution effect (decreasing ^{14}C abundance) was abruptly reversed by injection of ^{14}C from nuclear weapons-testing. How do these man-caused disturbances of the carbon-14 content of modern carbon affect the interpretation of analyses of ancient material? Is the carbon-14 method no longer valid? Can you think of a way to circumvent these difficulties? (*Hint:* Strictly speaking, does the material used as a standard of comparison need to be modern, or only of a known age?)

8 Refer to Figure 14-23. Dates that are appreciably older than about 2000 years are found to be systematically incorrect (too young). Did the samples formed earlier than that time inherit an abundance of ^{14}C that was larger or smaller than in modern material?

9 Have you heard of any theories of the causes of an ice age other than those mentioned in this chapter? In the light of geological data, do they appear to be plausible?

10 Refer to the Milankovitch curve, Figure 14-30b. Do times of the modern interglacial, "Wisconsin" glacial, and "Sangamon" interglacial agree with the interpretation of isotopic data from the Camp Century core seen in Figure 14-18b?

READINGS

Flint, Richard F., 1971: *Glacial and Quaternary Geology*, John Wiley & Sons, Inc., New York, 892 pp.

Matthews, William H., W. W. Kellogg, and G. D. Robinson (eds.), 1971: *Man's Impact on the Climate*, The MIT Press, Cambridge, Mass., 594 pp.

Post, Austin, and E. R. LaChapelle, 1971: *Glacier Ice*, University of Washington Press, Seattle, 110 pp.

Sharp, Robert P., 1960: *Glaciers*, University of Oregon Press, Eugene, 78 pp.

Turekian, Karl K. (ed.), 1971: *The Late Cenozoic Glacial Ages*, Yale University Press, New Haven, Conn., 606 pp.

chapter 15
Dilemma

Spaceship Earth—man's only habitation.
[*National Aeronautics and Space Administration
photograph.*]

Man is one of the most recent species to appear upon the earth. To a degree unparalleled in the history of living things, man has been able to exploit the resources of his surroundings. In preindustrial times, every society patterned its life in close conformity to the demands and generosity of the local environment—there was no other choice. For countless generations mankind survived uneasily in a world from which he required, and obtained, little. Even today, two-thirds of the earth's peoples have never known any existence outside their original environment.

But for those of us who dwell in the industrialized nations, how different life has become! The use of water illustrates the enormous difference in rate of consumption in the "have" and "have not" countries. A person drinks and uses in food preparation about 200 cubic meters* of water in a lifetime. Keeping clean, a luxury denied to millions of those who live in primitive conditions, might take as a minimum an equal quantity. Water in the industrial societies goes to irrigate crops, manufacture goods, fill swimming pools, etc., in addition to indispensable personal uses. In these nations the per capita consumption of water may approach a hundred times that in less affluent societies. Most other resources are being used at similar high rates.

Modern civilization has clearly begun to have a significant (and accelerating) impact on the environment. Our technological achievements have placed us in a perilous dilemma. While making it possible for world population to soar, they have empowered the individual to consume ever more rapidly. And yet, neither the population nor the rate of consumption can increase indefinitely. Underdeveloped nations are beginning to realize that already it may be too late for them to acquire their share of the world's resources. You will recall the conclusion of Malthus that "the power of population is indefinitely greater than the

* Equivalent to a cube about 20 feet on a side.

461

power in the earth to produce subsistence for man. . . . Population, when unchecked, increases in a geometrical ratio. Subsistence increases only in an arithmetical ratio."

Is the frightening impasse between population growth and a limited earth, predicted by Malthus, about to be fulfilled, and if so, when? Can creeping disaster be averted? What is the time scale of population increase and depletion of resources? How many untapped deposits of minerals and fuel remain in the earth? What percentage of the available resources has already been extracted?

Of course, these substances are still present even after being "consumed," but they become unusable. As fabricated metals serve their various purposes, they become widely dispersed; combusted hydrocarbons and coal enter the atmosphere in the form of CO_2 and water. Could we perhaps recycle these substances, thereby using them over and over? Here also is a dilemma, as more energy is needed to recombine the combustion products into fuel than the fuel in turn provides. Recycling of metals and of water will become routine in the future, but at high cost in energy.

Natural recycling processes in the earth are too slow. Some 300 tons of the evaporite mineral trona* are mined daily from Lake Magadi, Kenya. About 15 times that much trona is deposited in the lake each day. Lake Magadi is probably the only place on earth where nature is replenishing a deposit faster than it is being mined. Water pumped from the ground may be replaced in some areas in a matter of years. Man is extracting most other valuable substances millions of times faster than new concentrations are being created. Some of these resources (e.g., potassium, iron) are sufficiently abundant to last for centuries at present rates of consumption; others, such as mercury and silver, are already in critically short supply. Superabundant cheap power could alleviate the mineral shortage by making it possible to extract metals dissolved in minute concentrations in seawater, or dispersed throughout ordinary rock. Extremely large volumes of rock or water would have to be processed. In the meantime, further unacceptable "insults" would no doubt be delivered to the environment in the form of pollution.

If these comments sound both gloomy and hopeful, they only reflect the uncertainty of our future. It is a future whose somber prospect can be prevented only at great cost. An important task of geologists will be to search out and evaluate how the world's resources were formed and distributed. Let us examine some of the more important categories of these resources.

* Trona ($Na_2CO_3 \cdot NaHCO_3 \cdot 2H_2O$) is a raw material used in the manufacture of washing soda, etc.

GROUNDWATER

A fundamental characteristic of all resources, even the most plentiful, is their *sporadic distribution*. Three-quarters of the fresh water, for example, is tied up in glaciers far from the big cities (which are also unevenly distributed). In the United States, 95 percent of the needed water is drawn from streams and lakes, even though the volume of water stored in the upper kilometer of continental crust is about 40 times the amount contained on the surface. In the Ganges Valley of India, an example of a rural area densely populated with small villages, nearly all the water is pumped from wells. Water is by far the cheapest of all natural resources except air which, if not pure, is at least free.

Ancient philosophers had many odd notions about the origin of groundwater which betrayed their misunderstanding of the *hydrologic cycle* (Figure 15-1). This cycle is a summary of the pathways connecting the various reservoirs of H_2O. The ancients, fascinated by springs, thought that water was fed to them from the sea through subterranean channels. (We often hear mistaken references to "underground rivers" today.) Supposedly, salt was removed as the water filtered through the soil. Certainly the philosophers were convinced that precipitation was not sufficient to supply the springs, much less whole rivers. Not until the late 1600s was it finally established by the Frenchmen P. Perrault and E. Mariotte that the *overhead* return of water to the continent is a more plausible route than the *underground* path. Their measurements showed that annual discharge of the Seine River is less than one-sixth of the yearly rainfall over its drainage basin.

In fact, we may wonder where all the rest of that water goes. Take the situation in the well-studied United States as an example. Annual precipitation is equivalent to covering the land with a layer of water 76 centimeters deep*—enough to fill a medium-sized swimming pool every day for each citizen of the country. Most of this water is not available, however. Nearly 55 centimeters (72 percent) of the precipitation evaporates directly, or is *transpired* into the air through the leaves of vegetation. Of the usable 28 percent which becomes runoff, Americans withdraw one-third. After this water is used, part is returned to streams and the remainder evaporates. A branch of the runoff cycle includes water stored temporarily in the ground before entering a stream. Crude estimates suggest that on the average, a water molecule spends several hundred years in transit through the underground reservoir, but the residence time varies greatly in different localities. This time scale is short enough to make it possible to remedy situations of mismanaged groundwater. On the other hand, groundwater, though difficult to pollute

---

* Worldwide average precipitation ranges from near zero in some places to more than 26 meters per year in others.

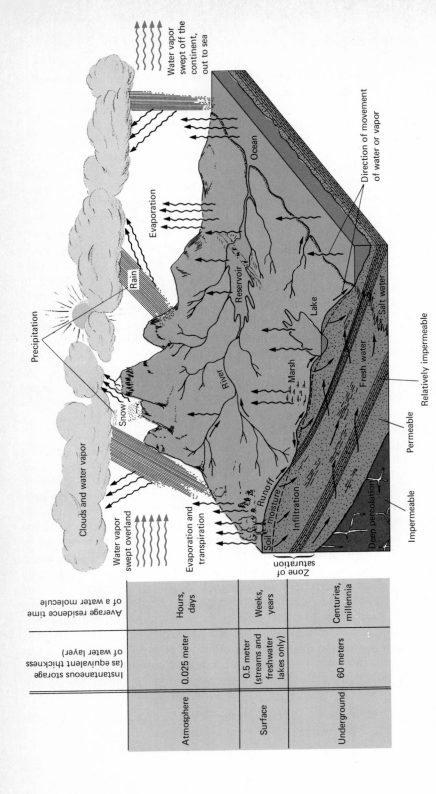

FIGURE 15-1 Most of the continental hydrologic cycle passes unnoticed because it involves solely clouds and vapor. Only 20 percent of the water vapor swept overland (colored arrows) is actually precipitated. In spite of the large mass of water handled by the atmosphere, its *instantaneous* storage capacity is relatively very small. [*Texas Water Development Board.*]

	Instantaneous storage (as equivalent thickness of water layer)	Average residence time of a water molecule
Atmosphere	0.025 meter	Hours, days
Surface	0.5 meter (streams and freshwater lakes only)	Weeks, years
Underground	60 meters	Centuries, millennia

in many areas, may be equally difficult to cleanse quickly. Many communities, supplied with water from streams, would be completely dependent on underground reserves if surface waters were poisoned by, say, a nuclear attack.

Geology of Groundwater

Open spaces (pores) must be present if a material is to contain water. The *porosity* (fraction of the volume occupied by pore space) varies considerably according to rock type (Figure 15-2). For example, the porosity of most granite is less than 1 percent, though the rock may contain numerous water-filled fractures. Porosity of coarse gravel is 20 to 30 percent and, surprisingly, the porosity of clay may be as high as 80 percent if the microscopic clay particles are not tightly packed together.

In order to transmit water, the pores must be interconnected; the rock must be *permeable*. Clay is quite impermeable because the pore water is tightly bound to the clay particles. Permeability is a fundamental property of a material, but temperature and viscosity also affect the ease by which a fluid can penetrate through a rock. (Water would move less readily than natural gas, but more easily than petroleum, through the same rock.) The term *hydraulic conductivity* summarizes the combined effect of permeability and other factors that influence the flow of the fluid.

FIGURE 15-2 Porosity is a characteristic feature of the texture of a rock. High porosity of a well-sorted sandstone (*a*) is diminished if the spaces among large grains are filled by smaller clastic particles (*b*), or by cement precipitated from solution (*c*). Porosity of limestone (*d*) is largely due to solution of channelways, and in a variety of igneous and metamorphic rocks (*e*), fractures are responsible for what little porosity there is. [*After O. E. Meinzer, "Ground Water,"* Physics of the Earth—IX: Hydrology, *Dover Publications, Inc., New York, 1942.*]

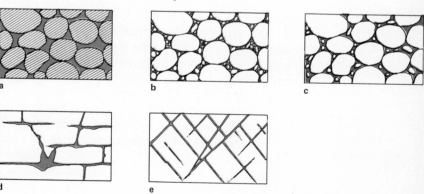

Material that is permeable is also porous. It is called an *aquifer* (from the Latin *aqua*: "water," and *ferre*: "to bear"). Only an aquifer can yield a sustained flow, but other kinds of material that are not good conductors of fluid can be induced to yield their water under special circumstances. For example, water for Houston, Texas, is pumped from buried delta deposits—sand aquifers situated amongst bodies of clay. At present, some 20 percent of the total supply comes out of the clay which compacts as it loses water. Not only is the clay a "once only" supplier, but the compaction has allowed large areas of Houston to sink. (The Terminal Island portion of Long Beach harbor, California, was subsiding into the Pacific because vast quantities of petroleum had been withdrawn from beneath the harbor. Since 1957 the rate of subsidence has been greatly reduced by injection of salt water in proportion to the quantity of petroleum removed.)

Darcy's law Modern studies of groundwater began in 1856 with the work of Henri Darcy, an engineer who was commissioned to develop a water-purification system for the city of Dijon, France. Darcy constructed a simple apparatus (Figure 15-3) to determine how water flows through a sand filter. He filled a tube whose cross-sectional area was A, with a length of sand l. Differences between coarse, permeable sand and finer, less permeable sand can be described by a factor K, later named the coefficient of hydraulic conductivity. Darcy opened the valve of an intake pipe, allowing water to enter at the top of the column, force its way through the sand, and drain into a measurement pan at the base. Connected at the top and bottom were open-ended, U-shaped tubes (manometers) partly filled with mercury. The pressure of the water displaced the mercury upward in the two manometers to elevations h_1 and h_2. By regulating pressure with the valve, Darcy could vary the "head" of water* $(h_1 - h_2)$ indicated by the manometers. Darcy's observations are described by a formula:

$$Q = KA \frac{h_1 - h_2}{l} \qquad (15\text{-}1)$$

This equation shows that the flow rate (Q) will increase if hydraulic conductivity (K), the cross-sectional area of the cylinder (A), or the head $(h_1 - h_2)$ were to increase, but Q will decrease if the length (l) increases. Intuitively we can see that increasing the area (A) or decreasing the length (l) decreases the amount of friction to be overcome by the percolating fluid. Or, to look at the formula another way, we can think of the quantity $(h_1 - h_2)/l$ as a slope or hydraulic gradient. As the gradient increases, flow rate of groundwater also increases just as the

---

* To obtain the water head, Darcy had to correct the readings to allow for the difference in density between water and mercury.

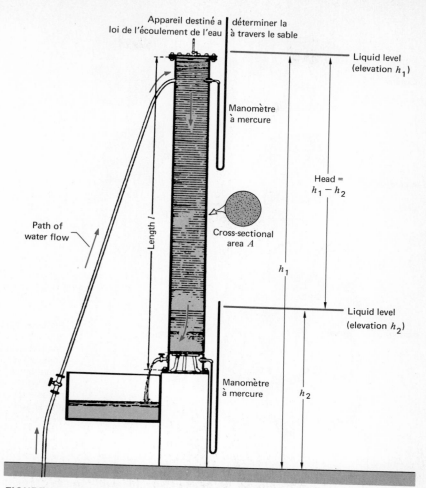

Appareil destiné a | déterminer la
loi de l'écoulement de l'eau | à travers le sable

Liquid level
(elevation h_1)

Manomètre
à mercure

Head =
$h_1 - h_2$

Path of
water flow

Length l

Cross-sectional
area A

h_1

Liquid level
(elevation h_2)

Manomètre
à mercure

h_2

FIGURE 15-3 A facsimile of Darcy's original drawing (with added explanation) shows what a simple device was needed to establish the most fundamental law of groundwater hydrology. Translated, Darcy's caption reads: "Apparatus designed to determine the law of leakage of water across the sand." [*After M. K. Hubbert, "Darcy's Law and the Field Equations of the Flow of Underground Fluids,"* Transactions, American Institute of Mining, Metallurgical, and Petroleum Engineers, *vol. 207, 1956.*]

velocity of a river would. Darcy's law has successfully described many situations in nature that bear little resemblance to a bed of sand in a pipe.

Unconfined aquifers There are forces of adhesion between water and its aquifer that, to a degree, oppose the force of gravity. Suppose that the fluid were allowed simply to drain out of the sand in Darcy's

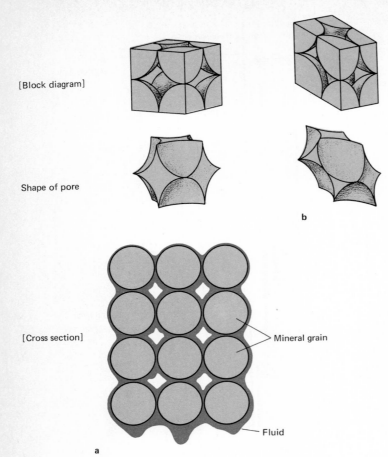

[Block diagram]

Shape of pore

b

[Cross section]

Mineral grain

Fluid

a

FIGURE 15-4 Recovery of water (or any other fluid) from a rock is never complete. These drawings show two possible ways to pack together equal-sized spherical particles. Even with simple grain textures, the pores may assume a variety of sizes and shapes. If the open packing (*a*) were reorganized into a more compact arrangement (*b*), the porosity would decrease.

apparatus. A film of liquid would remain behind, coating each sand grain (Figure 15-4). The *specific yield*, or fraction of the volume of the rock occupied by *drainable* fluid, may range from nearly zero to as high as 40 percent, depending upon porosity and the size and shape of the pore spaces.

If water is able to percolate freely down into an aquifer and to move in any direction, the aquifer is said to be *unconfined*. Just below the surface of the ground is a moist but unsaturated region where water is partly held in suspension, partly migrating downward. At the base of this zone is the *water table*, beneath which the aquifer is saturated. In most places, the surface of the water table imitates the topography of the land above it (Figure 15-5*a*). Downward and lateral motion of the groundwater tends to flatten the water table into a plane, but the next rainstorm elevates the water table and restores its irregularities once again.

As water is pumped from an aquifer, the water table sags sharply into a "cone of depression" centered about the well (Figure 15-5*b*).

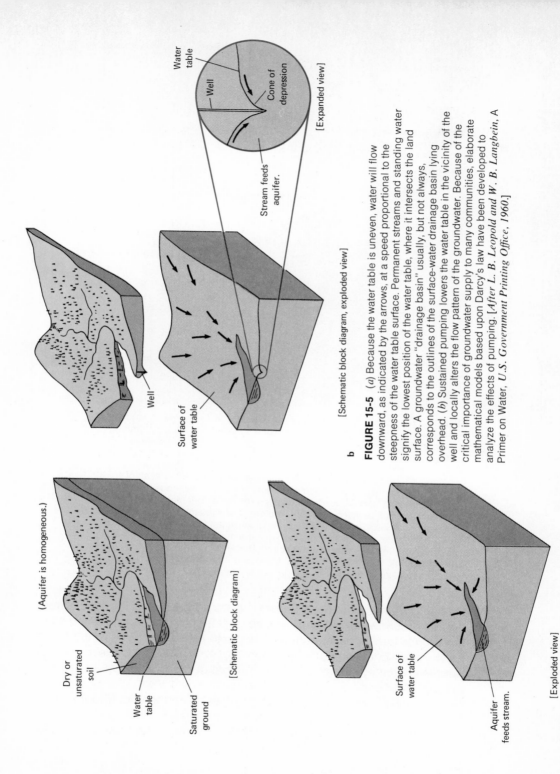

(Aquifer is homogeneous.)

Dry or unsaturated soil

Water table

Saturated ground

[Schematic block diagram]

Surface of water table

Aquifer feeds stream.

[Exploded view]

a

Well

Surface of water table

[Schematic block diagram]

Stream feeds aquifer.

[Schematic block diagram, exploded view]

Water table

Well

Cone of depression

[Expanded view]

b

FIGURE 15-5 (*a*) Because the water table is uneven, water will flow downward, as indicated by the arrows, at a speed proportional to the steepness of the water table surface. Permanent streams and standing water signify the lowest position of the water table, where it intersects the land surface. A groundwater "drainage basin" usually, but not always, corresponds to the outlines of the surface-water drainage basin lying overhead. (*b*) Sustained pumping lowers the water table in the vicinity of the well and locally alters the flow pattern of the groundwater. Because of the critical importance of groundwater supply to many communities, elaborate mathematical models based upon Darcy's law have been developed to analyze the effects of pumping. [*After L. B. Leopold and W. B. Langbein, A Primer on Water, U.S. Government Printing Office, 1960.*]

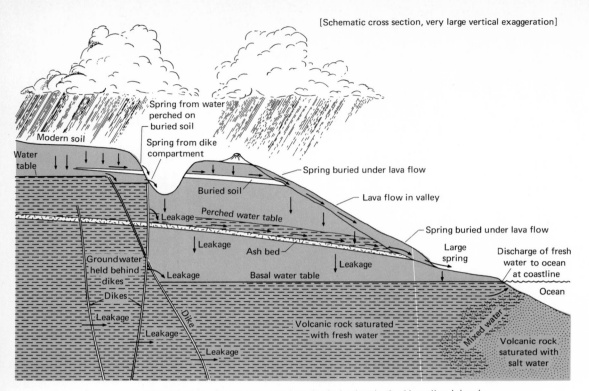

[Schematic cross section, very large vertical exaggeration]

Spring from water perched on buried soil

Spring from dike compartment

Modern soil

Water table

Spring buried under lava flow

Buried soil

Lava flow in valley

Leakage

Perched water table

Spring buried under lava flow

Leakage

Ash bed

Large spring

Discharge of fresh water to ocean at coastline

Groundwater held behind dikes

Leakage

Basal water table

Leakage

Ocean

Dikes

Dike

Leakage

Volcanic rock saturated with fresh water

Mixed water

Volcanic rock saturated with salt water

Leakage

Leakage

FIGURE 15-6 Challenges confront hydrologists in the Hawaiian Islands, which contain what is undoubtedly the world's most complicated plumbing system. The uneven distribution of rainfall and small catchment areas oblige the Hawaiians to make do with quantities of water that would seem meager even in the semiarid Western states. A three-dimensional network of impermeable dikes and sills divides the extremely permeable basalt of these islands into a mosaic of compartments. Water that overtops or filters through a vertical barrier may run laterally, creating a "perched" water table above a horizontal partition. Special problems are encountered at the coast (and along many coasts elsewhere). Here, fresh water floats above a lens of denser salty water that encroaches beneath the dry land. Depletion of fresh water by overpumping along the coast encourages the undesirable brackish water to rise into the well. [*After S. N. Davis and R. J. M. DeWiest,* Hydrogeology, *John Wiley & Sons, Inc., New York, 1966.*]

Pumping from a well located near a stream may cause the local direction of groundwater flow to reverse. Instead of receiving groundwater through its banks, the stream may begin to lose water. Where numerous wells have been steadily pumped for a long period, the cones of depression merge together into a regional depression that may assume giant proportions.

The southern High Plains of the Texas Panhandle (Figure 15-7) have become the site of a famous example of depletion by overpumping. In Pliocene times a vast sheet of clastic debris, known today

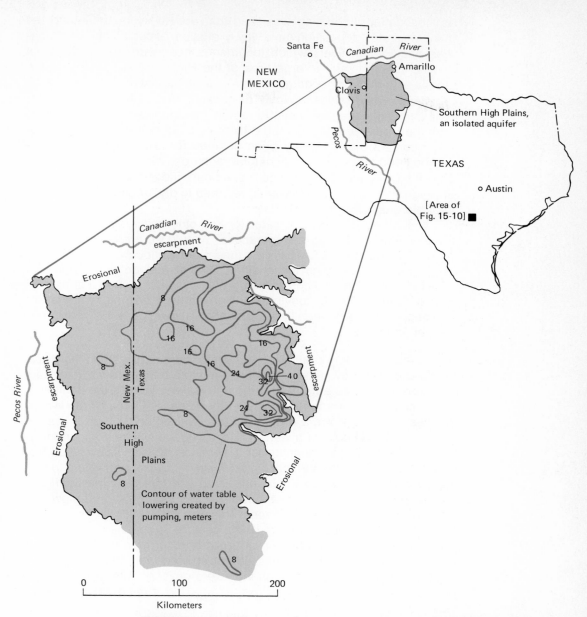

FIGURE 15-7 In three decades of heavy pumping between 1937 and 1967, the water table in the Ogallala aquifer dropped as much as 40 meters in places. More than 50,000 pumped wells are busy lowering the water table ever more rapidly. Already, some wells must descend 100 meters to reach water. [*Data from J. G. Cronin, "Ground Water in the Ogallala Formation in the Southern High Plains of Texas and New Mexico,"* U.S. Geological Survey Hydrologic Investigations Atlas HA-330, *1969.*]

as the Ogallala Formation, was shed from the Rocky Mountains across this area. Later, deposition gave way to erosion, especially by the Canadian and Pecos Rivers which lengthened headward toward the west and north. Eventually a large area of the High Plains, capped by the Ogallala Formation, became isolated as a high-standing, "perched" tableland. Early settlers quickly discovered that, although no permanent streams cross this semiarid region, abundant water was available from the Ogallala. More than 60 trillion liters (25 percent of the original water content) have already been pumped for irrigation, at a rate exceeding 100 times the rate of natural recharge. Within a few more decades the Ogallala aquifer will be "mined out"; indeed, parts of the southern High Plains have already been returned to dry farming and ranching capable of support by rainfall alone.

Is it wrong to mine this water (or any other exhaustible resource), thereby denying it to future generations? Should we *preserve* a resource (leave it untouched), or *conserve* it (use it up cautiously, with an eye to the future)? Solutions to these difficult moral and political questions depend in part on an understanding of geological factors. Farmers in the High Plains, having convinced the government that their water asset is disappearing, are permitted a depletion allowance (a tax reduction) on their earnings.

Confined aquifers Many aquifers are permeable strata *confined* above and below by more or less impermeable beds. Groundwater can recharge a confined aquifer only through leakage across the confining strata, or where the aquifer intersects the surface (Figure 15-8). Water that has migrated down a dipping aquifer will be under high pressure at depth. The concept of a water table, a surface whose contours may freely change, has no meaning for a confined aquifer. In its place we may define a pressure surface, or *piezometric surface*, equivalent to the elevation to which water would rise in a well that taps the confined aquifer at any given spot. Note that the land surface may lie below the piezometric surface in some places. This was the situation in the plains of South Dakota around the turn of the century. A confined aquifer, the Dakota Sandstone, extends from a region of outcrop in the Black Hills to underlie the plains farther east. Water spurted high into the air from some of the first wells as though they were geysers. After a number of years the piezometric surface dropped, making it necessary to pump these wells.

Behavior of other fluids in the ground is similar to that of water. Most reservoirs of oil and natural gas are confined, else the hydrocarbons would have leaked away long ago. A maximum recovery of petroleum, which on a unit-volume basis is worth hundreds of times as much as water, is obviously of great importance to the oil industry. *Primary* recovery includes the oil forced to the surface while gas pressure in the reservoir is still high, followed by oil that can be lifted out by

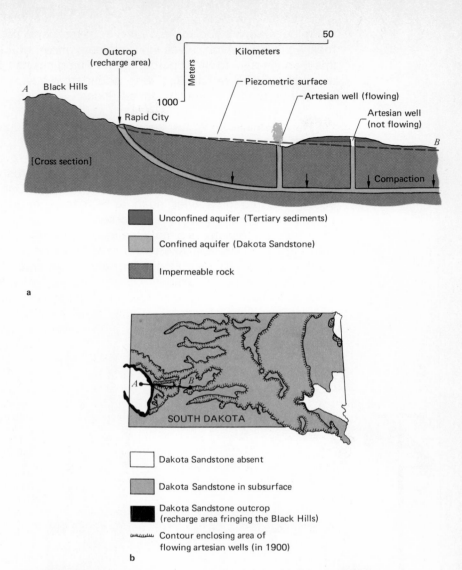

FIGURE 15-8 Most of South Dakota is underlain by the Cretaceous Dakota Sandstone, a highly prolific aquifer that receives groundwater where its upturned edge is exposed in the Black Hills. Under the Great Plains, the Dakota aquifer is sealed off above and below by impervious beds. Because the elevation of the sandstone steadily drops toward the east, water confined in the aquifer is under high pressure beneath the Great Plains. A hole drilled into the aquifer becomes an artesian well, so named after Artois, the northern French province where deep wells first encountered pressurized water. Water is forced upward into the artesian wells by compaction of the aquifer, not by flowage downward from the Black Hills recharge area, which would take many thousands of years. An unconfined aquifer, the superficial Tertiary sediments, is entirely independent of the confined water system of the Dakota Sandstone.

473 GROUNDWATER

pumping. One method of *secondary* recovery is to flood the reservoir rock with water which drives the petroleum ahead of it (Figure 15-9). In this manner, a yield of 20 percent (after pumping) can be increased to as much as 50 percent. By 1980, secondary recovery will account for half the oil production in the United States.

Flow velocities   Water seeps through most aquifers at speeds ranging from a meter or so per day to a few meters per year. Hydrologists have used the carbon-14 method to date groundwater containing dissolved CO_2. Water in the soil zone absorbs carbon 14 from the air and from decaying plants and soil bacteria. Once the water has descended into the aquifer, no more ^{14}C can be received from the soil and air reservoirs, while the radiocarbon already present decays away. The "age" of the water is simply the length of time it has spent en route from the outcrop (the intake point) to a given point farther within the aquifer. Flow velocities obtained by radiocarbon ages from the Carrizo Sand aquifer, in south Texas, are in good agreement with results calculated from Darcy's law (Figure 15-10).

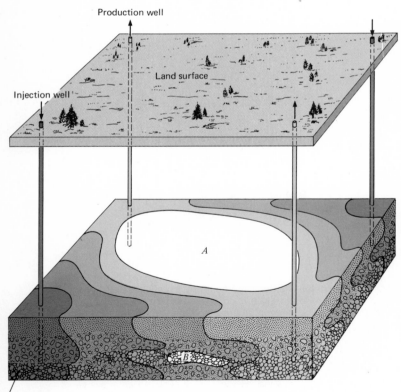

Production well

Land surface

Injection well

A

B

Petroleum-bearing rock

[Schematic block diagram]

FIGURE 15-9 Shaded contours trace the annual advance of the contact between oil and the water that is being injected into the reservoir rock. Because water is more fluid (less viscous) than oil, tongues of water tend to break through the oil and bypass regions (such as *A* and *B*), which remain stranded. [*After N. de Nevers, "The Secondary Recovery of Petroleum,"* Scientific American, *vol. 213, no. 1, 1965. Copyright © 1965 by Scientific American, Inc. All rights reserved.*]

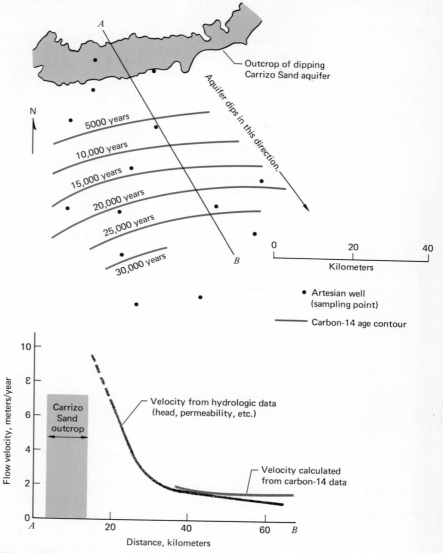

FIGURE 15-10 The Carrizo Sand, a confined aquifer, dips southeastward from its outcrop (recharge) area. Contours depict increasing carbon-14 ages as the water travels downdip. The velocity of the water—the distance between contours divided by the age difference—is in good agreement with velocity data obtained from Darcy's law. [*After F. J. Pearson and D. E. White, "Carbon 14 Ages and Flow Rates of Water in Carrizo Sand, Atascosa County, Texas,"* Water Resources Research, *vol. 3, no. 1, pp. 260, 261, 1967. Copyright by American Geophysical Union.*]

475 GROUNDWATER

ORE DEPOSITS

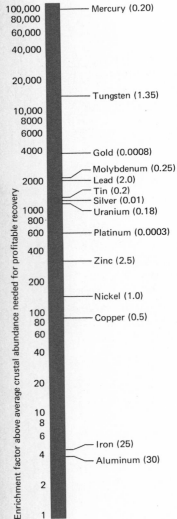

Despite the amazing diversity of the earth's crust, from an economic standpoint all but a very small part can best be regarded as "ordinary, useless rock." In certain rare environments, geologic processes have produced unusually high concentrations of some economically valuable metal, gemstone, or fuel. Generally speaking, an *ore* is a rock whose metal content can be recovered at a profit. Sometimes nonmetals (for example, sulfur) are called ore, but never stone, salt, coal, or hydrocarbon deposits. Since the definition of an ore depends partially on fluctuations of the market, a rock body may be ore one day but not the next. Obviously, the more concentrated a metal is, the more profitable it is to mine the ore. For major rock-forming elements such as iron and aluminum, a deposit enriched in the metal about 5 times above the average crustal abundance qualifies as an ore (Figure 15-11). Other elements, ordinarily present in trace quantities, can be mined profitably only if they are enriched hundreds or thousands of times above the average crustal abundances.

Even so, the total quantity of these metals in common rocks far exceeds the amount in ore deposits. For example, the crust contains about 15 parts per million of the element lead (Pb). Even at this low abundance, the amount of lead in just 25 cubic kilometers of average crust—a modest-sized mountain—is equal to that in the Broken Hill district, Australia, one of the world's greatest lead-zinc deposits. It is impossible to pinpoint a "source rock" for the metal in most ore bodies. A very large volume of source rock may have suffered the loss of an undetectably small fraction of the metal originally present in it.

Although ore deposits have been vitally important to civilization for thousands of years, the question of their origins continues to be one of the most frustrating and controversial in geology. Part of the problem is due to the rarity of ore deposits; for some types, only one, or perhaps a half-dozen, are known in the entire world. Geologists disagree about the origin of even the largest and most intensively studied ore bodies. Nearly 70 percent of the world's annual production of gold is extracted from ancient conglomerates of the Witwatersrand district, near Johannesburg, South Africa. Some geologists affirm that the tiny flecks of gold were simply laid down as sedimentary particles along with the rest of the conglomerate. Others insist that hot aqueous solutions introduced the gold at a much later time. A third group combines these

FIGURE 15-11 Numbers in parentheses are minimum, or cutoff, percentages of metal that can be profitably mined. For example, if mercury is enriched by a factor of 100,000, it forms an ore body in which the concentration is 0.2 percent. [*After B. J. Skinner,* Earth Resources, *Prentice-Hall, Inc., Englewood Cliffs, N.J., 1969.*]

ideas: They say that the gold, originally deposited as clastic particles, was later dissolved and reprecipitated after moving short distances.

Sudbury, Ontario, site of the world's largest known deposit of nickel,* is the subject of another sharp debate. Some geologists believe that the source of the nickel was a nickel-iron meteorite that struck the earth. Others say that the energy of impact would have caused the meteorite to evaporate, but that the impact fractured the earth, permitting nickel-rich magma to rise from great depth. A third group of geologists accept the magmatic origin of the nickel sulfide, but they deny the existence of the meteorite. Yet a fourth group appeals to hot aqueous fluids as a source of nickel. Proponents of each viewpoint have assembled a wealth of evidence to support their case, but the situation is so complex that the opposition can always reinterpret contradictory evidence in favor of what *they* believe to be true.

What difference does it make whether we know the origin of an ore deposit? Gold is where you find it, goes the old saying. Although some deposits have been found by luck, well-established ideas about the origin of ore have long served as a guide to explorations. No one looks for coal, a material of sedimentary origin, in the midst of a granite intrusion. Kimberlite pipes (Chapter 10) are a good place to search for diamonds, another form of carbon created only under high pressure. Our ability to distinguish the origins of coal and diamonds is useful, but now that the obvious mineral resources are rapidly being discovered, exploration techniques must become more sophisticated.

For example, copper minerals in an important type of deposit appear to have been precipitated from hot gases. Some geologists postulate that deposition took place through volcanic eruptions on the sea floor. Others favor a deposition at high pressure, beneath thousands of meters of rock overburden. These geologists point out that because pressure at the bottom of a shallow sea is not very great, the density of the volcanic gas, hence its ability to transport dissolved copper compounds, would be too small in the submarine environment. Results of exploration seem to confirm the latter viewpoint; copper deposits of this type are found only in rocks that once were deeply buried. Subtle indications such as this will be important in the continued search for ore.

Economic Geology

A specialized discipline known as *economic geology* is devoted to the study of these valuable deposits (Table 1-1). The challenges of economic geology are almost as varied as the scope of geology itself. Prospecting is a major task for geologists concerned with fossil fuels (Chapter 13) or ore deposits of the minor metals (tin, copper, silver, etc.). Even some very large deposits of abundant metals have escaped notice until recently. Not until 1963 was the Hamersley Range, West-

---

* As the iron-nickel sulfide, pentlandite $(Fe,Ni)_9S_8$.

TABLE 15-1

Classification of Earth Resources

Metallic	Nonmetallic
Abundant metals*: iron, aluminum, nickel, manganese, chromium	Fresh water
	Building materials: sand, gravel, gypsum, cut stone, etc.
Scarce metals*: copper, zinc, cobalt, lead, uranium, titanium, (magnesium†)	Fossil fuel: coal, crude oil, natural gas, oil shale (undeveloped)
Trace metals: silver, gold, tin, beryllium, molybdenum, platinum, etc.	Specialty products: salt, phosphate rock, asbestos, sulfur, mica, diamond, fluorspar, potash, etc.

* In order of quantity of ore (*not* quantity of metal) mined each year. Trace metals are not conveniently rated according to ore tonnages. For example, much of the world's silver is obtained as a by-product from ore mined primarily for some other metal.
† Ocean water is a major "ore" of magnesium.

ern Australia, recognized as containing the world's largest known reserve of iron ore. (Mining has just begun.) Other classes of resources include water, building materials (sand, cement, gypsum, etc.), and special-purpose substances such as fertilizer and salt. Such an enormous bulk of these latter materials is consumed that the problem is to find deposits near the areas in which they are to be used. Cost of transporting a carload of crushed limestone from a distant quarry to the city may be more than the cost of the limestone itself. Restoring the land scarred by giant strip mines and quarries has become a serious environmental problem.

Sedimentary ores Let us see how a selected assortment of ore types illustrates the geological processes that have concentrated the metal. The origins of *sedimentary* ores are perhaps easiest to understand. In size, sedimentary ores far outweigh the sum total of all other kinds. One variety called *placer* (PLASS-er) ore contains clastic particles of heavy metals such as gold or platinum. Because of their high density, these tiny grains slowly work down to the base of the sedimentary fill of a stream channel. Their softness protects them from being shattered by impact as the stream picks up and deposits its bedload again and again. Since the transporting ability of flowing water diminishes as turbulence decreases, rich "ore streaks" are commonly stranded on the inside bank of old meander loops (Figure 15-12).

Other dense minerals such as cassiterite (tin oxide: SnO_2), magnetite (Fe_3O_4), and diamonds also occur as placers. In some deposits, pounding waves substitute for flowing stream water as an energy source. The Soviets and Japanese are mining beach placers—black sand containing less than 4 percent iron in the form of magnetite grains. Although far leaner than most commercial iron deposits, this

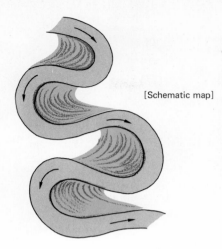

[Schematic map]

FIGURE 15-12 Knowledge of depositional processes in a meandering stream (Chapter 13) is important in the search for placer gold. A clever prospector recognizes that heavy, sandsized gold particles may concentrate in old meander scars, not only in the active meander belt but also in distant, abandoned meander belts (compare with Figures 13-10 and 13-12). Placer bonanzas were the object of a fervent search during the Gold Rush days in California and Alaska. The old prospectors quickly learned that large pieces—nuggets—are more likely to be found upstream near the bedrock source of gold. The largest nugget ever discovered would have fetched nearly $400,000 at today's price.

placer is economically valuable because the ore need not be ground up, and magnetite is readily pulled out of the sand with magnets.

The largest accumulations of iron oxide originate as a chemical precipitate, perhaps a colloidal product of the life processes of iron-secreting bacteria. Though agreeing that the ore is of sedimentary origin, at least six competing theories suggest different sources and environments of deposition of the iron. A handicap to our understanding is that nearly all these prodigious ore bodies formed during the interval between 3.2 and 1.7 billion years ago, under circumstances greatly different from today's. Some geologists believe that free oxygen, first appearing in the atmosphere at about this time, combined with the iron to precipitate insoluble iron oxides.* In the early years of iron mining in the Upper Great Lakes district, the only material considered to be ore was iron oxide from which most of the silica originally precipitated with it had been leached out (Figure 15-13). Fortunately, by the time the high-grade ores were depleted, technological advances made it possible to use the low-grade (30 percent Fe) ore containing silica. In fact, the low-grade ore, after treatment to enrich the iron content somewhat, was found actually to make better blast-furnace feed than pure iron oxide.

Weathering, another process allied to the sedimentary cycle, is responsible for concentrating the ores of aluminum and some kinds of nickel and manganese ores. You recall that aluminum, the third most abundant element in the earth's crust, is a prominent constituent of feldspar. Weathering in the wet tropics may go far beyond the point of making clay minerals out of feldspar. Severe leaching decomposes even the clay minerals; silica is dissolved, leaving only hydrated

* See Chapter 5 for further discussion of sedimentary iron ores.

FIGURE 15-13 This photograph of ancient sedimentary iron ore from Africa shows contorted layers of iron oxide finely interbedded with chert. (See Chapter 5 for a discussion of chert.) In some of these deposits, weak metamorphism has caused the iron oxide to react with the chert, forming iron silicate minerals. [*From C. F. Park and R. A. MacDiarmid,* Ore Deposits, *W. H. Freeman and Company, San Francisco, 1970.*]

oxides of aluminum, the most insoluble materials of all. Because these aluminum-rich *bauxite** deposits are residues left upon the earth's surface, they are highly vulnerable to erosion. Most of them formed no earlier than a few tens of millions of years ago. The extent of tropical weathering seemed all the more impressive when geologists realized that the source rock for some bauxites is limestone with just a little clay impurity. A very large amount of rock indeed must have been removed, since the calcium carbonate part of the limestone makes no contribution to the accumulation of bauxite.

By similar means, solution weathering has left residues of nickel silicate in Cuba and in New Caledonia (a Pacific island near Australia). Nickel in the underlying source rock is found in the mineral olivine where it substitutes for atoms of magnesium and iron. Brazil contains a large deposit of manganese oxide created by weathering of metamorphic rock (schist). Garnet in the schist is the source of manganese.

Magmatic ores Experimental studies over the years have helped us to visualize what happens as a magma crystallizes deep underground. Most intrusive igneous rocks are composed of minerals that crystallized at different times. Usually an early-crystallizing mineral grain does not

* Bauxite is the rarer, aluminum-oxide equivalent of laterite (hydrated iron oxides). See Chapter 5.

have the same density as the still molten material surrounding it. In 1915 the American petrologist N. L. Bowen performed an interesting experiment in which he kept a silicate melt, containing a few crystals of olivine, for a few minutes at high temperature. Then he suddenly cooled the melt, "freezing" the olivine crystals in their positions. Examination of the quenched melt showed that much of the dense olivine had sunk to the bottom of the crucible (Figure 15-14a).

This same process has operated on a gigantic scale in igneous intrusions in many places. It has created ore deposits in the Bushveld Intrusion of South Africa, an enormous saucer-shaped mass (largest of its kind in the world) of basaltic composition. Evidently the magma was injected into the host rock quickly, before crystallization had progressed very far. Soon after the very earliest crystals had formed and settled, dense chromite grains ($FeCr_2O_4$), a valuable concentrated source of chromium, began to crystallize. Astonishingly uniform layers of chromite rained down upon the bottom of the magma chamber (Figure 15-14b); one stratum about 15 centimeters thick has been traced for a distance of at least 250 kilometers! Here is an igneous "sediment" that is more continuous than most water-laid sediments.

Since early crystals settling to the floor of the Bushveld Intrusion were of ferromagnesian minerals, it follows that the remaining liquid became enriched in the elements *other* than iron and magnesium.

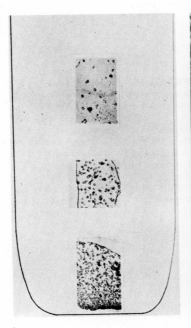

a

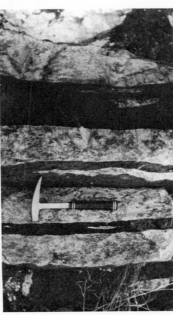

b

FIGURE 15-14 (*a*) Bowen cut three slices of the quenched silicate melt for microscopic examination. Photographs of the slices, in restored position, show dense olivine crystals embedded in a matrix of clear glass. [*After N. L. Bowen, "Crystallization-Differentiation in Silicate Liquids,"* American Journal of Science, *vol. 189, no. 230, 1915.*] (*b*) Lower parts of the Bushveld Intrusion contain black, chromite-rich layers interbedded with light-colored, feldspar-rich strata. The resemblance between this rock and deep-ocean sediments is uncanny. Extremely widespread and continuous layers actually show graded bedding and other evidence for turbidity currents, but the Bushveld rock is igneous! [*Courtesy Stephen Clabaugh.*]

During this natural sorting process, water also was concentrated into the final residue of magma. Indeed, the Bushveld today resembles a layer cake some 7 kilometers thick and 65,000 square kilometers in area, in which dark-colored, dense rocks at the base grade upward into rocks that look more like granite.

Other intrusions are composed of granitic magma at the outset. In this case, the final watery magma may crystallize not only into the usual quartz, feldspar, and mica, but sometimes also into minerals containing the metals lithium, beryllium, tantalum, and others that are normally present only in trace quantities in rocks. The final result, a *pegmatite*, is a spectacular granitic rock composed of enormous crystals. Masses of pure quartz the size of a large house, and sheets of muscovite* a meter or two across, are commonly encountered. Some pegmatitic minerals are exceedingly rare; for example, the Harding pegmatite, in northern New Mexico, is studded with giant crystals of spodumene [$LiAl(SiO_3)_2$], a source of lithium (Figure 15-15).

Hydrothermal ores   So far, we have seen that well-known depositional or igneous processes can create ore deposits. Another class of mineral resources that includes the biggest reserves of silver, copper, lead, and zinc is of more uncertain, and very controversial, origin. Seemingly the variations of composition and form of these ores are

---

* The term *muscovite* comes to us from the days of the czars when wealthy citizens of Moscow had windowpanes made of this transparent variety of mica (Muscovy glass).

FIGURE 15-15 Big "logs" of spodumene, a lithium-bearing mineral of the pyroxene family, lie strewn about in the Harding pegmatite. Natural cleavage of spodumene even mimics the splintery appearance of wood. The largest single crystal ever discovered measured about 12.7 meters by 1.7 meters, and weighed more than 16 metric tons. Crystals of other pegmatitic minerals are of similar gigantic size. [*Courtesy William Muehlberger.*]

endless, but they share several important features in common. In some deposits the rock is peppered with tiny specks or shot through with countless small veins and stringers of metal sulfide. Along these channelways the shattered, mineralized rock may be intensely altered—impregnated with silica, or its feldspar turned into clay. Complex textures indicate that early mineral grains were dissolved and replaced (sometimes again and again) by fresh surges of incoming mineral material. On a larger scale, mineralization in an entire mining district may be arranged in definite zones (Figure 15-17). These observations are best explained if we postulate that the ore was dissolved in hot water when introduced into the rock; it is of *hydrothermal* origin.

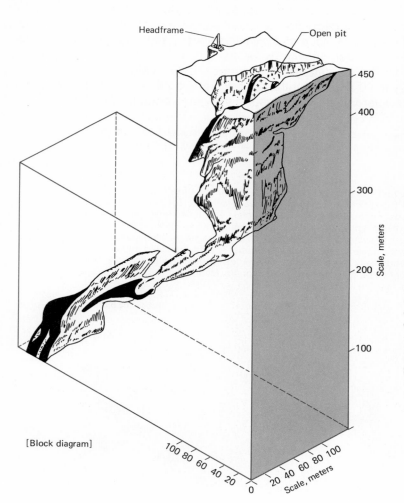

[Block diagram]

FIGURE 15-16 Ascending hydrothermal fluids deposited the contorted ore body of the Tsumeb lead-zinc-copper mine, South West Africa. Like many other mines, this one began as an open pit, later enlarged by underground workings. A scale drawing of the headframe, a building several stories tall that houses a cable drum and lift machinery, emphasizes the impressive size of the mine. Complexly shaped mineral deposits are an unceasing source of legal disputes over ownership. Is the landowner entitled to surface rights only, or to everything beneath the surface in a wedge-shaped column extending to the center of the earth? What happens if the underground continuation of a mine extends beneath someone else's property? From the legal viewpoint, is an ore body that is interrupted and offset by faults a separate deposit, or the same deposit? [*After H. Schneiderhöhn, "Das Otavi-Bergland und seine Erzlagerstätten," Zeitschrift für Praktische Geologie, vol. 37, Sonderheft zum XV Internationalen Geologen-Kongress in Südafrika, 1929.*]

A pegmatite perched on the side of a large granite body was almost certainly derived from the granitic magma. Sources of fluid responsible for many hydrothermal deposits are not so obvious. In some cases, there are no igneous rocks anywhere near some of the largest hydrothermal deposits. Others, such as the porphyry copper* deposits prominent in the western United States, are always closely associated with igneous intrusions, but even here the field relationships are unclear. Were hydrothermal fluids expelled from the igneous body, or did they penetrate it from the outside? What rock was the source of copper? Evidence from stable isotopes has suggested a partial answer to this difficult problem. We saw (Chapter 14) that as storm clouds sweep inland, the hydrogen and oxygen in the precipitation become more and more enriched in the light isotopes (^1H and ^{16}O). In some porphyry copper deposits the clay minerals are "labeled" with distinctively light hydrogen and oxygen isotopes, suggesting that the clay formed in equilibrium with water that entered the ground as local precipitation. Isotope data from the deposit at Butte, Montana, demand that some 50 to 90 percent of the hydrothermal fluid be of groundwater origin, not water initially contained in the granitic magma.

* So named because the copper minerals are dispersed throughout huge volumes of porphyry, an igneous rock consisting of large feldspar crystals set in fine-grained matrix.

FIGURE 15-17 A granite batholith (Chapter 4) underlies southwestern England. Cupolas of granite project upward to the surface, forming a string of outcrops (middle drawing). Hydrothermal fluids have deposited metals in a series of concentric zones in the granite and adjacent host rock (right-hand drawing). Tin oxide (cassiterite) occurs nearest the granite; farther away, mineralization is mostly copper sulfide, then lead and zinc sulfide, and finally iron oxide (hematite). These zones have provided a convenient guide to further exploration for ore. Mining in this region has continued ever since 600 B.C.

[Schematic block diagram]

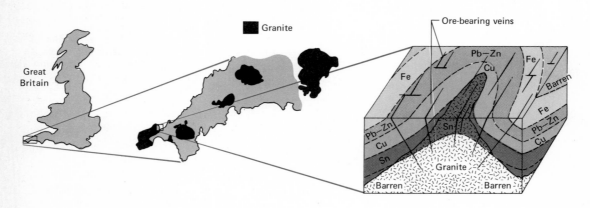

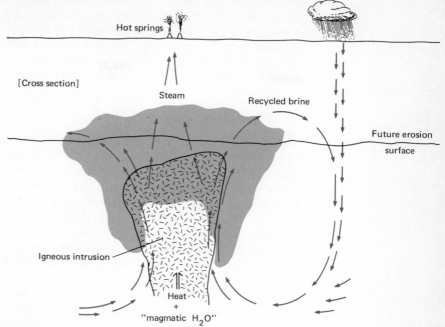

[Cross section]

Hot springs

Steam

Recycled brine

Future erosion surface

Igneous intrusion

Heat
+
"magmatic H_2O"

FIGURE 15-18 Hot copper-bearing brines move upward through a highly fractured igneous intrusion. Copper sulfide minerals are precipitated as a result of chemical reaction between the solutions and the rocks. Millions of years later, when erosion has uncapped the copper deposit, permitting atmospheric oxygen to reach the sulfides, another process of concentration may occur. Copper sulfide is oxidized to soluble copper sulfate which trickles down to the water table. There it reprecipitates as secondary, insoluble sulfide. Groundwater has thereby concentrated a very large, diffuse ore deposit into a smaller, highly enriched deposit.

With the benefit of recent findings, we may begin to assemble a picture of how hydrothermal ore deposits form. Today, about 40 major geothermal areas are scattered about in active volcanic regions, which significantly are also near the edges of the great tectonic plates (Chapter 12). (Examples are the spectacular hot springs and geyser field of Yellowstone National Park, and a geothermal region that supplies steam to heat buildings in Reykjavík, Iceland.) Quite likely, large igneous intrusions, very slowly cooled, lie hidden at depth beneath these areas. (The Steamboat Springs in Nevada has discharged boiling water for at least a million years.) We surmise that near a cooling intrusion, groundwater slowly descends on a journey that may take as long as 100,000 years or so to complete (Figure 15-18). Water approaching the intrusion is strongly heated, and its power to dissolve ions out of the host rock or the intrusion is increased. As the brine ascends from the intrusion, it precipitates ore minerals in a series of zones as temperature continues to decrease. The most volatile metal compounds are not precipitated until the fluid emerges in hot springs. (About 5 metric tons of arsenic and 1 ton of antimony are transported to the surface by the Steamboat Springs each year.) Other hot springs, modern and long inactive, are prominent sources of mercury minerals.

ECONOMICS OF GEOLOGY

The value of raw mineral and energy resources is only 2 percent of the GNP* of the United States—a rather insignificant part of the national wealth. Many economists have assumed that the ingenuity of the engineers and exploration geologists will continue, as always, to provide the necessary materials as we have need of them. After all, new discoveries are made every day, while lower and lower grades of ore are mined at a profit (Figure 15-19). Nowadays many a handsome return is obtained by *re*processing old mine tailings (waste heaps) from which earlier smelting operations had failed to extract all the metal. The same economists argue that because scarcity of raw materials will drive prices up, economic pressure alone will cause deposits that are currently below the cutoff grade to be reclassified as ore.

Worldwide Reserves

Are these assertions an accurate description of future supply and demand, or are they dangerous half-truths? Even if mineral resources are a small part of the GNP, they have been indispensable to the devel-

* Gross national product, the total market value of all goods and services created by a nation in a year.

FIGURE 15-19 Indians in the Upper Great Lakes region once made hand weapons and tools from pieces of pure copper metal found locally in volcanic rocks. Early mining operations by advanced techniques were centered in secondarily enriched copper deposits (see Figure 15-18). As time has gone on, the grade of ore that could be mined profitably has decreased rather steadily. One-third of all the money spent in the United States in search of new ore deposits is used in looking for copper ore. [*After T. S. Lovering, "Mineral Resources from the Land,"* Resources and Man, *Publication 1703, Committee on Resources and Man, National Academy of Sciences–National Research Council, W. H. Freeman and Company, San Francisco, 1969.*]

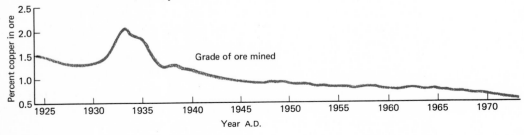

opment of any nation, modern and ancient. A sobering indication of what the future holds is to be found in the immediate past. In 1962, President John F. Kennedy reported to Congress that in the preceding 30 years the people of the United States had consumed more mineral products than all the world's peoples had previously consumed since the dawn of history. Production data emphasize the usage of materials having little or no practical value less than a century ago. Since 1882, when the United States first began to report mineral statistics, the annual production (and consumption) of mineral resources has risen to staggering levels. Compared to production in 1882, domestic production of petroleum in 1972 was up by a factor of 115; cement was up 5300 times; sulfur production had risen by a factor of 17,000! Not only that, but technology-based societies are making economic use of argon, dysprosium, europium, fluorine, germanium, hafnium, helium, krypton, lutetium, neon, radium, radon, rhenium, and xenon, none of which was even known in 1882, and of perhaps 40 other chemical elements that at the time were regarded as mere laboratory curiosities.

Another way to predict the future of supply and demand is through the *reserve-production index:* the ratio

$$\frac{\text{Known reserves of a resource (metric tons)}}{\text{Annual production rate (metric tons per year)}}$$

This index is simply an estimate of how long the known but untapped quantity of a resource would last at the present rate of extraction. Since more deposits are frequently being discovered, and since production rates continue to spiral upward year by year, the index is only a rough guide, but informative. Barring some radical new developments, approximately a dozen of these vital materials will be depleted within a generation (Figure 15-20). We are running out of earth and out of the earth's resources. Reserves of mercury, for example, are so short that for some years a cartel* has been able to control the market price (Figure 15-21). As a consequence, there is no simple relationship between price and consumption, in violation of the customary economic cause-and-effect responses. No doubt the price of other scarce metals will soon begin to fluctuate violently as the supply nears exhaustion.

In view of the current 10-year doubling period for the demand of energy, reserves of fossil fuel are in serious jeopardy. They are quite accurately evaluated, since fossil fuels are located in the earth's superficial cover of sedimentary rocks. The American geologist M. K. Hubbert has developed a mathematical model to predict the future of crude oil production. His calculations take into account not only the past history of consumption, but also proven reserves, the rate of discovery of new oil fields, and the "lead time" needed to develop production in a new field. Rate of withdrawal of any exhaustible resource must start from

* *Cartel:* a combination of independent business organizations formed to regulate the production and pricing of goods.

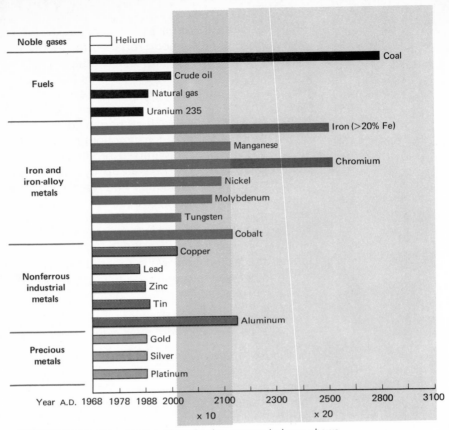

FIGURE 15-20 This diagram of production-reserve indexes shows apparent lifetimes of mineral reserves recoverable by today's technology. Note changes of scale on the time axis (horizontal). [*After S. F. Singer (ed.),* Is there an Optimum Level of Population?, *McGraw-Hill Book Company, New York, 1971.*]

zero, proceed through one or more maximum points, and return to zero. Hubbert's curve for the production rate in the United States reaches a peak in the late 1960s (Figure 15-22); for world production, the maximum appears only a few years later. Even the oil field at Prudhoe Bay, Alaska, which promises to be twice as big as the largest American field previously found, would supply the needs of the United States for only about 3 years. Drilling records are another measure of success in petroleum exploration. From 1860 to 1920, when oil was easy to find, wells produced an average of 100 cubic meters of petroleum per meter of well drilled. Today, with the benefit of advanced techniques, the rate has dropped to only 18 cubic meters of oil per meter of well drilled.

Knowledge of ore reserves is more uncertain, partially because so many reserves presumably lie hidden beneath the sedimentary cover. Moreover, several years (typically, 7 or more) of heavy investment go

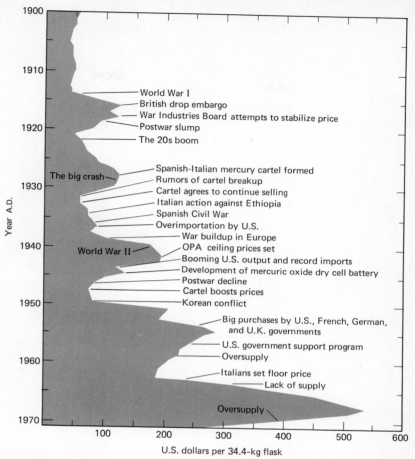

FIGURE 15-21 "Mercurial," meaning quick and changeable in character, is an apt description of the behavior of liquid mercury (quicksilver) and of its market price. Numerous fluctuations are superimposed upon a general inflationary price increase over the years. [*After R. A. Pendergast (ed.), "Mercury," E. & M. J. Metal and Mineral Markets, Jan. 25, 1965.*]

into proving and developing an ore deposit in advance of full mine production, whereas a successful exploratory well instantly becomes the conduit through which oil is made available to the market.

The American geologist F. E. Ingerson has estimated an upper limit to worldwide ore reserves obtainable by orthodox methods. He reasoned that a large area of bedrock obscured by superficial cover should contain about the same amount of ore that is found in an equal area of exposed bedrock. More accurately, because ore deposits may be stacked one above another, equal *volumes* of a given kind of rock should be weighted equally. (In Missouri, for example, lead and zinc sulfides are mined from Paleozoic sedimentary strata that overlie Precambrian basement with iron deposits.) Let us suppose that all these

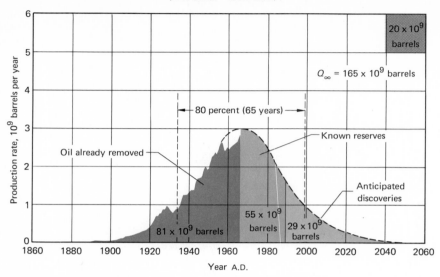

FIGURE 15-22 A graph indicates the crude oil production rate in the United States (including continental shelves but excluding Alaska) *vs.* time. Area under the bell-shaped curve equals (quantity of oil produced per unit time, times time), or ultimate total production (Q_∞). The curve suggests that approximately half the oil that will ever be discovered has already been removed from the ground. Statistics of production agree quite well with this theoretical model; production rates have leveled off, beginning in the early 1970s. [*After M. K. Hubbert, "Energy Resources,"* Resources and Man, *Publication 1703, Committee on Resources and Man, National Academy of Sciences–National Research Council, W. H. Freeman and Company, San Francisco, 1969.*]

hidden deposits could be exploited—actually an impossible dream when we consider that more than one-fourth of them are masked by polar ice sheets or submerged beneath the continental shelves. In all, these types of cover, plus lakes, recent volcanic fields, tropical lateritic soil, glacial debris, tundra, and desert dunes, conceal about twice as much area of ore-bearing rock as the area that is exposed. Hence the total potential may equal about 3 times the amount of ore that can be discovered from surface outcrops. Suppose we somewhat arbitrarily multiply this figure again by 3 to allow for full exhaustion of stacked deposits to great depth.

According to mining statistics, production rates have doubled approximately every 15 years since 1870. If, for example, we use one unit of ore by the early 1980s (when most of the currently known reserves will be depleted), how much longer would eight additional units last in view of this doubling rate? What if the estimate of reserves falls short by a factor of 2? We have gained only one additional 15-year grace period! Production rates of ore minerals, another exhaustible resource, must also decline in the future.

Could we postpone the inevitable by mining some unconventional deposits? For example, could we mine the ocean, which occupies 70 percent of the earth's surface? With few exceptions, geologic data from both the ocean water and the crust beneath are not encouraging. Seawater is already a major source of magnesium, bromine, and common salt. Significant yields of most other elements would require that unreasonably large volumes of water be processed. Although the ocean contains 10 billion metric tons of gold, the amount so far extracted after much effort by various researchers is less than 0.0001 gram. Certain restricted water bodies have abnormally high concentrations of dissolved ions. Cadmium, for instance, is 10,000 times more abundant in the Red Sea than in the open ocean.

Oceanic crust is also an unpromising mineral prospect. Exploration and geochemical calculations suggest that as much as 25 percent of the ocean floor is coated with nodules or with solid pavements of slowly deposited manganese oxide. Once the mining and smelting techniques are perfected, this sedimentary ore could supply the world demand for manganese for centuries. Oceanic basalt, on the other hand, is a very primitive rock derived directly from the mantle. Trace elements in the basalt probably have not been sorted and concentrated by the various processes that make ore deposits.

Yet another proposal to ease the mineral crisis is to extend mining operations into lower-grade ore. Economists have traditionally assumed that as the grade of ore diminishes *arithmetically*, the volume of ore increases *geometrically*—a sort of happy reversal of the "dismal theorem" of Malthus. Unfortunately, the geologic data give little basis for their optimism. An arithmetic-geometric relationship may hold for the major crust-forming elements such as iron and aluminum, but trace elements appear to be highly concentrated into tiny, widely scattered spots in the midst of practically barren terrain (Figure 15-23).

Chemical elements have been vertically separated into different zones to a remarkable degree. Exploration at greater depth will not reveal important concentrations of the upward-migrating, volatile elements lead, zinc, cadmium, and mercury. These metals will be found near the surface, or not at all. Geothermal heat at depth decomposes the hydrocarbon fuels into a useless carbonaceous residue. Moreover, deeply buried reservoir rocks tend to have low porosity (storage capacity).

Doomsday?

What consequences does this somber picture of diminishing reserves have for national policy and for our daily lives? Because mineral distribution is so sporadic, the very large nations are more likely to have inherited the mixture of resources necessary for economic development. Not even the biggest country, nor even a continent, is self-suf-

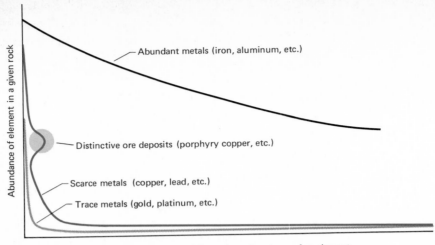

FIGURE 15-23 An abundance diagram represents the distributions of "abundant," "scarce," and "trace" metals (see Table 15-1). The area under a curve is proportional to the total quantity of the metal in the earth's crust. Occurrence of certain common types of ore body (such as porphyry copper) may result in a small kink in the abundance curve. Only a very small, sharply distinct volume of rock is highly mineralized with trace metals. Precise shapes of these curves are potentially of great importance to economic geologists. These shapes are currently unknown, hence no scales can be attached to the axes of the graph.

ficient, though.* North America is rich in molybdenum but poor in tin, tungsten, and manganese; Asia is the opposite. More than 97 percent of the world's accessible coal is located in the three Northern Hemisphere continents (Figure 15-24). For every unit of coal allotted to an inhabitant of Latin America, Africa, and Australia, the quota for each American is 24 units, each Soviet citizen obtains 60 units, and each Canadian inherits 95 units. Our southern neighbors must regard such an unequal distribution as one of nature's cruelest jokes.

While resources dwindle, world population continues to grow at a net rate of two persons per second, equivalent to creating an additional United States every 3 years (Figure 15-25). On the average, population increase is about 2 percent per year, but a growth rate nearly twice as high is found in some of the less developed countries. Imagine that world population suddenly were to stabilize at the modern value of about 3.7 billion. Suppose that metals were entirely recycled after use. To match the per capita consumption in the United States, the world population would have to keep in steady-state circulation about 350 million metric tons of zinc, 500 million tons of lead (not including lead for car batteries), and 50 million tons of tin. These quantities exceed the known and inferred estimates of reserves by factors of 3, 5, and 10 times. Without their quota of metal and energy, the less developed coun-

* Among all the nations, the U.S.S.R. is most nearly self-sufficient.

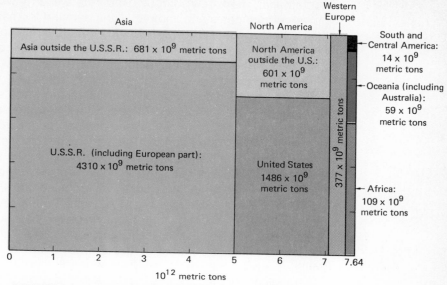

FIGURE 15-24 All earth resources, rare and abundant alike, are unevenly distributed. Except for Antarctic coal, the world inventory, diagrammed here, is well known. [*After M. K. Hubbert, "Energy Resources,"* Resources and Man, *Publication 1703, Committee on Resources and Man, National Academy of Sciences–National Research Council, W. H. Freeman and Company, San Francisco, 1969.*]

tries will never reach the high standard of living practiced by more affluent nations. We have *already* run out of earth.

Some geologists, deeply alarmed and seeing no realizable solutions to these problems, have become prophets of doom. Others are

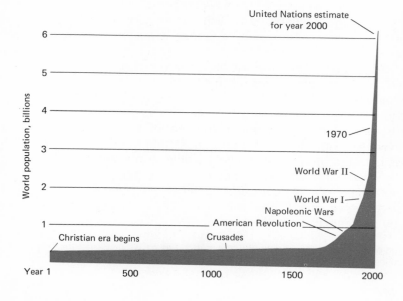

FIGURE 15-25 Current doubling time for world population is 35 years, or slightly longer than one generation. Wars, famine, and disease have had a negligible effect in reducing population growth over the past 2000 years. At present, the rate of world population increase is 1000 times greater than it was during the centuries preceding A.D. 1700.

hopeful that, soon, new sources of abundant energy will satisfy both the power requirements and, indirectly, the demand for materials. Following the earliest successful experiment in controlled fission, development of atomic power advanced in just 12 years to the first commercial power plant. Conventional nuclear reactors are extremely wasteful and inefficient users of the uranium fuel, as they consume only the rare* isotope, ^{235}U. At projected future consumption rates, the world's high-grade uranium ores will be exhausted at about the same time as the petroleum reserves. Another class of nuclear devices under intensive development will "breed" the abundant but nonfissionable isotopes of uranium and thorium by bombarding them with neutrons supplied from ^{235}U starter fuel. After a series of radioactive decays, uranium 238 is converted into fissionable plutonium 239, and thorium 232 becomes fissionable uranium 233. Once breeder reactors are a reality, certain ordinary rocks will become a huge fuel resource. For example, under an area of eastern Tennessee, the Chattanooga Shale contains about 60 parts per million of uranium;† the Conway Granite, New Hampshire, contains about the same concentration of thorium. Quarries a few square kilometers in size in either rock would provide for breeder reactors an amount of fuel equivalent to the initial supply of petroleum in the United States.

Another, more distant possibility is energy from the controlled fusion of deuterium atoms (^{2}H), or of deuterium with tritium (^{3}H). Lithium is needed to promote the deuterium-tritium fusion, the more promising of the two types of reaction. Low abundance of lithium would limit the energy supply from this source to roughly the equivalent amount available from fossil fuel. For the deuterium-deuterium reaction, ocean water is a limitless "ore." The amount of solar energy—yet another possibility—falling upon the earth in about 2 minutes equals the amount consumed by all mankind in a year. Like fusion power, abundant solar power could be made available only with much more research and development. Geothermal energy can be tapped in selected areas, but at best it could supply only a tiny fraction of the total need.

At least temporarily, a serious imbalance will open in the near future between the huge demand for resources and shortage of supply. Wealthy nations will continue to import minerals and fossil fuel at higher cost, but at an ever increasing pace as long as political stability makes free trade possible. In the appalling event of a general nuclear war, the survivors, unable to draw upon concentrated earth resources as our forefathers did, may never again be able to construct a technological civilization. If relative peace continues, the gap in standards of living in affluent and less developed countries will become greater and greater Whatever the future, everyone, rich and poor, will be confronted by the limitations of the earth.

* The proportion of atoms ^{235}U/^{238}U in natural uranium is 1:138.
† Roughly 1 pound of uranium per 8 tons of rock.

FURTHER QUESTIONS

1 Groundwater from cavernous limestone or fractured igneous rocks (Figure 15-2*d* and *e*) may not be pure enough for drinking (the bacterial counts are too high). Can you postulate why?

2 Refer to Figure 15-11. What is the average concentration of mercury in the crust? Of lead? Of iron?

3 Some placer ores are not easy to trace to their source. Recently discovered placer diamonds on the coast of South West Africa are unlike the diamonds found in other parts of Africa, but are similar to diamonds mined on the other side of the Atlantic in Brazil. Can you postulate an explanation for this?

4 The average concentration of gold in a placer deposit may be far higher than the concentration in the source bedrock. Why?

5 Do you have any suggested solutions to the mineral- and energy-shortage dilemma? Is an adequate political solution possible? Are the technological innovations that would be demanded likely to appear, or even to be possible? Will human ingenuity be able to cope with the crisis that lies ahead?

6 Historians have argued that mineral resources (or lack of them) were crucial to the rise and fall of the Roman Empire. Research this subject, and criticize it. On this basis, what do you predict for the future well-being of Great Britain, of South Africa, of the Soviet Union, of Brazil, of the United States?

READINGS

* Committee on Resources and Man, 1969: *Resources and Man,* National Academy of Sciences–National Research Council, W. H. Freeman and Company, San Francisco, 259 pp.

* LaPorte, Leo F., and others, 1972: *The Earth and Human Affairs,* Canfield Press, San Francisco, 142 pp.

Leopold, Luna B., and W. B. Langbein, 1960: *A Primer on Water,* U.S. Government Printing Office, 50 pp.

Singer, S. Fred (ed.), 1971: *Is There an Optimum Level of Population?,* McGraw-Hill Book Company, New York, 426 pp.

* Skinner, Brian J., 1969: *Earth Resources,* Prentice-Hall, Inc., Englewood Cliffs, N.J., 149 pp.

* Various authors, 1968: "Limitations of the Earth: A Compelling Focus for Geology," *Texas Quarterly,* vol. 11, no. 2, 154 pp.

* Available in paperback.

English and Metric Units Compared

All physical measurements can be defined in terms of combinations of fundamental units of time, mass, and length. Fortunately, the second is already universally accepted as the basic unit of time. Here are common English units of mass and of length (and area, or length squared), compared pictorially to metric units used in this book.

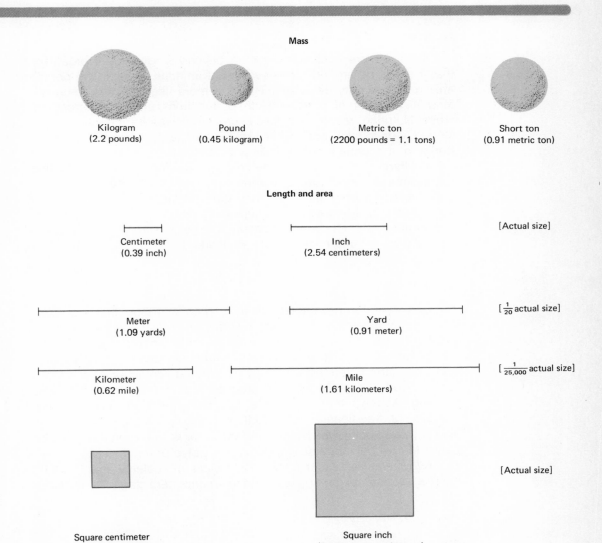

Mass

Kilogram
(2.2 pounds)

Pound
(0.45 kilogram)

Metric ton
(2200 pounds = 1.1 tons)

Short ton
(0.91 metric ton)

Length and area

Centimeter
(0.39 inch)

Inch
(2.54 centimeters)

[Actual size]

Meter
(1.09 yards)

Yard
(0.91 meter)

$[\frac{1}{20}$ actual size]

Kilometer
(0.62 mile)

Mile
(1.61 kilometers)

$[\frac{1}{25,000}$ actual size]

Square centimeter
(0.15 square inch)

Square inch
(6.45 square centimeters)

[Actual size]

Fossil Record of Animal Phyla

Porifera

Sponges are such a short step beyond single-celled organisms that it could be argued that they are not animals. At best, the porifers are an evolutionary dead end. They are more like cell colonies exhibiting just a hint of specialization of functions among different cell types. Nutrition, respiration, and excretion of waste products are performed individually, cell by cell, but water is circulated through the pores* of a sponge along well-directed pathways.

Porifers first appeared before the Cambrian Period. Modern species may be freshwater or marine. Sponges may be bowl-shaped, or cup-shaped, or irregular. Their hard parts (if there are any) consist of tiny spines, or spicules, of calcium carbonate or silica (Figure B-1). Glassy sponge spicules are probably a major source of silica that can readily dissolve and reprecipitate as chert.

Coelenterata

Coelenterates have three distinct types of tissue that are much more highly diversified than that of sponges. Many coelenterates have a rather complex life cycle. At one stage, the animal is a polyp, an attached form. The polyp is essentially a tube, closed off at the attached end and surrounded by an array of stinging tentacles at the open mouth. Food and waste material proceed in and out through the same opening. A polyp may develop a bud that breaks off to become a medusa, or free-floating individual. A medusa can release eggs or sperm (never both) which unite to form a larva that grows into a new polyp. Many coelenterates skip either the polyp or medusa stage.

Jellyfish are an exclusively marine class of coelenterates in which the medusa form of the animal is the more important stage. A jellyfish is

---

* Hence the name *porifera*.

498

a

b

FIGURE B-1 (*a*) These vase-shaped fossil sponges from the Burgess Shale are exceptionally well preserved. [*American Museum of Natural History photograph.*] (*b*) Tiny sponge spicules may assume a variety of shapes.

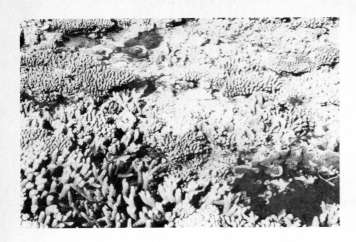

FIGURE B-2 This scene in Heron Island, in Australia's Great Barrier, the world's mightiest living reef (see Figure 5-12), shows masses of colonial coral. Each colony contains hundreds to thousands of tiny individual animals. [*Courtesy Joan Lang.*]

hardly more than a gelatinous, umbrella-shaped body fitted with tentacles and rudimentary sensory and digestive systems. Adults may be from 1 centimeter to 2 meters across, with tentacles up to 10 meters long. Though we would hardly expect to find many jellyfish in fossil form, there do exist some impressions made by jellyfish that were washed up on a beach and quickly covered by sand, as at the Ediacara site (see Chapter 8). Jellyfish were among the first true animals to appear, sometime before the Cambrian Period.

A vastly more important fossil is *coral*, another class of marine coelenterates. Coral never produces a medusa. Varieties of soft coral—polyps with no hard parts, or with flesh reinforced by spicules—are little known from fossils. Stony corals, that build a rigid skeleton of calcium carbonate, are a major constituent of many limestones, and are a familiar sight in modern shallow tropical waters (Figure B-2). (A few species can tolerate deep, frigid water.) Coral grows today in only a few localities, but in past times, when there were widespread shallow seas, this animal was both abundant and represented by more species than are living today.

Some "solitary" corals, now extinct, built cone-shaped skeletons resembling a horn of plenty (see Figure 2-22). Other corals reproduced asexually by developing buds that grew into new adults. The colonies thus formed could accumulate into giant reefs featuring all manner of fantastic shapes: fans, compact masses, treelike branches, and others.

Brachiopoda

Most brachiopods are attached filter-feeders that superficially resemble clams, and indeed were once classified in the same phylum

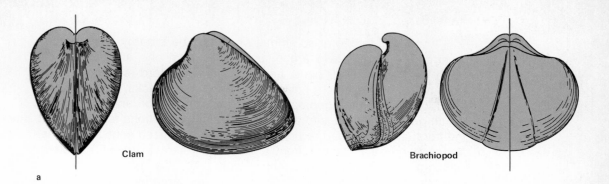

Clam   Brachiopod

a

b

FIGURE B-3 (*a*) The symmetry of clams is between the valves. The line of symmetry of brachiopods lies across the valves. (*b*) The ciliated feeding structure inside many brachiopods is arranged in a complicated series of loops. The literature describing fossil brachiopods has grown to bewildering size; about 700 terms have been used just to describe the shape of the shell.

with clams. Both organisms have bilateral symmetry, but the line dividing the equal halves is different in each case (Figure B-3*a*). The valves (shells) of a brachiopod, composed of calcium carbonate or calcium phosphate, are bound together, and attached to the sea bottom via a muscular stalk. Inside is a complicated set of fleshy coils coated with tiny beating filaments. This feeding structure traps food particles as water is moved along by the filaments waving to and fro. Sometimes details of the calcified support of the feeding structure are preserved (Figure B-3*b*). Every so often, the animal clears itself of sediment and waste material by clapping its valves violently together. The valves of fossil species range in size from 1 millimeter to almost 40 centimeters in some gigantic forms. Brachiopods are marine organisms that can tolerate brackish* water only for short periods.

Obscure and little understood, the modern brachiopods have definitely been displaced by a more successful group of animals, the mollusks. Brachiopods are the only phylum for which the known fossil species (30,000) far exceed the number of modern, living species (220).

* With salinity between that of marine and freshwater.

501   FOSSIL RECORD OF ANIMAL PHYLA

Annelida

About half the 20 or so animal phyla can be broadly described as worms. The worm phyla are distinguished from one another entirely upon the basis of soft parts, hence the fossil record is absent aside from a few impressions, horny jaws, and trails, tracks, and burrows.

Most significant geologically is the phylum of annelid worms, which are segmented into a number of similar sections terminated at the front end by a head region containing specialized sensory organs. Annelids range in length from a few millimeters to about 3 meters. Although annelid worms appeared before the Cambrian Period, the most celebrated collection site is the Middle Cambrian Burgess Shale (see Figure 8-11).

Mollusca

From general appearance, mollusks do not seem to be related to annelid worms, and for that matter, neither do snails, clams, and octopuses (all mollusks) resemble one another closely. The correct assignment of the relationship of these diverse organisms has been a small triumph of taxonomy. What do the mollusks share in common? These animals (excepting clams) possess a well-developed head with tentacles and eyes. A mollusk also has a single muscular foot (sometimes greatly modified) that enables it to "spud into" soft sediment, and to creep or glide along. In addition, mollusks have a sheety tissue called a mantle that drapes downward to enclose the animal's guts. One of the several functions of the mantle is to secrete a protective shell of calcium carbonate or of chitin (a flexible horny substance). The octopus has an internal shell, and in certain rare species, shells are entirely lacking.

The origin of mollusks was debated for many years. And then, in 1957, several living specimens of a primitive mollusk (*Neopilina*), dredged from 3500 meters' depth in the East Pacific, shed new light on the question. The soft parts of these living fossils (their nearest relatives had become extinct about 400 million years ago) are definitely arranged in segments. Although most of the mollusks have lost their segmentation, it appears likely that long ago this phylum split off from the annelid worms.

Gastropods—snails and their kin—are the most diverse mollusks; they live in all types of water from marine to fresh, and they are the only group of mollusks to have invaded the dry land. Different species graze upon algae, scavenge dead organisms, or actively pursue a carnivorous diet. Meat-eating snails can bore through the shell of an unlucky clam or some other supposedly secure victim. The body of a snail becomes severely twisted as it matures toward adulthood, and usually so does its shell. An astonishing variety of snail-shell shapes are

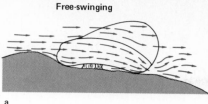

Free-swinging

a

Cemented to
sediment surface

b

Free-living

c

Burrowing

FIGURE B-4 By making careful comparisons with modern species, paleontologists have been highly successful in interpreting the living habits of extinct species of marine pelecypods from the form of the fossil shell.

Rock-boring

Fused
siphons

Mantle
edge

Foot

d

e

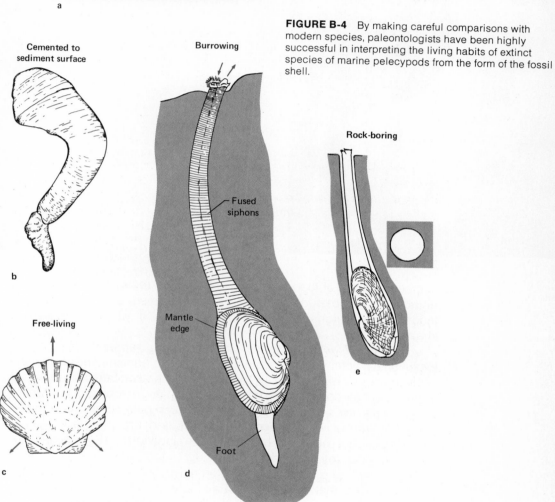

preserved in the fossil record, but unfortunately they tell us little about the life habits of the animal.

Pelecypods, another class of mollusks, include oysters, clams, and scallops. Species of these animals are adapted to life in almost any aqueous environment, where they may crawl, or swim, or burrow through loose sand and mud, or even bore a living space out of solid rocks (Figure B-4). Most pelecypods are filter-feeders that strain out

FIGURE B-5 A nautilus shell sliced parallel to the plane of symmetry reveals a succession of more than 30 chambers culminated by the large end chamber currently occupied by the soft parts of the animal at the time of death. [*Judy Camps.*]

microscopic bits of food with tiny beating hairs (cilia). Their shells may be from 1 millimeter long to as much as 1.5 meters across. The largest modern clam weighs more than 270 kilograms, of which 95 percent is shell material.

The *cephalopods*, a marine class, probably represent the highest evolutionary potentiality of any unsegmented organism. These animals evolved more rapidly, and in more directions, than almost any other group. Cephalopods are represented today by octopuses (umbrella-like), squids and cuttlefish (cigar-shaped), and by a single coiled-shell genus, *Nautilus*. The deep-ocean squid is undoubtedly the largest of all animals that have no backbone. An individual 16 meters long, including tentacles, was once caught.

The fossil record suggests that primitive cephalopods began as a snaillike animal that deposited, not a single chamber as snails do, but a series of chambers separated by thin partitions. If extinct species were like the modern pearly nautilus, only the most forward of these chambers was occupied by the fleshy part of the animal (Figure B-5). Compare with the exterior of a coiled cephalopod (Figure 7-20). Pearly nautilus is a swimming carnivore, but extinct species probably were filter-feeders, grazers, floaters, bottom-dwelling forms, etc., in various combinations.

Arthropoda

The arthropod ("jointed feet") phylum accommodates far more living species (perhaps more than 8 million) than all other phyla combined. Actually, not only the feet, but the entire body of an arthropod, are divided into many jointed segments. An affinity of arthropods to their probable ancestors, the annelid worms, is thus obvious. One of the Burgess Shale fossils is an impression of an organism that looks strikingly like both segmented worms and certain arthropods such as the millipede. Arthropods evolved early in the Cambrian Period, and in fact some geologists use the appearance of the trilobite genus *Olenellus* to define the base of the Cambrian System.*

The arthropod skeleton, an external, tough, laminated crust secreted by skin cells, serves both as support for the soft tissue inside and as protective armor. However, the platy covering of an arthropod cannot grow along with the rest of the individual. Arthropods solve the space problem by molting, or shedding off their armor from time to time. Until they can grow new protection, they are quite vulnerable to being eaten.

An extinct class of arthropods, the *trilobites,* is exceedingly abundant in the Paleozoic record. The name trilobite, or "three-lobed one," refers not to the animal's segments (which were numerous in some

* Since *Olenellus* is an "advanced" animal, perhaps arthropod evolution had already been in progress for some time before the beginning of the Cambrian Period.

FIGURE B-6 The three-part plan, one central and two side sections (all segmented), is evident in these fossil trilobites from eastern Nevada. [*Courtesy James Sprinkle.*]

1 centimeter

species), but rather to its division lengthwise into three parts separated by grooves (Figure B-6). Basically, trilobites were marine scavengers that swam, skimmed through the surface of the mud, or burrowed into it. Adults of various species were as little as 0.5 centimeter to as much as 75 centimeters long.

Another arthropod class represented by abundant fossils is composed of the *crustaceans* (lobsters, crabs, shrimp, and barnacles). Crustaceans populate the ocean by the uncounted trillions; they are the most abundant animals on earth on the basis of number of individuals. Adult crustaceans may be as much as 3.5 meters across, but most species are an inconspicuous 0.5 millimeter or so in length. The 40,000 species of these animals have adapted to land habitats and to all types of water from concentrated brine to fresh water.

Insects are enormously successful arthropods that have invaded a great variety of environments, but regrettably their fragile skeletons made of chitin do not fossilize readily. Insect fossils may be flattened impressions in fine-grained sediment, or delicately preserved entire bodies trapped in lumps of amber—fossil tree resin (Figure B-7).

Echinodermata

The echinoderms are a large and important marine phylum, so plentiful that thick layers of some limestones are composed of echino-

FIGURE B-7 This primitive ant, preserved in amber, is the most ancient known fossil of a "social insect." Features of its anatomy strongly indicate that ants and wasps had a common ancestry. The specimen was discovered in Cretaceous sediments on the New Jersey coast in 1965. [*From E. O. Wilson and others, "The First Mesozoic Ants,"* Science, *vol. 157, no. 3729, 1967. Copyright 1967 by the American Association for the Advancement of Science.*]

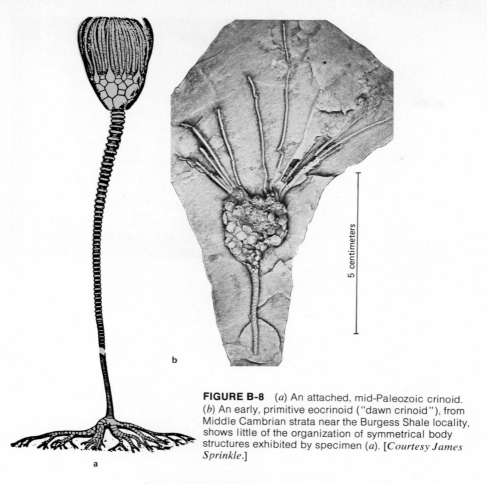

5 centimeters

FIGURE B-8 (*a*) An attached, mid-Paleozoic crinoid. (*b*) An early, primitive eocrinoid ("dawn crinoid"), from Middle Cambrian strata near the Burgess Shale locality, shows little of the organization of symmetrical body structures exhibited by specimen (*a*). [*Courtesy James Sprinkle.*]

derm skeletal fragments. Echinoderm fossils first appeared in Cambrian rocks. Here we have the only phylum whose adult symmetry is five-sided, like a pentagon. The skin of most of these animals is a tough, leathery covering in which just a few to as many as hundreds of calcium carbonate plates are embedded. Since the plates are not fused together, each plate can enlarge quite simply as the animal grows. The disconnected plates do, however, tend to get scattered after the animal dies.

Another structure common to this phylum is a set of canals (tubes or grooves) called the water vascular system. The animal uses its hydraulic network to inflate one area while relaxing another to accomplish a rather stiff and awkward movement.

Most adult *crinoids* and *blastoids* are attached forms of echinoderms, and extremely abundant as fossils. A typical crinoid is fastened to the sea bottom by a long stalk consisting of little disks attached like buttons on a string (Figure B-8). At its base are hold-fasts that in some

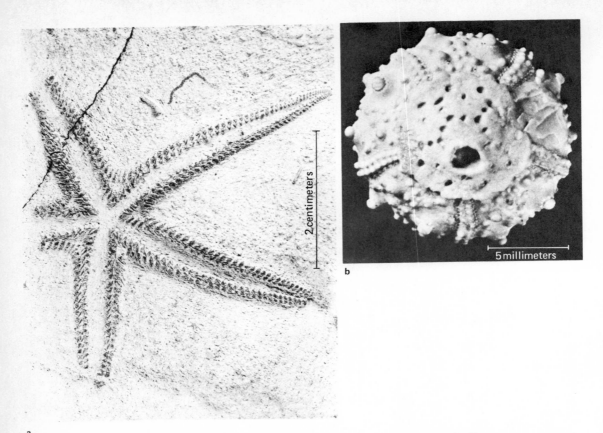

FIGURE B-9 (*a*) This slab of limestone from a locality in Austin, Texas, contains an exceptional, nearly intact fossil starfish. (*b*) Spines are generally not found associated with the central body of a fossilized spiny echinoid. They were formerly attached to the many small knobs that adorn the surface of this specimen. Note the echinoid's five-fold symmetry. [*Courtesy James Sprinkle.*]

species even developed into a system of plantlike rootlets. At the top is a cup from which there sprouts a thick sheaf of waving arms. The arms are not tentacles—they do not seize large prey that is forcibly shoved into the mouth. Instead, the arms are outfitted with beating cilia that waft microscopic particles down to the digestive system located in the central cup.

If the dense beds of extinct crinoids were anything like their brilliantly colored modern descendants, they must have looked just like animated underwater flower gardens. Fossil crinoid heads are a rarity compared to the abundance of stalks. Perhaps the great number of flattened, crushing-type sharks' teeth in crinoid beds is a clue that the fish considered crinoid heads (hard plates and all) to be a tempting meal.

Starfish and *sea urchins* are mobile echinoderms. Some starfish are filter-feeders, like the crinoids, but a number are voracious carnivores that devour clams, corals, and other prey. Fossil starfish are rare but instructive (Figure B-9*a*). Sometimes hollow molds are found into which the paleontologist can pour a preparation of latex. The rubber cast that comes out is a perfect replica; it can even be flexed gently to show how the hard parts of the starfish mesh together.

The sea urchin, or spiny echinoid, is the porcupine of the marine world. Its central body, which may be anything from spherical to a flattened disk, is armored with an imposing set of sharp-pointed spines. Not only are the spines protection against predators, but they help the animal shift about, dig burrows (even in rock), and collect food. Fossil sea urchins are only occasionally accompanied by their spines, which are easily loosed and scattered (Figure B-9*b*).

Chordata

Every individual in the chordates has a rod called a notochord running along its back (at least during the embryo stage of development). A subphylum called *vertebrata* includes species in which the embryonic notochord is later replaced by a jointed structure, the backbone. Chapter 9 takes a detailed look at vertebrate evolution.

Fossil Record of Plant Divisions

Psilophyta

Psilophytes are unquestionably the most primitive plants, though not necessarily ancestral to other groups. Psilophytes are, in fact, about as simple as vascular plants can be. The most famous, and one of the oldest, psilophyte fossil localities is a Devonian chert bed (a former peat bog) near the village of Rhynie, in eastern Scotland.

A restoration of the petrified Rhynie psilophytes (Figure C-1) shows naked upright stems connected to short horizontal stems that crept along at the ground level or slightly below it. Nutrients and water entered the horizontal stem through tiny hairs. The plant stems divided by branching into two equal-sized parts. This style, called dichotomous branching, is considered to be a primitive attribute.

Microphyllophyta

Numerous, but tiny and simple, nonvascular, leaflike structures are the distinguishing feature of this plant division. Club mosses and spike mosses (technically not true mosses) are modern creeping forms of microphyllophyta, but extinct representatives reached tree size, approaching 50-meter heights on occasion. This division appeared in the Devonian Period but attained fullest development during the Mississippian and Pennsylvanian Periods, when it was prominent in extensive coastal swamplands of the Northern Hemisphere. Entire logs of the genus *Lepidodendron* are often encountered in coal mines. The microphyllophytes were probably the first to experiment (so to speak) with the production of roots, "leaves," and a treelike growth habit (Figure C-2).

FIGURE C-1 Stems of some species of the Rhynie flora (above) were naked; stems of other species were densely clothed with scalelike emergences (not true leaves). Knobs at the tips of the branches are spore-bearing reproductive organs. [*From "Reconstruction of Ancient Vegetation by the Late Professor Paul Bertrand,"* The Paleobotanist, *vol. 1, 1952.*]

FIGURE C-2 Dichotomous branching of the crown of *Lepidodendron* (right) caused its growth to be self-limiting. An analogy would be a thick bundle of wires that can be divided and divided until eventually the frayed ends of individual wires are separated. Similarly, the strands of vascular tissue in *Lepidodendron* divided until the tree was "used up." Spirally arranged rows of scars on the trunk are points of former attachment of leaves. [*After D. A. Eggert, "The Ontogeny of Carboniferous Arborescent Lycopsida,"* Paleontographica, *Band 108, Abt. B, 1961.*]

511

FIGURE C-3 Fossils of a Late Paleozoic treelike genus show the hollow, jointed stem and delicate leaf whorls typical of the arthrophytes. [*From H. N. Andrews*, Ancient Plants, *Comstock Publishing Associates, a division of Cornell University Press, Ithaca, N.Y., 1947.*]

Arthrophyta

The arthrophytes, represented today by horsetails (branched species) and scouring rushes (unbranched), provide an interesting variation on the plant theme. In this division, the stems, both above and below ground, completely dominate the structure. They are hollow and interrupted by many joints, and their cell walls are heavily charged with gritty silica. In most species the leaves, of minor importance at best, are attached at intervals along the branches as radiating leaf-bursts, or "whorls" (Figure C-3). The arthrophytes also made their first appearance in the Devonian, reaching full climax late in the Paleozoic Era.

Pterophyta

The highly successful ferns took the opposite tack—they developed leaves as the dominant organ. The lovely fern leaves, familiar to most people, are complexly branched with generally a large number of leaflets (see Figure 8-7). Unlike the plant divisions mentioned above, ferns have continued to flourish as a large number of species to this day.

Coniferophyta

The conifers are represented by pine, sequoia, cedar, yew, etc. These richly branched plants produce distinctive cones as part of the reproductive cycle, and are endowed with numerous needlelike or straplike leaves. Conifers and all the divisions described below are seed-bearing. Conifers appeared in the Paleozoic Era and probably reached their peak during the Mesozoic Era.

Cycadophyta

Modern cycads (SIGH-cads) are a rather obscure group that look like a cross between ferns and palms, but cycads differ from both ferns and palms in the texture of the stem and mode of reproduction.

An extinct group of cycadophytes known as seed ferns points up one of the hazards of an attempt to imply too much from fragmentary fossils. For many years, species of supposed fern leaves were known which consistently failed to reveal any associated spore-bearing organs. Then, in 1905, seeds were discovered actually attached to unmistakable fernlike foliage (Figure C-4). Inasmuch as seeds are never produced by true ferns, these mysterious fossils had to be reassigned to another plant division, the cycadophytes.

Once it became apparent that the fernlike leaves are an important and recurring type, much interest was generated in learning their origin. One popular theory takes the simple, all-stem plants (such as psilophytes) as a starting point. As evolution continued, the stems became increasingly branched and flattened into a single plane (Figure C-5). Then a webbing of vegetation developed to connect the strands of vascular tissue, forming a fern-type leaf with branching veins. The extinct seed ferns support this theory (Figure C-6).

FIGURE C-4 In spite of diligent searching, paleobotanists have found seeds attached to fernlike foliage only a dozen or so times. This Permian seed fern is actually allied more closely to cycads than to true ferns. [*After H. N. Andrews*, Ancient Plants, *Comstock Publishing Associates, a division of Cornell University Press, Ithaca, N.Y., 1947.*]

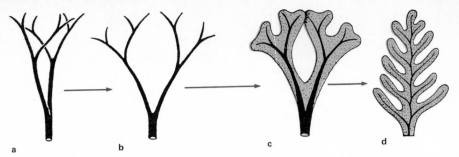

a   b   c   d

FIGURE C-5 The progression of forms, (*a*) to (*d*), represents the evolutionary development of leaves. [*After W. N. Stewart, "An Upward Look in Plant Morphology,"* Phytomorphology, *vol. 14, no. 1, 1964.*]

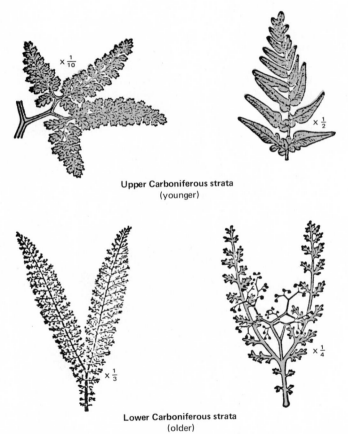

Upper Carboniferous strata
(younger)

Lower Carboniferous strata
(older)

FIGURE C-6 In accord with theory, the lower (more ancient) leaves look more like stages (*a*) and (*b*) of Figure C-5, whereas the upper leaves look more like stages (*c*) and (*d*). [*After K. R. Sporne*, The Morphology of Gymnosperms, *Hutchinson & Co. (Publishers), Ltd., London, 1965.*]

514   FOSSIL RECORD OF PLANT DIVISIONS

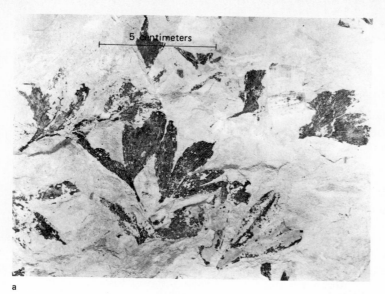

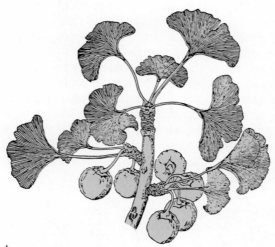

FIGURE C-7 (*a*) Leaves of a certain extinct ginkgo species are fan-shaped, like leaves of the modern plant (*b*), but the fossil leaves are more deeply indented. Thus it appears probable that the ginkgo leaf, like those of the ferns, gradually fused into more continuous bladelike forms. [(*a*) *from T. Delevoryas*, Plant Diversification, *Holt, Rinehart & Winston, Inc., New York, 1966.* (*b*) *after H. N. Andrews*, Ancient Plants, *Comstock Publishing Associates, a division of Cornell University Press, Ithaca, N.Y., 1947.*]

Ginkgophyta

The modern ginkgo (hard "g") is truly a living fossil. Long thought to be extinct, a group of living ginkgo trees was discovered by Western-ers in a small area of China in the late 1600s. The slow-growing ginkgo has been exported to many parts of the world where it survives hand-somely. The one extant ginkgo species forms a stately tree covered by distinctive fan-shaped leaves (Figure C-7). Although it superficially resembles the flowering trees, the ginkgo makeup differs from that of angiosperms in numerous technical details. Accordingly, ginkgo is placed in a separate division.

Anthophyta

The flowering plants beggar description because they are so abundant and diverse. Aside from the presence of flowers, other features that set angiosperms apart are well known to botanists but not apt to be fossilized. This enormous and successful division includes plant species that have invaded nearly every land environment in which life is possible.

Flowering plants have undoubtedly taken over some of the niches previously held by the still abundant but less widespread conifers. Two factors could contribute to the rapid evolution, hence relative success, of the anthophytes. Many of these plants have short reproductive cycles, in contrast with the long periods needed by most conifers to reach maturity. Another factor is that many anthophytes require insects to spread their pollen, whereas the pollen of conifers simply blows about indiscriminately. The evolution of insect-pollinated plants is firmly locked into the evolution of the insects themselves. This interdependence also caused rapid speciation.

Specialists have studied the tough, resistant fossil angiosperm pollen grains as sensitive indicators of ancient climate (Figure C-8). Extensive changes took place as a previously moderate world climate shifted into its present sharply stratified zonation in which polar ice sheets, temperate forests, deserts, and tropical rain forests are known. Pollen research has received great impetus from the invention of the scanning electron microscope. This instrument pictures the pollen surfaces in vivid detail never before witnessed.

50 microns

FIGURE C-8 Details of angiosperm pollen are studied with the scanning electron microscope. Modern pollen, pictured here, are strikingly similar to their ancient fossil counterparts. [*Courtesy Thomas Taylor.*]

Index

This Index also incorporates a Glossary in which some of the more troublesome words are defined. These include highly technical terms, common words used in a technical sense, and similar terms that must be carefully distinguished. The customary page references follow the glossary definitions. Italicized page numbers refer to information in figures or in figure captions; *n.* stands for footnote; *t.* stands for table. For further reading see Gary, Margaret, Robert McAfee, Jr., and Carol L. Wolf (eds.),1972: *Glossary of Geology,* American Geological Institute, Washington, D.C., 805 pp.

Abundance of chemical elements:
  cosmic, 13–14
  in earth's crust, 60, *492*
  of rare gases, 41–42
Abyssal hill, 336
Abyssal plain, 300, *380*, 412
Actualism (*see* Uniformitarianism)
Adaptive radiation, 192, *193*
  of mammals, 265
Airy, G. B., 307–309
Alanine, *168*, 169
Albedo. The fraction of sunlight or other electromagnetic radiation that is reflected off a surface. 452–457
Algae, 113, 115, 175–176, 197, *198*, 502
  (*See also* Blue-green algae)
Alps, *322*, 340–348, 350, 436
Aluminum:
  ore, 479
  retention in moon, 43
  in silicate minerals, 61, 63, 65, 68
Amber, 506
Amino acid, 167–169, 172, 179, 189
Ammonia, 16, 19, 30*n.*, 105, 168
  (*See also* Methane)
Amniote egg, 252–254, 256
Amphibian, 233*t.*, *235*, 247–251
Amphibole, 63–65, 69*t.*
Analogous organs. Body structures that are similar in form or in function, but not in evolutionary origin (*See also* Homologous organs). *191*, 192
Andalusite, 120, *121*
Andes Mountains, 340, 350, 351
Angiosperm (*see* Anthophyta)
Angular momentum:
  of earth-moon system, 46
  of planets, 16–17, 20
Anhydrite, *113*, 343–344, 416
Animal phyla, *200*, 204*t.*, 498–509
Annelida, 204*t.*, 207, *208*, 236–238, 502
Anthophyta, 209*t.*, 210, 516
Aquifer, 466–475

Aquifer:
  confined, 472–475
  Dakota Sandstone, 472, *473*
  Ogallala, *471*, 472
  unconfined, 467–472, *473*
Archaeopteryx, 262, *263*
Arithmetic-geometric relationship, 181, 491
Artesian well, 472, *473*
Arthrophyta, 209*t.*, 512
Arthropoda, 204*t.*, 205, *208*, 236–238, 505–506
Asteroid, 16, 19, 27
Atmosphere:
  origin of, 41–42, *162*
  oxygen buildup in, 174–175
  in polar regions, 440
  role in Ice Ages, 452–457
Atoll, 337–338, *367*
  (*See also* Guyot)
Atom, 10–12
  (*See also* Matter, structure of)
Atomic structure:
  of kaolinite, *110*
  of NaCl, 58–59
  of nonsilicate minerals, 70–73
  of silicate minerals, 62–70

Bacon, Sir Francis, 355, 356
Bacteria:
  infected by virus, 189
  as oldest fossils, 149
  as prokaryotes, 197–198
  role in origin of life, 166, 169–172, 175
  as sulfate-reducers, 416, *417*
  in weathering and sedimentation, 107, 111, 479
Bahama Islands, *116*, 117, *118*
Barrier island, 410, *412*
Barringer meteorite crater, 29
Barrow, George, 127, 128
Basalt. A dark-colored, fine-grained, dense extrusive igneous rock in which calcium plagioclase and pyroxene are

Basalt:
  the chief minerals.
  crosscutting relationship with granite, *95*
  description of, 86–87
  in Hawaii, *470*
  on ocean floor, 300, 364, 366, 374, 491
  stratification of, 107
Baselevel, 388, *389*
Basement rock:
  in Alps, 341–342
  isotopic ages of, 154–158, 359, *361*
  in North America, 151–158
  ore deposits in, 489
Batholith. A large, generally discordant mass of coarse-grained intrusive igneous rock of granite to granodiorite composition. It is deeply eroded, with a large surface exposure (*See also* Granite).
  associated with ore deposits, 484
  contact-metamorphism around, *124*
  description, origin, 94–98
  at plate boundaries, *368*, 370
  slow cooling of, 145
Bauxite, 480
  (*See also* Lateritic soil)
Beach:
  depositional processes on, 408–410, *411*, *412*
  as a geologic formation, *215*, 216
  (*See also* Water wave action)
Becquerel, Henri, 140
Bedload of stream, 388–394, 399, 478
Bernal, J. D., 172
Bestiary, 227
Big Bang, 9–10, 14
Biotite:
  atomic structure of, 65, *66*
  in granite, 93
  as a hydrous mineral, 42, 119
  in K-Ar isotopic age method, 144

517

Crater:
  volcanic. A basinlike depression, usually at the summit of a volcanic cone, whose diameter is approximately that of the vent (*See also* Caldera, volcanic). *79*, 91
Crevasse:
  in glacier ice, 202n., 429, *430*, 435
  in natural levee, *397*
Crick, F. H. C., 170
Crinoid, 507–508
Crosscutting relationships, law of, 88, *95*, 155, *156*
  (*See also* Superposition, law of)
Crossopterygian, 247–248, *250, 251*
  (*See also* Fish)
Crust, 39, 476
  chemical abundances in, 60, 70
  continental, 300–301, 303–309, 311, 363
  oceanic, 300–301, 303–309, 324n., 363, 367, 369, *376*
Crustacean, 506
Crystal. A three-dimensional atomic, ionic, or molecular structure in which the constituent atoms are located in a repeating pattern (*See also* Glass).
  atomic structure of, 53–56
  crystalline state of matter, 25, 30
  external form of, 51–53
Curie temperature, 311, 315
Curvature of space, 5–6, 9
Cuvier, Georges, *196*, 216, 218
Cycadophyta, 209t., 513, *514*
  (*See also* Seed fern)

Darcy, Henri, 466
Darcy's law, 466–469, *475*
Darwin, Charles, 47, 163–165, 176–178, 180–188, 210, 216, 218, 338
Darwin, George, 47
Darwin's finches, 186, *187*
Daughter atom, 12, 140–142
  (*See also* Parent atom)
Davis, W. M., 390
Dead Sea, 414–415
Deformation:
  of entire earth, 276–277, 373
  of fold mountains, 309, 341–350
  at plate boundaries, 363, 367–370
  of rocks, 119, 121–127, *480*
  of salt, 415–417
  of snow and ice, 425–432
Delta, 215–216, 394–404, 408
  (*See also* Depositional environment)
Delta cycle, 399, *404*

Delta lobe, 396–399
Dendritic drainage, *382, 448*
Density:
  of core, mantle, and crust, 39, *299*
  of igneous rocks, 86, 94
  of matter in space, 9–10, 17, *18*, 19
  of moon versus earth, 47
  beneath ocean trenches, 336
  role in isostasy, 307–308
  of silicate minerals, 69t., 299
  of weathering products, 102
Deoxyribonucleic acid (DNA):
  as distinguishing prokaryotes and eukaryotes, 197–198
  as distinguishing species, 178, 188
  ingredient of gene, 183
  mutations of, *190*
  structure of, 167, 170–172
Depositional environment, 379–420
  coastal, 404–410
  continental shelf, 412–416
  deepwater, 416–418
  delta, 394–404
  fluvial, 382–394
Deuterium, 440, 494
Diamond:
  chemical bonds in, 60
  mineral, 74
  origin of, 300, 477
  placer, 478
Dichotomous branching, 510, *511, 514*
Differentiation. Process by which the components of a mixture are separated from one another.
  age of, in earth, 149
  of basaltic magma, 481–482
  of earth, 40–42
  heat of, 40
  of moon, 43
Dike, 88–90, 364, *470*
  (*See also* Sill)
Dimetrodon, 228, 229, 256
Dinosaur, 257–260
  extinction of, 259, 261
  skeleton of, *258, 259*
Distributive net, 382–384, *398, 402*
  (*See also* Contributive net)
Dollo's law, 261
Dolomite, 115, *116*
  (*See also* Calcite; Carbonate mineral)
Donn, W., 455
Doppler effect, 7–8
Dust, atmospheric, 85, 91, 106, 454, *455*
Du Toit, A. L., 355, 356, 360

Earth:
  ages of, 147–149

Earth:
  average rigidity, 292, 294–295
  core of, 294–296
  crust of, 300–301
  internal layering of, 39–40
  mantle of, 298–300
  moment of inertia of, 292, *293*
  shape of, 274–277, 303–305
  in space, *460*
Earth-moon system:
  origin of, 47–49
  tidal interactions of, 44–47
Earthquake:
  characteristics of, 277–280
  deep-focus, 291–292, 336, *337*, 367–369
  elastic rebound theory of, 279, *280*
  first motion of, *280, 371*
  global distribution of, 362–363
  intensity of, *281*
  locating distant epicenters of, 287–291
  due to meteorite impact, 30
  ringing action of, 41, *287*
  (*See also* Seismic wave)
Echinodermata, 204t., 236–238, 506–509
Echinoid, 237, *508, 509*
Economic geology, 477–486
Ediacara fauna, 206–207
  (*See also* Burgess Shale, fossils in)
Einstein, Albert, 5, 6, 9
Electron:
  in chemical bonds, 57–59
  description of, 10–11
  in magnetic minerals, 311
  shells of, 58–59
Element (*see* Chemical element)
Elevation of earth's surface, *339*
Endemic species. A species that is restricted to a particular region or ecological environment (*See also* Cosmopolitan species). 219
English units of measurement, 496–497
Epeirogeny. Broadscale crustal movements, dominantly vertical, that create large, open structures such as basins and plateaus (*See also* Orogeny). 150, 151, *154*, 159
Epicontinental sea, 117, 159, 343
Eratosthenes (lunar crater), *32*, 37–38
Eratosthenes (man), 274–275
Erosion:
  in creating unconformities, 136–137
  exposing granite, 95, *96*
  by ice sheets, 432–439

Erosion:
of lunar rille, 37
of lunar surface, 34, 36
of meteorite craters, 31
of mountains, *310*, 339–348,
351–352
by rivers, 382–385
of sedimentary rocks, *100*, *217*
of a volcanic landscape, *88*
by waves, 406–410
Esker, *435*
Estuary, 410, *412*, *413*
Eukaryote. An organism whose
cells contain internal membranes and organelles (specialized structures including chromosomes, nucleus, mi-
tochondria, etc.). Includes all living organisms except bacteria
and blue-green algae (*See also*
Prokaryote). 197–200
Evaporite sediment, 112, 343,
414–415, 462
Evolution (*see* Organic evolution)
Ewing, M., 455
Exfoliation dome, 102, *439*
Extinction, causes of: by continuing
evolution, 192, *193*, 242
by Divine catastrophes, 216
during the Pleistocene, 266
of small populations, 236
due to unadaptability, 185, 268
Extrusive igneous rock (*see* Volcanic
rock)

Fault:
accompanying metamorphism,
121–123, *127*, 150
in deformation of mountains, *346*,
350
in ocean basins, 332
of ore deposits, *483*
Faunal succession, law of.
Faunas (and floras) succeed
one another in a definite, recognizable order; each geologic
formation contains a fossil assemblage that is different from
those in the formations below
and above it, thus making it possible to determine the passage
of geologic time. 216–221
Feldspar:
alteration of, 483
atomic structure of, 67–68, 69*t*.
in igneous rock, 80, 83, 86, 92, 93,
97, *98*
in metamorphic rock, 127, 128
radioactive isotopes in, 142
in sedimentary rock, 108
weathering of, 103, 104, 1-10, 479

Ferromagnesian mineral. A mineral in which magnesium and
ferrous iron are major constituents.
description of, 62–67
in igneous rock, 83, 92
in metamorphic rock, 126
in ore deposits, 481–482
in weathering processes, 103
Fig Tree Chert, 172, *173*
Firn, *426*, *427*, *428*
Fish, 233*t*., *235*
bony and cartilaginous, 244–247
jawless, 240–242
origin of, 238–240
placoderms, 242–244
Fold mountain (*see* Mountain belt)
Foraminiferan, 113, *114*, 197, 201,
219, *220*, 448, *449*
Force field, 273–274
Formation, 215–216
Fossil record:
animal, 204–207
definition of, 164
evidence for continental drift, 361,
362
how made, 200–204
imperfections in, 210–212,
218–219, 222
in metamorphic rocks, 126
plant, 207–210
of vertebrates, 230–231
Fracture. Uneven breaking of a
mineral along surfaces other
than cleavage planes (*See also*
Cleavage).
Fracture zone, oceanic, *331*, 332,
334, 335, *369*, 370
Frequency distribution curve,
183–185
Fungi, *200*

Galápagos Islands, 185–186, *187*
Galaxy:
light spectra of, 6–9
structure of, 4–5
Galileo, 31, 34
Garnet, 126, 127, 300, 480
Gastropod (*see* Snail)
Gene, 170, 182–184, 190, 231, 236
Genesis, Book of, 132–134
Genetic code, 170, 189
Genetic drift, 236
Geoid. The figure of the earth that
corresponds with sea level, or
with sea level projected continu-
ously through the continents. It
is everywhere a horizontal sur-
face (locally perpendicular to
the plumb bob) (*See also*
Spheroid). 303–309

Geologic map, *79*, *217*, *223*, *342*,
380
Geologic time:
history of thought about, 131–140
hourglass methods of telling,
138–140
isotopic methods of telling,
140–147
Geologic time scale, 213–224
Geophysics, geophysical in-
struments, *272*, 273–274
geophysical fish, 336
Geotherm, *298*
Geothermal area, 485
Geothermal energy, 494
Geyser, 472, 485
Gilbert, William, 310, 311
Ginkgophyta, 209*t*., 515–516
(*See also* Seed fern)
Glacial. An interval during an Ice
Age when the ice sheets are
greatly expanded (*See also* Interglacial).
Glacial erosion, deposition,
432–439
Glaciated landscape, 436–437, *438*,
439
Glaciation (*see* Ice Age)
Glacier ice:
accumulation, wastage, 428–432,
435
motion dynamics, 428–432
warm (temperate) versus cold,
425–432
Glacier surge, 432, *433*, 457
Glass. A state of matter intermediate between the highly organized array of atoms in a crystal
and the highly disordered arrangement of atoms in a gas
(*See also* Crystal). 30–31
volcanic, 85
Gneiss. A regionally metamorphosed rock in which layers
containing granular (equidimen-
sional) minerals alternate with
layers rich in flaky or prismatic
minerals (*See also* Schist).
126–127, 150, *480*
Gold, 74, 150, 476–479
Goldich, Samuel, 104
Graded bedding, 416, *481*
Granite:
associated with ore deposits, 477,
482, 484
contact relationships, 95–96, *124*
description of, 93–95
origin of, 96–98
weathering of, 101–104
(*See also* Intrusive igneous rock)
Granodiorite, 93–94, 301, 307
(*See also* Granite)

Gravity, force of:
between earth and moon, 44–47
on lunar surface, *34*, 43
role in isostasy, 306–309, *310*
in shaping the earth, 275–277
in solar system, 15, 19, 20
in universe, 6, 9
variations of, 301–305, 335–336
Gravity meter, *272*, 302, *303*
Gross National Product (GNP), 486, 487
Groundwater:
aquifers, 466–474
Darcy's law, 466–467
flow velocities, 474, *475*
hydrologic cycle, 463–465
in origin of ore deposits, 484–486
in origin of sedimentary rocks, 111, 118
Gulf of Mexico, 150–154, 379–381, 414–418, *419*
Gutenberg, Beno, 294
Guyot. A flat-topped seamount (*See also* Atoll; Seamount). 338–339, *367*
Gypsum, 72, 478

H_2O, physical properties of, 423–424, *425*
Habitat. Ecologic environment (*See also* Niche).
Haldane, J. B. S., 167, 172
Half-life:
of erosion of mountains, *310*
of radioactive isotope, 140–142, *446*
Halite, 72, *73*, *113*
(*See also* Salt; Sodium chloride)
Halley, Edmund, 138
Haüy, René, *52*, 53, 56
Hawaiian Islands, 83, 86–90, 186, 374, *375*, *470*
Heat in earth:
to power drifting tectonic plates, 372–374
sources of, 40
from tidal flexing, 46
Heat flow, 372–374
Helium:
escape from planets, 43
as a resource, *488*
in universe, *13*, 14
Hematite, 71, *72*, 103, 311
(*See also* Magnetite)
Herodotus, 394
High-pressure, high-temperature equipment, *81*
Himalaya Mountains, 340, 343, 350, 351, 360, 367
Homologous organs. Body structures that are similar, but not

Homologous organs:
identical, whose evolutionary development in different organisms proceeded from the same ancestral organ (*See also* Analogous organs). *191*, 192
Hornblende, 64, 93, 119, 142, 144
Horse evolution, 268–270
Hot spot, 374, *375*
Hot spring, 42, *485*, 486
Hoyle, Fred, 20
Hubbert, M. K., 487
Hutton, James, 135–138, 154, 155, 381
Huxley, T. H., 270
Hydraulic conductivity, 465, 466 (*See also* Permeability; Porosity)
Hydrogen, hydrogen ion:
as energy source, 494
escape from planets, 43
in hydrous minerals, 42
in origin of life, 167
in universe, *13*, 14, 16
in weathering, 103
Hydrograph, 386–388
Hydrologic cycle, 463–465
Hydrothermal ore, 482–486
Hydrous mineral, 42, 43, 67, 72, 102, 103, 119, 120*t.*, 127, 300

Ice Age, 423
ancient glaciations, 447, 452, *453*
Cenozoic multiple glaciations, 447–450, 452–457
Iceberg, *424*, 432
Iceland, *76*, 83, 86, 87*n.*
Ice sheet:
erosion by, *438*, *439*
location of, 423, 424
motion dynamics of, 430–432
origin of, 440–443
Pleistocene advances of, 266
Ice shelf, 432
Ichthyosaur, *192*, 211, 261, 270
Ichthyostega, *248*, *251*
Igneous rock:
definition of, 77, 78
field and laboratory interpretations of, 78–82
intrusive, 93–98
volcanic, 82–92
Ignimbrite. A volcanic rock formed from explosive, far-traveled, avalanches of hot particles that usually have welded together (*See also* Tuff). 92, 107, *222*
Impression. The form, or indentation, of a fossil made on a sedimentary surface (*See also* Compression). 203, *206*
Inert gas:
electron shells in, 58

Inert gas:
terrestrial abundances of, 41–42
Ingerson, F. E., 489
Inheritance of acquired characters, 183
Insect, *191*, 249, 254, 506, 516
Interfacial angles, law of, *52*, 53
Interglacial. An interval during an Ice Age when the climate is relatively mild, and the ice sheets are contracted (*See also* Glacial). *442*, 447, *448*
Intrusive igneous rock:
description of, 93–98
role in ore deposits, 480–482, *484*, 485–486
Iodine-xenon age, 148
Ion, 11, 58
Ionic bonding. A chemical bond in which negatively charged ions (anions) are bound by electrostatic forces to positively charged ions (cations) (*See also* Covalent bonding). 58–59
Iron, iron mineral:
in early earth history, 174, *175*
in earth's core, 295
in ferromagnesian minerals, 62–67
in meteorites, 25–27
in nonsilicate minerals, 71, 72, *73*, 103
in universe, *13*, 14–15
in weathered rocks, 103, 106, 107
Ironstone sedimentary deposit, 111, 174, 478–479, *480*
Island arc, *324*, 335–336
Isoseismal, *281*
Isostasy. Flotational equilibrium of relatively brittle surficial part of the earth above relatively plastic interior material. 306–309, *310*, 351, 436, *437*
Isotope. Different species of atom in which the number of protons is the same, but the number of neutrons differ.
creation of, 13–15, 148
definition, 11
stable and radioactive, 12
Isotopic age:
of basement rocks, 149–159, *361*
of the creation of chemical elements, 148
of earliest life, 149
of earth's differentiation, 149
as evidence for continental drift, 359, *361*
of geologic time scale, 221–223
of igneous rocks, 144–145
of metamorphic rocks, 145–147
of ocean sediment, 326, *328*

Model:
ficult to observe directly.
Mohorovičić, A. 296
Mohorovičić discontinuity (Moho),
296–298, 301, *330*
Mollusca, 204*t*., 236–238, 254,
501–504
Monera, *200*
Monophyletic evolution. Evolving
from a single ancestral stock
(*See also* Phylogeny; Polyphyletic evolution). 254, *255*
Moon:
atmosphere, surface water of,
41–43, 101
comparison with earth, 38–44
craters on, 31–36
earth-moon system, 44–49
highlands (terrae) on, 31, 33
maria on, 31–33, *34*
meteorite impacts on, 44
moonquakes in, 41
origin of, 47–49
recession of, 46–47
rilles on, 36–37
size, density of, 38–39
in space, *22*
stratigraphy of, 37–38
Moonquake, 41
Moraine, *433, 451*
Mosaic evolution, 270
Mountain belt, 339–353, 367, *368*
Alps, 340–348
deformation of rocks in, 349–350
generalizations about, 350–353
Multivariate analysis, 179, *180*
Murchison, R. I., 220, 221, 224
Muscovite:
atomic structure of, 65, *66*
in igneous rocks, 93
in metamorphic rocks, 126
in pegmatites, *67*, 482
in sedimentary rocks, 108
(*See also* Biotite)
Mutation. Imperfect replication of
genetic material (*See also* Natural selection; Recombination).
188–190, 236

Nappe, 345–348
Natural levee, 393, *397, 398,* 399,
401, 402
Natural selection. The process
whereby organisms survive to
the age of reproductive maturity
if they are adapted to their environment, or do not survive if they
are not sufficiently well adapted
(*See also* Mutation; Recombination). 181–182, 236
Nautilus, 189, 504

Neutron:
description of, 10–11
in synthesis of chemical elements,
15
Newton, Isaac, 302
Niche. The position of an organism
or population in an environment,
including where it lives, and
how it lives (*See also* Habitat).
Nickel, nickel mineral, 71, 150
ores of, 477, 479–480
in universe, 14
Nickel-iron alloy:
in earth's core, 39–40, 295
in meteorites, 25–27
Nile delta, 395, 400–401
Nonsilicate mineral, 70–74
Notochord, 231, *237,* 238, *239,* 240,
509
Nuclear force, 10, 140
Nucleus of atom, 10–11, 140

Ocean, 323–339
age of, 149
birth to death, 368*t*.
origin of, 41–43
sediments in, 115, *116,* 325–328,
335, 369, 416–419, 447–449,
481, 491
small ocean basins, 324–325
as a source of atomic energy, 494
as a source of earth materials, 491
topography of floor of, 325–339
Ocean-floor spreading, 366–367,
369, 376
Ocean ridge system:
description of, 330–335
earthquakes in, *363*
role in plate tectonics, 364–367,
371, 372, 373, *374, 376*
Ocean sediment core, *328,* 447–448,
449
Ocean trench system:
description of, 335–336
earthquakes in, *363*
location of, *331*
role in plate tectonics, 367–372
Oceanus Procellarum, *32*
Octopus, 502–504
Oldest thing on earth:
fossil, 149
living organism, *445*
rock, *130,* 149
sediments, 149
Oldham, R. D., 294
Olivine:
atomic structure of, 62–63, 69*t*.
in igneous rocks, 82, 86, 87
in mantle, 300
in ore deposits, 480, 481
in weathered rocks, 104

Oolite, 118
Oparin, A. I., 167, 172
Orbit:
of artificial earth satellites, 275,
304, *305*
of earth, 454, *456*
of lunar satellites, 33, *34*
of moon, *45,* 49
of planets, 15
Ore deposit:
cutoff percentages of metal in, *476*
hydrothermal, 482–486
legal disputes over, *483*
magmatic, 480–482
mine tailings from, 486
sedimentary, 478–480
source rock of, 476
sporadic distribution of, 491–493
worldwide reserves of, 486–491
Organelle. A specialized structure,
carrying out definite functions,
within a eukaryotic cell.
Organic evolution. The theory that
life on earth has gradually developed from simple organisms
to more complex organisms. All
organisms are related to one
another historically. 163–193
of vertebrates, 227–270
Organic molecule, 164–174
Organic reef, 115–117, 207, 338, 500
Organisms, classification of,
178–181, 197–200
Organization of life, 165*t*.
Orogeny. The process of formation
of mountains by folding and
faulting, and accompanied by
metamorphism and intrusion of
igneous rocks. The latter two
processes do not always accompany orogeny.
ages of, 158–159
of Alps, 341–348
generalizations about, 350–353
in Greece, *223,* 224
in North America, 149–158
at plate boundaries, 367–370
Orthogenesis, 268
Oxbow lake, 393, *394, 396, 397*
Oxide mineral:
in basalt, 86
in bauxite, 479–480
description of, 71, *72*
in mantle, 300
on ocean floor, 491
Oxygen:
in atmosphere, 107, 174–175, 479
covalent bonding of, 59
crustal abundance of, 60, 70
in oxide minerals, 71, *72*
in silicate minerals, 27, 30, 61–62
stable isotopes of, 11, 440–443,